MW01274808

Stability by Fixed Point Theory for Functional Differential Equations

Stability by Fixed Point Theory for Functional Differential Equations

T. A. BURTON
Department of Mathematics
Southern Illinois University
Carbondale, Illinois

DOVER PUBLICATIONS, INC.
Mineola, New York

Bibliographical Note

Stability by Fixed Point Theory for Functional Differential Equations is a new work, first published by Dover Publications, Inc., in 2006.

Library of Congress Cataloging-in-Publication Data

Burton, T. A. (Theodore Allen), 1935–
Stability by fixed point theory for functional differential equations / T.A. Burton.
p. cm.
Includes bibliographical references and index.
ISBN 0-486-45330-8 (pbk.)
1. Functional differential equations—Numerical solutions. 2. Stability. 3. Fixed point theory. I. Title.

QA372.B879 2006
518'.6—dc22

2006047426

Manufactured in the United States of America
Dover Publications, Inc., 31 East 2nd Street, Mineola, N.Y. 11501

To Professor V. Lakshmikantham

Preface

This book is the first general introduction to stability of ordinary and functional differential equations by means of fixed point techniques. It contains a massive collection of new and classical examples worked in detail and presented in a most elementary way. More than three fourths of the work relies mainly on three principles: a complete metric space, the contraction mapping principle, and an elementary variation of parameters formula. The material is highly accessible to upper division undergraduate students in the mathematical sciences, as well as working biologists, chemists, economists, engineers, mathematicians, physicists, and other scientists using differential equations. It also introduces many research problems which should be of interest for a long time.

For more than one hundred years Liapunov's direct method has been the primary technique for dealing with stability problems in ordinary and functional differential equations. We have studied and contributed to it for more than forty years. Yet, a variety of difficulties persist which lead us to consider other methods in conjunction with the direct method. The direct method usually requires pointwise conditions, while so many real-world problems call for averages. There is the difficulty of constructing Liapunov functions and functionals. When a suitable Liapunov function is found, there remain significant problems with ascertaining limit sets when the equation becomes unbounded or the derivative is not definite. Many of these difficulties vanish when fixed point theory is used. On the other hand, this book strives to combine fixed point techniques and Liapunov techniques whenever it can be fruitfully done.

There is one more central issue which makes fixed point methods so very desirable in the study of stability properties of differential equations. Students and users of differential equations frequently do not have an extensive background in the theory of differential equations which is so often required in Liapunov theory. Indeed, to even apply Liapunov's direct method, one must independently verify that a solution exists. In one step, a fixed point argument yields existence, uniqueness, and stability. Section 0.2 contains a lecture we have polished over several years which shows some of the outstanding features of fixed point methods in stability.

The aforementioned matter of conditions involving averages is discussed throughout the book and is important for describing real-world problems. John Appleby studied these results and has shown that, in addition to the averages, the same fixed point technique also yields stability under stochastic perturbations, making it doubly useful for real-world problems. Chapter 7 is entirely his contribution. He gives an introduction to stochastic definitions, problems, and techniques. Then he applies the fixed point technique to study stability properties of two classical functional differential equations. His work is particularly exciting for those, such as the author, who are seasoned investigators of stability for deterministic problems, but total novices when it comes to stochastic problems. In Section 7.3 he shows how to split a stochastic differential equation into two equations, one being a fixed stochastic problem and the other a deterministic stability problem with forcing function. The first is fixed for all problems of the class and he solves it. The second equation varies depending on the original problem, but is deterministic and can be directly attacked by the fixed point methods developed throughout this book. Clearly, it is worth the study to discover the intricacies. References are given to his work along the lines introduced in Chapter 7. This is a most inviting area for further research.

Chapter 1 discusses half-linear problems for which a fixed point mapping is readily obtained. Section 2.1 introduces a paper by K. Cooke and J. Yorke which leads us to the idea of harmless perturbations whereby a functional can be replaced by a pointwise function. In effect, the functional differential equation is often replaced by an ordinary differential equation with a harmless perturbation. Many papers have followed the Cooke-Yorke paper and it is our conjecture that every one of those papers will lead us to additional harmless perturbations, thereby multiplying the results contained in this book many-fold. Thus, we feel that this will be a very fruitful area for mathematical research for many years to come.

Thanks goes to John Appleby, Leigh Becker, Tetsuo Furumochi, V. Lakshmikantham, Adrian Petrusel, Laszlo Hatvani, Marcel Serban, Alfredo Somolinos, Christian Vladimirescu, and especially to Bo Zhang for reading portions or all of the manuscript and for their helpful comments.

I am pleased beyond measure to see the book published by Dover. It is a company I have admired during my entire career for making important mathematical work available at prices affordable to people throughout the world. A special thanks goes to the editor, John Grafton, for his interest in the book. It is always a pleasure to work with him.

Finally, I wish to dedicate this book to Professor V. Lakshmikantham who, for about fifty years, has played a leading role in the investigation of differential equations both by his own work and his constant encouragement of others.

T. A. Burton
Northwest Research Institute
Port Angeles, Washington
February 10, 2006

Contents

Chapter 0

Introduction and Overview

0.1 The Origin of the Work

It has long been our view that Liapunov's direct method is the leading technique for dealing with stability problems in many areas of differential equations. We have contributed to that theory for more than forty years and continue to do so. Yet, numerous difficulties with the theory and application to specific problems persist and it does seem that new methods are needed to address those difficulties. There is, of course, the problem of constructing appropriate Liapunov functionals and in carrying out various calculations such as is seen in the annulus argument. These are things which we feel we can overcome with work and imagination. But there are also problems with the types of conditions which are typically imposed on the functions in the differential equations. These conditions are virtually always precise pointwise requirements. Real-world problems with all their uncertainties call for conditions which are averages. In this book we explore the use of fixed point theory in meeting some of those problems.

Not only do the fixed point conditions emerge as averages, but very recently John Appleby has shown that the fixed point methods will also admit stochastic perturbations. In Section 7.3 he discusses how these perturbations relate to real-world applications. Chapter 7 is entirely his contribution. He introduces basic stochastic theory and shows how the fixed point methods yield stability results under stochastic perturbations. Thus, in two ways the fixed point methods seem more applicable to real-world problems than does Liapunov's method.

There is an added benefit to studying stability by fixed point methods. In the next section we illustrate how very complete, simple, and rigorous stability analysis of a highly sophisticated problem can be given using only a complete metric space, the contraction mapping theorem, and the elementary variation of parameters formula. In one step we prove existence, uniqueness, and boundedness of solutions of a problem which has challenged investigators for decades. This is important for students and investigators who are not primarily stability analysts, but whose work leads them to sophisticated stability problems. Increasingly, stability problems emerge in biology, economics, and unlikely places. These problems can often be solved using fixed point theory in an elementary and mathematically honest manner. What is required is imagination in setting up a fixed point mapping. This book is a guide to the process.

For a long time we have been interested in the use of fixed point theory in conjunction with Liapunov's direct method. Fixed point theory can be used directly to study a differential equation when solutions are being considered on a finite interval. But when we are discussing solutions on an infinite interval, some technique seems to be necessary to keep control of the solution. In Burton (1984) we use Liapunov functionals in conjunction with Browder's fixed point theorem to obtain periodic solutions of nonlinear functional differential equations. In Burton (1985) a large part of the monograph is devoted to the use of Liapunov functionals with Horn's fixed point theorem to obtain periodic solutions. In Burton (1994) we use mildly unbounded Liapunov functionals with Schaefer's fixed point theorem to prove global existence for a variety of equations. In a parallel way Seiji (1989) and others have used Liapunov functions with fixed point theory to prove stability. In a different vein, Serban (2001) has used fixed point theory to prove asymptotic stability for a difference equation.

In 1997 several investigators began a systematic study toward a comprehensive stability method based on fixed point theory. As it developed it became more of an investigation into what could be said about classical problems which had offered resistance to Liapunov's direct method and what kinds of absolutely new properties can be discovered. The accent has been on using averaging techniques associated with variation of parameters to avoid pointwise conditions generated by the direct method. In several places we use Liapunov's direct method, but in different ways than described above. First, it is used as a contrast with fixed point methods. Next, we frequently use the direct method to prove that solutions can be continued for all future time. Finally, in one of our sections fixed point methods are used to identify harmless perturbations, while Liapunov functionals are then used to obtain new stability results.

In this book we offer a large collection of examples worked in great detail which clearly establish fixed point theory as a viable tool in stability theory. We also present stability results of types not seen before in the literature which emerge because of the fixed point theory. There are other avenues for investigation and we believe that this can be a major area of research for many years. Moreover, there is such a wide class of problems that this book should serve as a handbook for investigation into the stability properties of specific problems confronting biologists, chemists, economists, engineers, mathematicians, physicists, and other scientists.

0.2 An Introductory Lecture

During the past year we have had the opportunity to lecture in several countries on fixed point methods in stability. The following material has emerged as what we believe is an informative introduction to the need for the method, the basic techniques, and some convincing examples.

Functional differential equations are often used to model real-world problems in which there are great uncertainties. Yet, our conditions frequently demand excessively precise and detailed conditions which are totally unverifiable. A concrete example of this begins with Volterra (1928) who sought to model a certain biological problem by means of the equation

$$x'(t) = -\int_{t-L}^{t} a(t-s)g(x(s))\,ds \tag{0.2.1}$$

in which $xg(x) > 0$ if $x \neq 0$, $L > 0$, and $a(t)$ is twice differentiable with $a''(t)$ not identically zero. He was interested in conditions to ensure that solutions tend to zero as $t \to \infty$. At the time he suggested an idea for constructing a Liapunov functional which might yield some stability properties. While that construction was a long time in coming, interest in the formulation of the problem continues to this day. Ergen (1954) and others used the equation to model a circulating fuel nuclear reactor. It is used as a model of one-dimensional viscoelasticity. And it has been used to model neural networks as well as many other problems. Levin (1963) followed Volterra's suggestion and successfully constructed a Liapunov functional for a companion equation; thus, we call (0.2.1) the Volterra-Levin equation. The next year, Levin and Nohel (1964) extended that Liapunov functional to the case of (0.2.1), itself. Here is their result.

Let $L > 0$, $xg(x) > 0$ when $x \neq 0$, $a : [0, L] \to R$ and consider the IVP

$$x'(t) = -\int_{t-L}^{t} a(t-s)g(x(s))\,ds, \ \psi : [-L, 0] \to R. \tag{0.2.2}$$

We need an initial function, ψ, because we must define $x'(0)$:

$$x'(0) = -\int_{-L}^{0} a(-s)g(\psi(s))\,ds.$$

The resulting solution is $x(t,0,\psi)$ where $x(t,0,\psi)=\psi(t)$ for $-L\le t\le 0$.

Theorem Levin-Nohel. *Let*

$$a(L)=0,\ a'(t)\le 0,\ a''(t)\ge 0,\ a''(t) \text{ not } \equiv 0,$$

$$\int_0^x g(s)ds\to\infty \text{ as } x\to\pm\infty.$$

For a given ψ and for $x(t)=x(t,0,\psi)$ the functional

$$V(t)=\int_0^x g(u)du-(1/2)\int_{t-L}^t a'(t-v)\left[\int_v^t g(x(s))ds\right]^2 dv\ge 0$$

satisfies

$$2V'(t)=a'(L)[\int_{t-L}^t g(x(s))\,ds]^2-\int_{t-L}^t a''(t-s)[\int_s^t g(x(v)\,dv]^2\,ds$$

along the solution of (0.2.2) and $x^{(j)}(t)\to 0$ as $t\to\infty$ for $j=0,1,2$.

It can readily be argued that this is one of the most beautiful results in all of stability theory by Liapunov's direct method. But when we look at the precise and detailed conditions on $a(t)$ we are compelled to complain that never can these conditions be well verified in a real-world problem. Other solutions to these kinds of problems do exist.

The following set of steps (S_1, S_2, S_3) represents the way in which we establish stability of the zero solution of the functional differential equation $x'=f(t,x_t)$ by fixed point methods in this book.

S_1. An examination of the differential equation reveals that for a given t_0 there is an initial interval E_{t_0} and we require an initial function $\psi: E_{t_0}\to R^n$. We then must determine a set $\mathcal{M}$ of functions $\phi: E_{t_0}\cup[t_0,\infty)\to R^n$ with $\phi(t)=\psi(t)$ on E_{t_0} which would serve as acceptable solutions. Usually, this means that we would ask ϕ bounded and, sometimes, $\phi(t)\to 0$ as $t\to\infty$.

S_2. Next, shrewdly invert $x'=f(t,x_t)$ obtaining an integral equation

$$x(t)=a(t)+\int_{t_0}^t G(t,s,x(\cdot),\psi)ds$$

so that for the mapping P defined by $\phi\in M$ implies

$$(P\phi)(t)=a(t)+\int_{t_0}^t G(t,s,\phi(\cdot),\psi)ds$$

then $P:\mathcal{M}\to\mathcal{M}$ and a fixed point solves the initial value problem, $x'=f(t,x_t)$ and $x(t)=\psi(t)$ on E_{t_0}. The critical point here is that the inversion must be so shrewdly accomplished that the implied mapping does map bounded sets into bounded sets. Virtually never will a simple integration of the differential equation accomplish this.

S_3. Finally, select a fixed point theorem which will show that P has a fixed point in $\mathcal{M}$. Since the fixed point is in $\mathcal{M}$, it is an acceptable solution.

We will now give a sequence of examples and techniques leading us to a solution of the Volterra-Levin equation. Our first example is a half-linear equation in which we can directly use the variation of parameters formula to invert the differential equation and obtain the mapping. The contraction mapping principle is given in the next chapter, as well as considerably more detail than we offer here in this brief introduction.

Example 0.2.1. Let a, b, r be continuous, $r(t) \geq 0$,

$$x' = -a(t)x(t) + b(t)x(t - r(t)), \tag{0.2.3}$$

where

$$\int_0^t e^{-\int_s^t a(u)du}|b(s)|\, ds \leq \alpha < 1. \tag{0.2.4}$$

Theorem 0.2.1. *Let $-\int_0^t a(s)ds$ be bounded above and let (0.2.4) hold. If $\psi : (-\infty, 0] \to R$ is bounded and continuous, then the solution $x(t, 0, \psi)$ of (0.2.3) exists, is unique, and is bounded for $t \geq 0$.*

Notice that on average $a(t)$ dominates $b(t)$. There are no bounds on a, b, r, or r'. The function a can have positive and negative values. Also, $b(t)$ can be large when $a(t)$ is negative. All of these properties would cause substantial difficulties with Liapunov's direct method.

Proof. Identify $b(t)x(t - r(t))$ as an inhomogeneous term and apply the variation of parameters formula to obtain

$$x(t) = e^{-\int_0^t a(s)\, ds}\psi(0) + \int_0^t e^{-\int_s^t a(u)du} b(s)x(s - r(s))\, ds.$$

This is the integral equation which will define the mapping and it will be easy to see that it is a contraction. We will use the supremum metric and define the complete metric space $(\mathcal{M}, \|\cdot\|)$ by

$$\mathcal{M} = \{\phi : (-\infty, \infty) \to R \,|\, \phi \in \mathcal{C}, \phi(t) = \psi(t) \text{ for } t \in (-\infty, 0],\ \phi \text{ bounded}\}.$$

For $\phi \in \mathcal{M}$ define $(P\phi)(t) = \psi(t)$ if $t \leq 0$ and for $t \geq 0$ then

$$(P\phi)(t) = e^{-\int_0^t a(s)\,ds}\psi(0) + \int_0^t e^{-\int_s^t a(u)du} b(s)\phi(s - r(s))\,ds.$$

Then $P\phi$ is bounded and continuous. If $\phi, \eta \in \mathcal{M}$ then

$$\begin{aligned}
&|(P\phi)(t) - (P\eta)(t)| \leq \\
&\int_0^t e^{-\int_s^t a(u)du}|b(s)||\phi(s - r(s)) - \eta(s - r(s))|\,ds \\
&\leq \alpha\|\phi - \eta\|.
\end{aligned}$$

Since the mapping is a contraction there is a unique fixed point residing in $\mathcal{M}$. It is bounded, continuous, satisfies the initial condition, and satisfies the differential equation.

Notice that the process is simple, fast, rigorous, and only requires knowledge of the elementary variation of parameters formula. In one step we obtain existence, uniqueness, and boundedness of solutions.

We can readily add small conditions implying solutions tend to zero. This is done in a later section. Next, we come to a linear equation which offers challenges.

Example 0.2.2. Let $a : [0, \infty) \to R$ be bounded and continuous, let r be a positive constant, and let

$$x' = -a(t)x(t - r). \tag{0.2.5}$$

Given a continuous $\psi : [-r, 0] \to R$, we can write the equation as

$$x' = -a(t + r)x(t) + (d/dt)\int_{t-r}^t a(s + r)x(s)\,ds$$

so that by the variation of parameters formula, followed by integration by parts, we have

$$\begin{aligned}
x(t) &= x(0)e^{-\int_0^t a(s+r)\,ds} + \int_{t-r}^t a(u + r)x(u)\,du \\
&\quad - e^{-\int_0^t a(u+r)\,du}\int_{-r}^0 a(u + r)\psi(u)\,du \\
&\quad - \int_0^t a(s + r)e^{-\int_s^t a(u+r)\,du}\int_{s-r}^s a(u + r)x(u)\,du\,ds.
\end{aligned}$$

With $\mathcal{M}$ as before we will define a contraction mapping from this formula for $x(t)$ if there is a constant $\alpha < 1$ with

$$\int_{t-r}^{t} |a(u+r)|\,du + \int_{0}^{t} |a(s+r)|e^{-\int_s^t a(u+r)du} \int_{s-r}^{s} |a(u+r)|\,du\,ds \le \alpha. \tag{0.2.6}$$

Notice again that (0.2.6) represents averages; there is no bound, either above or below, on $a(t)$. But if $a(t) \ge 0$, then (0.2.6) can be replaced by

$$2\int_{t-r}^{t} a(u+r)\,du \le \alpha < 1.$$

As we are interested in asymptotic stability we will need

$$\int_{0}^{t} a(s+r)\,ds \to \infty \text{ as } t \to \infty.$$

Theorem 0.2.2. *Let $\int_0^t a(s+r)ds \to \infty$ as $t \to \infty$ and let (0.2.6) hold. Then for every continuous initial function $\psi : [-r, 0] \to R$ the solution $x(t, 0, \psi)$ is bounded and tends to zero as $t \to \infty$.*

Proof. Let $(\mathcal{B}, \|\cdot\|)$ be the Banach space of bounded and continuous functions $\phi : [-r, \infty) \to R$ with the supremum norm. Let $(\mathcal{S}, \|\cdot\|)$ be the complete metric space with supremum metric consisting of functions $\phi \in \mathcal{B}$ such that $\phi(t) = \psi(t)$ on $[-r, 0]$ and $\phi(t) \to 0$ as $t \to \infty$.

Define $P : \mathcal{S} \to \mathcal{S}$ by $(P\phi)(t) = \psi(t)$ on $[-r, 0]$ and for $t \ge 0$ then

$$\begin{aligned}(P\phi)(t) &= \psi(0)e^{-\int_0^t a(s+r)\,ds} + \int_{t-r}^{t} a(u+r)\phi(u)\,du \\ &\quad - e^{-\int_0^t a(u+r)\,du} \int_{-r}^{0} a(u+r)\psi(u)\,du \\ &\quad - \int_{0}^{t} a(s+r)e^{-\int_s^t a(u+r)\,du} \int_{s-r}^{s} a(u+r)\phi(u)\,du\,ds.\end{aligned}$$

Now $P\phi$ is continuous, and from (0.2.6) it follows that $P\phi$ is bounded. P is a contraction by (0.2.6).

We can show that the last term of $P\phi$ tends to zero by using the classical proof that the convolution of an L^1–function with a function tending to zero, does also tend to zero. There is a fixed point in $\mathcal{M}$, a solution of the equation.

We return now to the Volterra-Levin problem in which we ask more of g, but conditions on $a(t)$ which are far more verifiable in real-world problems.

Example 0.2.3. Suppose that there is a function $g : R \to R$ satisfying:

$$|g(x) - g(y)| \leq K|x - y| \tag{0.2.7}$$

for some $K > 0$ and all $x, y \in R$,

$$\frac{g(x)}{x} \geq 0 \quad \text{and} \quad \lim_{x \to 0} \frac{g(x)}{x} \quad \text{exits,} \tag{0.2.8}$$

and sometimes

$$\frac{g(x)}{x} \geq \beta \tag{0.2.9}$$

for some $\beta > 0$.

Consider the scalar equation

$$x' = -\int_{t-L}^{t} p(s-t)g(x(s))\, ds \tag{0.2.10}$$

with $L > 0$, p continuous,

$$\int_{-L}^{0} p(s)\, ds = 1, \tag{0.2.11}$$

(This is notation; we only need $\int_{-L}^{0} p(s)\, ds > 0$.) and for the K of (0.2.7) let

$$2K \int_{-L}^{0} |p(v)v|\, dv =: \alpha < 1. \tag{0.2.12}$$

Theorem 0.2.3. *If (0.2.7), (0.2.8), (0.2.11), and (0.2.12) hold, then every solution of (0.2.10) is bounded. If (0.2.9) also holds, then every solution and its derivative tends to zero.*

Proof. Let $\psi : [-L, 0] \to R$ be a given continuous initial function and let $x_1(t) := x(t, 0, \psi)$ be the unique resulting solution. By the growth condition on g, $x_1(t)$ exists on $[0,\infty)$. If we add and subtract $g(x)$ we can write the equation as

$$x' = -g(x) + \frac{d}{dt}\int_{-L}^{0} p(s)\int_{t+s}^{t} g(x(u))\,du\,ds.$$

Define a continuous non-negative function $a : [0,\infty) \to [0,\infty)$ by

$$a(t) := \frac{g(x_1(t))}{x_1(t)}.$$

Since a is the quotient of continuous functions it is continuous when assigned the limit at $x_1(t) = 0$, if such a point exists.

Thus, for the fixed solution, our equation is

$$x' = -a(t)x + \frac{d}{dt}\int_{-L}^{0} p(s)\int_{t+s}^{t} g(x(u))\,du\,ds$$

which, by the variation of parameters formula, followed by integration by parts, can then be written as

$$\begin{aligned}
x(t) &= \psi(0)e^{-\int_0^t a(s)\,ds} \\
&\quad + \int_0^t e^{-\int_v^t a(s)ds}\frac{d}{dv}\int_{-L}^{0} p(s)\int_{v+s}^{v} g(x(u))\,du\,ds\,dv \\
&= \psi(0)e^{-\int_0^t a(s)\,ds} + e^{-\int_v^t a(s)\,ds}\int_{-L}^{0} p(s)\int_{v+s}^{v} g(x(u))\,du\,ds\Big|_0^t \\
&\quad - \int_0^t a(v)e^{-\int_v^t a(s)\,ds}\int_{-L}^{0} p(s)\int_{v+s}^{v} g(x(u))\,du\,ds dv \\
&= \psi(0)e^{-\int_0^t a(s)\,ds} + \int_{-L}^{0} p(s)\int_{t+s}^{t} g(x(u))\,du\,ds \\
&\quad - e^{-\int_0^t a(s)\,ds}\int_{-L}^{0} p(s)\int_{s}^{0} g(\psi(u))\,du\,ds \\
&\quad - \int_0^t e^{-\int_v^t a(s)ds}a(v)\int_{-L}^{0} p(s)\int_{v+s}^{v} g(x(u))\,du\,ds\,dv.
\end{aligned}$$

With the notation $\phi_0 = \psi$ meaning that $\phi(t) = \psi(t)$ for $-L \le t \le 0$, define

$$\mathcal{M} = \{\phi : [-L,\infty) \to R \,|\, \phi_0 = \psi,\ \phi \in \mathcal{C},\ \phi \text{ bounded}\}$$

and define $P : \mathcal{M} \to \mathcal{M}$ using the above equation in $x(t)$. For $\phi \in \mathcal{M}$ define $(P\phi)(t) = \psi(t)$ if $-L \le t \le 0$. If $t \ge 0$, then define

$$\begin{aligned}(P\phi)(t) &= \psi(0)e^{-\int_0^t a(s)\,ds} + \int_{-L}^{0} p(s) \int_{t+s}^{t} g(\phi(u))\,du\,ds \\ &\quad - e^{-\int_0^t a(s)ds} \int_{-L}^{0} p(s) \int_{s}^{0} g(\psi(u))\,du\,ds \\ &\quad - \int_0^t e^{-\int_v^t a(s)\,ds} a(v) \int_{-L}^{0} p(s) \int_{v+s}^{v} g(\phi(u))\,du\,ds\,dv.\end{aligned}$$

To see that P is a contraction, if $\phi, \eta \in \mathcal{M}$ then

$$\begin{aligned}&|(P\phi)(t) - (P\eta)(t)| \\ &\le \int_{-L}^{0} |p(s)| \int_{t+s}^{t} |g(\phi(u)) - g(\eta(u))|\,du\,ds \\ &+ \int_0^t e^{-\int_v^t a(s)ds} a(v) \int_{-L}^{0} |p(s)| \int_{v+s}^{v} |g(\phi(u)) - g(\eta(u))|\,du\,ds\,dv \\ &\le 2K\|\phi - \eta\| \int_{-L}^{0} |p(s)s|\,ds \\ &\le \alpha\|\phi - \eta\|.\end{aligned}$$

There is a unique fixed point, a bounded solution.

If $\frac{g(x)}{x} \ge \beta > 0$, then add to $\mathcal{M}$ the condition that $\phi(t) \to 0$ as $t \to \infty$. We can show $(P\phi)(t) \to 0$ when $\phi(t) \to 0$ so the fixed point tends to 0. From (0.2.10) we see that $x'(t)$ also tends to zero. With assumptions on boundedness of the derivative of p we could claim that $x''(t)$ also tends to zero, as did Levin and Nohel.

Fixed point theory can answer some of our most fundamental questions about the behavior of solutions of differential equations. Our first encounters with differential equations show us that all solutions of the constant coefficient equation

$$x'' + ax' + bx = 0$$

tend to zero as $t \to \infty$ whenever a and b are positive constants; in fact, they do so exponentially, along with their derivatives. But we can prove using ideal theory that we can never solve the simple equation

$$x'' + tx = 0$$

with anything like elementary functions. Thus, we know that variable coefficients present real difficulties. We look at

$$x'' + h(t)x' + k^2 x = 0$$

with $k^2 > 0$ and $h(t) \geq h_0$, where h_0 is a positive constant and we conjecture that solutions will tend to zero. R. A. Smith (1961) proved a beautiful result showing that the zero solution of that equation is asymptotically stable if and only if

$$\int_0^\infty e^{-H(t)} \int_0^t e^{H(s)}\, ds\, dt = \infty$$

where $H(t) := \int_0^t h(s)ds$. It is not hard to show that the condition holds for $h(t) = h_0 + t$, but fails for $h(t) = h_0 + t^2$. These functions are crucial benchmarks in our thinking about differential equations. They tell us when our intuition works and when it fails. Volumes have been written about the behavior of solutions of

$$x'' + f(t,x)x' + g(x) = 0$$

but we have very little information about solutions of

$$x'' + f(t,x,x')x' + b(t)g(x(t-L)) = 0$$

in which $b : [0,\infty) \to [0,\infty)$ is continuous and bounded, $g : R \to R$ and $f : R \times R \times R \to [0,\infty)$ is continuous. We would be so interested in knowing what limits on these functions will preserve the property that all solutions tend to zero. In Section 2.4 we study that problem by means of transformations much like those in the proof of Theorem 0.2.3 and with contraction mappings. We show that for appropriate bounds on b and L, the benchmark of R. A. Smith holds in that f can increase as fast as t and still solutions will tend to zero.

Not only should we be able to attack classical problems and obtain better solutions, but new methods should present entirely new kinds of solutions. Fixed point theory does exactly that. Results of the following type are given in Section 3.1. We can use methods similar to those in the proof of Theorem 0.2.3 to obtain stability results for

$$x'(t) = -x^{2n+1}(t-r)$$

for r a positive constant. We have no idea how to prove a stability result for

$$x'(t) = -(1 - 2\sin t)x^{2n+1}(t-r).$$

Is it possible that we could use the higher order term in

$$x'(t) = -(1 - 2\sin t)x^{2n+1}(t - r) - 2\sin t x^{2n+3}(t - r)$$

to stabilize the equation? In fact, we can borrow the coefficient of the higher order term, add it to the coefficient of the lower order term, and use the techniques in the proof of Theorem 0.2.3 to show that the equation is stable. We do not believe that similar results have ever been conjectured or proved in any other way.

Chapter 1

Half-linear Equations

1.1 Statement of the problem

This section is an elementary introduction to the formulations of fixed point problems in differential equations. These formulations serve two purposes. First, they give a brief introduction to the kinds of problems we will be considering. But, more importantly, they offer motivation for the properties and results which we present in the next section; those properties are fundamental in the study of stability by fixed point theory.

Many different kinds of problems can be solved by means of fixed point theory. Generally, to solve a problem with fixed point theory is to find:

(a) a set $\mathcal{S}$ consisting of points which would be acceptable solutions;

(b) a mapping $P : \mathcal{S} \to \mathcal{S}$ with the property that a fixed point solves the problem;

(c) a fixed point theorem stating that this mapping on this set will have a fixed point.

We will be primarily interested in functional differential equations, but we begin with an ordinary differential equation

$$x'(t) = g(t, x(t)) \tag{1.1.1}$$

where $g : [0, \infty) \times R^n \to R^n$ is continuous. To start us on our way we will discuss problems which are central to the study and motivate the contents of future sections. Several concepts may be used here which will be more fully defined and discussed later.

Example 1.1.1 An existence theorem. Perhaps the most basic problem concerning (1.1.1) is to find a solution through a given point $(t_0, x_0) \in [0, \infty) \times R^n$ defined on some interval $[t_0, t_0 + \gamma]$ and satisfying (1.1.1) on that interval.

For this problem, our first guess would be that the set $\mathcal{S}$ should consist of differentiable functions $\phi : [t_0, t_0 + \gamma] \to R^n$ with $\phi(t_0) = x_0$. Next, the simplest way to find a mapping is to formally integrate (1.1.1) and obtain

$$x(t) = x_0 + \int_{t_0}^{t} g(s, x(s))ds$$

so that the mapping P on $\mathcal{S}$ is defined by

$$(P\phi)(t) = x_0 + \int_{t_0}^{t} g(s, \phi(s))ds.$$

A fixed point will certainly satisfy the equation. Since our mapping is given by an integral, our second approximation to $\mathcal{S}$ is the set of continuous functions; differentiability will be automatic. There is now a vast array of fixed point theorems which will yield a fixed point of that mapping and satisfy our initial value problem. This is typical of fixed point theory. The big step is in finding a suitable mapping; once that is done investigators may work for decades using that mapping to get progressively better results. We will begin with one of the simplest fixed point theorems.

Theorem 1.1.1 Contraction Mapping Principle. *Let $(\mathcal{S}, \rho)$ be a complete metric space and let $P : \mathcal{S} \to \mathcal{S}$. If there is a constant $\alpha < 1$ such that for each pair $\phi_1, \phi_2 \in \mathcal{S}$ we have*

$$\rho(P\phi_1, P\phi_2) \leq \alpha\rho(\phi_1, \phi_2),$$

then there is one and only one point $\phi \in \mathcal{S}$ with $P\phi = \phi$.

We frequently use the supremum metric denoted by $\|\cdot\|$.

For our illustration here, it is easiest to complete the solution by asking that g satisfy a global Lipschitz condition of the form

$$|g(t, x) - g(t, y)| \leq K|x - y|$$

for $t \geq t_0$ and for all $x, y \in R^n$, where $|\cdot|$ is any norm on R^n. This will allow us to give a contraction mapping argument. For any fixed interval $[t_0, t_0 + \gamma]$, our set $\mathcal{S}$ with the supremum metric is a complete metric space and $P : \mathcal{S} \to \mathcal{S}$. Checking our contraction requirement, we have

$$|(P\phi_1)(t) - (P\phi_2)(t))| \leq \int_{t_0}^{t} K|\phi_1(s) - \phi_2(s)|ds \leq K\gamma\|\phi_1 - \phi_2\|$$

so that if $\alpha := \gamma K < 1$ then P is a contraction with unique fixed point ϕ, a solution of our differential equation and it satisfies the initial condition.

So much more can be done. By adopting a different metric, we can increase the length of the interval of existence. By using Schaefer's fixed point theorem we can get existence on $[t_0, \infty)$. By using Schauder's theorem we can drop the Lipschitz condition and obtain existence without uniqueness. But everything begins with a suitable mapping; that is the central problem and it is the one on which this entire book centers. It can be relatively easy to state and prove theorems, once we have a proper mapping, but the real problem is in constructing the mapping.

Seldom will we see a problem in which it is so easy to find a suitable mapping as the one we just finished. We are concerned with a treatment of stability by means of fixed point theory. The main classical method of studying stability is Liapunov's direct method. In that study we must always begin by finding a Liapunov function which is a type of generalized distance function measuring the distance from a solution to the origin. It is a major problem to find an appropriate Liapunov function. But once a suitable Liapunov function is found, investigators may continue for decades deriving more and more information from that Liapunov function; we will later discuss the Liénard equation and Levin's equation as two such examples. In precisely the same way, our major problem is to find a suitable mapping. Once that mapping is found, investigators may find the mapping an endless source of results. This study began with the idea that difficulties encountered in Liapunov theory might be circumvented by fixed point theory. Thus, we will frequently refer back to classical difficulties with Liapunov theory.

Example 1.1.2 A bounded solution. Stability concerns a special kind of boundedness of solutions. Thus, we next ask how we can show that solutions of (1.1.1) are bounded. We would pick the set S as follows. For a given $(t_0, x_0) \in [0, \infty) \times R^n$, let S be the set of differentiable functions $\phi : [t_0, \infty) \to R^n$ which are bounded and satisfy $\phi(t_0) = x_0$. Next, we must select an appropriate mapping. If we try to write

$$(P\phi)(t) = x_0 + \int_{t_0}^{t} g(s, \phi(s))\, ds$$

then we immediately have difficulty. If ϕ is bounded, we have no simple way of establishing that $P\phi$ is bounded. That mapping will map a given function ϕ right out of the set and there is no hope of proving that there is a fixed point. And that gives rise to the name of this chapter.

Suppose that our problem is "half linear" in the sense that it can be written as

$$x' = Ax + f(t, x)$$

where A is an $n \times n$ real constant matrix, all of whose characteristic roots have negative real parts, while $f : [0, \infty) \times R^n \to R^n$ is continuous and $f(t, x)$ is bounded for x bounded. In truth, if g is bounded for x bounded then this is no assumption at all since we can add and subtract any convenient Ax. While that is an ancient technique it has its pitfalls and a later example will focus on just how we may avoid them. Formally, write $p(t) = f(t, x(t))$ so that our equation is "linear:"

$$x' = Ax + p(t).$$

For a given (t_0, x_0) use the variation of parameters formula to write

$$x(t) = e^{A(t-t_0)}x_0 + \int_{t_0}^{t} e^{A(t-s)}p(s)\, ds.$$

Recall that our goal is to show that for each (t_0, x_0) a solution $x(t) = x(t, t_0, x_0)$ is bounded. Let $(\mathcal{S}, \|\cdot\|)$ be the complete metric space of bounded continuous $\phi : [t_0, \infty) \to R^n$ with the supremum metric and satisfying $\phi(t_0) = x_0$. By the assumption on A we can find $L > 0$ and $\beta > 0$ with

$$|e^{At}| \leq Le^{-\beta t}$$

for $t \geq 0$. As $f(t, x)$ is bounded for bounded x it is now clear that $P : \mathcal{S} \to \mathcal{S}$. If f satisfies a global Lipschitz condition with constant K then

$$|(P\phi_1)(t) - (P\phi_2)(t)| \leq \int_{t_0}^{t} Le^{-\beta(t-s)}K|\phi_1(s) - \phi_2(s)|\, ds,$$

a contraction if $LK/\beta < 1$, with unique fixed point ϕ, a bounded solution of our problem.

It may be far too much to ask for a global Lipschitz condition and it may be out of the question to ask that $LK/\beta < 1$. There are interesting alternatives. But we can not restrict the t-interval as we did before. If we can construct a closed bounded set $\mathcal{S}$ such that $P : \mathcal{S} \to \mathcal{S}$, then the contraction condition need only hold in $\mathcal{S}$. But if $LK/\beta < 1$ fails to hold in S, there is still a simple way out. The fact that the mapping is given by an integral means that it is virtually always possible to define a weighted metric so that a Lipschitz map becomes a contraction. Moreover, if the Lipschitz condition itself fails in $\mathcal{S}$ then this weighted metric may permit us to use Schauder's theorem (Theorem 1.2.5) to establish the bounds.

Example 1.1.3 An acquired linear term. As noted above, asking that $x' = g(t,x)$ be written as

$$x' = Ax + f(t,x)$$

is no assumption at all since we can write

$$x' = Ax + [g(t,x) - Ax].$$

In the last example we wanted $f(t,x)$ to satisfy (at least a local) contraction condition. Suppose our equation is scalar and can be written as

$$x' = -x^3 + h(t,x)$$

where $h(t,x)$ satisfies a smallness condition. Note that x^3 will satisfy a local (near $x = 0$) contraction condition with constant as small as we please. If we write

$$x' = -x + (x - x^3) + h(t,x)$$

and then use the variation of parameters formula to obtain

$$x(t) = x_0 e^{-t} + \int_0^t e^{-(t-s)}[x(s) - x^3(s) + h(s,x(s))]\,ds,$$

we see that the integrand loses its contraction property as $x \to 0$. Indeed, for $x^2 + y^2 \leq 1/2$ we have

$$|x - x^3 - y + y^3| \leq |x - y|[1 - (1/2)(x^2 + y^2)]$$

and the contraction constant tends to one as $x^2 + y^2 \to 0$. The contraction mapping theorem fails. But in the next section we offer an improved contraction mapping theorem which saves the argument. A combination contraction-Schauder theorem then yields the desired boundedness. This is the motivation for Definition 1.2.6, Theorem 1.2.4, Theorem 1.2.7, Theorem 1.2.8, and Example 1.2.7. Later, we will get around the difficulty using a concept called exact linearization.

1.2 Background and definitions

This section contains an elementary set of definitions, theorems, and examples which were motivated by the examples in the last section and were formulated to aid us in deciding which fixed point theorem to use and which stability properties to prove.

Every stability problem we discuss is formulated in a complete metric space. Let it be noted at the beginning that in our work in this book there is usually a Banach space $(\mathcal{B}, \|\cdot\|)$ in the background. A subset $\mathcal{S}$ of $\mathcal{B}$ is selected and $(\mathcal{S}, \|\cdot\|)$ is the complete metric space in which we work, where the metric on $\mathcal{S}$ is defined by the norm inherited from the Banach space. Thus, the notation almost always suggests a norm $\|\cdot\|$ or $|\cdot|_h$ instead of a metric ρ. Even if the zero function, say θ, is not in $\mathcal{S}$, then for $\phi \in \mathcal{S}$, $\|\phi\|$ is interpreted as $\rho(\phi, \theta) = \|\phi - \theta\|$.

Definition 1.2.1. *A pair* $(\mathcal{S}, \rho)$ *is a metric space if* $\mathcal{S}$ *is a set and* $\rho : \mathcal{S} \times \mathcal{S} \to [0, \infty)$ *such that when* y, z, *and* u *are in* $\mathcal{S}$ *then*

(a) $\rho(y, z) \geq 0$, $\rho(y, y) = 0$, *and* $\rho(y, z) = 0$ *implies* $y = z$,
(b) $\rho(y, z) = \rho(z, y)$, *and*
(c) $\rho(y, z) \leq \rho(y, u) + \rho(u, z)$.

The metric space is *complete* if every Cauchy sequence in $(\mathcal{S}, \rho)$ has a limit in that space. A sequence $\{x_n\} \subset \mathcal{S}$ is a *Cauchy sequence* if for each $\varepsilon > 0$ there exists N such that n, $m > N$ imply $\rho(x_n, x_m) < \varepsilon$.

There are four spaces which occupy much of our time.

Example 1.2.1. Let $\phi : [a, b] \to R^n$ be continuous and let $\mathcal{S}$ be the set of bounded continuous functions $f : [a, \infty) \to R^n$ with $f(t) = \phi(t)$ for $a \leq t \leq b$. For $f, g \in \mathcal{S}$ define $\rho(f, g) = \sup_{a \leq t < \infty} |f(t) - g(t)| =: \|f - g\|$. Then $(\mathcal{S}, \|\cdot\|)$ is a complete metric space.

Let

$$\mathcal{M} = \{\phi : [0, \infty) \to R \mid \phi \in C, |\phi(t)| \leq 1, \phi(t) \to 0 \quad \text{as} \quad t \to \infty\}$$

and

$$\mathcal{Q} = \{\phi : [0, \infty) \to R \mid \phi \in C, |\phi(t)| \leq 1\}$$

Let $\|\cdot\|$ be the supremum metric and let $|\cdot|_h$ be a weighted metric defined by $h : [0, \infty) \to [1, \infty), h(0) = 1, h(t) \to \infty$ monotonically, and for $\phi \in \mathcal{M}$ or $\mathcal{Q}$ then

$$|\phi|_h = \sup_{t \geq 0} |\phi(t)|/h(t).$$

Example 1.2.2. $(\mathcal{M}, \|\cdot\|)$ is a complete metric space.

This example will cause us to choose contraction mappings in many problems. It is to be contrasted with Example 1.2.6. Contraction mappings only require a complete metric space, while many other common fixed point theorems require compactness.

Example 1.2.3. $(\mathcal{Q}, |\cdot|_h)$ is a complete metric space.

Proof. Suppose that $\{\phi_n\}$ is a Cauchy sequence in the space. If we restrict $\{\phi_n\}$ to the interval $[0,k]$ then the metric is essentially the supremum metric and the Cauchy sequence has a unique continuous limit on that interval. This is true for $k = 1, 2, 3, 4, ...$so we get a continuous limit on $[0,\infty)$ which is clearly in $\mathcal{Q}$.

Example 1.2.4. $(\mathcal{M}, |\cdot|_h)$ is not a complete metric space.

Proof. Let $\{\phi_n\}$ be the sequence of functions such that
$\phi_n(t) = 1, \quad 0 \le t \le n,$
$\phi_n(t) = 0, \quad n+1 \le t < \infty,$
ϕ_n is linear and continuous on $[n, n+1]$.

Claim. *$\{\phi_n\}$ is a Cauchy sequence in this space.*

Proof. Let $\varepsilon > 0$ be given. We must find N such that $n, m \ge N, t \in R$ imply that

$$|\phi_n(t) - \phi_m(t)| < \varepsilon h(t). \qquad (*)$$

Find T such that $\varepsilon h(T) > 2$. Clearly, for all n, m and for all $t \ge T$ we have (*). Find $N > T$. For $n, m \ge N$ and $0 \le t \le N$ we have $\phi_n(t) = \phi_m(t) = 1$ so (*) holds. This proves that it is a Cauchy sequence. But for any fixed t, $\phi_n(t) \to 1$ as $n \to \infty$, and 1 is not in $\mathcal{M}$. This completes the proof.

This will turn out to be a counterexample to several proposed theorems which we will encounter.

If the functions in a differential equation of interest all satisfy a local Lipschitz condition, then contraction mappings will usually be our first choice in our stability investigations. If the functions are not Lipschitz then we will turn to fixed point theorems of general Schauder or Krasnoselskii type. These require compactness instead of completeness and are formulated in complete normed spaces.

Definition 1.2.2. *A set L in a metric space $(\mathcal{S}, \rho)$ is* compact *if each sequence $\{x_n\} \subset L$ has a subsequence with limit in L.*

Definition 1.2.3. *Let U be an interval on R and let $\{f_n\}$ be a sequence of functions with $f_n : U \to R^d$. Denote by $|\cdot|$ any norm on R^d.*

(a) $\{f_n\}$ is uniformly bounded *on U if there exists $M > 0$ such that $|f_n(t)| \le M$ for all n and all $t \in U$.*

(b) $\{f_n\}$ is equicontinuous *if for any $\varepsilon > 0$ there exists $\delta > 0$ such that $t_1, t_2 \in U$ and $|t_1 - t_2| < \delta$ imply $|f_n(t_1) - f_n(t_2)| < \varepsilon$ for all n.*

This definition provides our main method for proving compactness. A proof of the next result can be found in any text on real variables or in Burton (1985; p. 165).

Theorem 1.2.1. Ascoli-Arzela. *If $\{f_n(t)\}$ is a uniformly bounded and equicontinuous sequence of real functions on an interval $[a, b]$, then there is a subsequence which converges uniformly on $[a, b]$ to a continuous function.*

Since our t–intervals are possibly infinite, we need an extension of that result. We can extend it either by using a weighted norm or as given below (cf. Burton and Furumochi (2001b; pp. 73-82)).

Theorem 1.2.2. *Let $R^+ = [0, \infty)$ and let $q : R^+ \to R^+$ be a continuous function such that $q(t) \to 0$ as $t \to \infty$. If $\{\phi_k(t)\}$ is an equicontinuous sequence of R^d-valued functions on R^+ with $|\phi_k(t)| \leq q(t)$ for $t \in R^+$, then there is a subsequence that converges uniformly on R^+ to a continuous function $\phi(t)$ with $|\phi(t)| \leq q(t)$ for $t \in R^+$, where $|\cdot|$ denotes the Euclidean norm on R^d.*

Proof. It is clear that the set of functions $\{\phi_k(t)\}$ is uniformly bounded on R^+. Thus, considering intervals $[0, n]$, n a positive integer, and using a diagonalization process there is a subsequence, say $\{\phi_k(t)\}$ again, converging uniformly on any compact subset of R^+ to some continuous function $\phi(t)$ with $|\phi(t)| \leq q(t)$ for $t \in R^+$. Because $\phi(t) \to 0$ as $t \to \infty$, it will now be possible to show that $\|\phi_k - \phi\| \to 0$ as $k \to \infty$, where $\|\cdot\|$ denotes the supremum metric on R^+. From the definition of $q(t)$, for any $\varepsilon > 0$ there is a $T > 0$ with

$$q(t) < \varepsilon/2 \text{ if } t \geq T,$$

which yields

$$|\phi_k(t) - \phi(t)| \leq 2q(t) < \varepsilon \text{ if } k \in N \text{ and } t \geq T, \tag{**}$$

where N denotes the set of positive integers. On the other hand, since $\{\phi_k(t)\}$ converges to $\phi(t)$ uniformly on $[0, T]$ as $k \to \infty$, for the ε there is a $\kappa \in N$ with

$$|\phi_k(t) - \phi(t)| < \varepsilon \text{ if } k \geq \kappa \text{ and } 0 \leq t \leq T,$$

which together with (**), implies that $\|\phi_k - \phi\| < \varepsilon$ if $k \geq \kappa$. This shows that $\|\phi_k - \phi\| \to 0$ as $k \to \infty$.

Definition 1.2.4. *A vector space $(\mathcal{V}, +, \cdot)$ is a normed space if for each $x, y \in \mathcal{V}$ there is a nonnegative real number $\|x\|$, called the norm of x, such that*

(i) $\|x\| = 0$ *if and only if* $x = 0$,

(ii) $\|\alpha x\| = |\alpha|\|x\|$ *for each* $\alpha \in R$, *and*

(iii) $\|x + y\| \leq \|x\| + \|y\|$.

A normed space is a vector space and it is a metric space with $\rho(x, y) = \|x - y\|$. But a vector space with a metric is not always a normed space.

Definition 1.2.5. *A* Banach space *is a complete normed space.*

Example 1.2.5. We list three examples.

(a) The space $\mathcal{C}([a, b], R^n)$ consisting of all continuous function $f : [a, b] \to R^n$ is a vector space over the reals.

(b) If $\|f\| = \max_{a \leq t \leq b} |f(t)|$, where $|\cdot|$ is a norm in R^n, then $(\mathcal{C}, \|\cdot\|)$ is a Banach space.

(c) For a given pair of positive constants M and K, the set

$$L = \{f \in \mathcal{C}([a, b], R^n) \mid \|f\| \leq M, |f(u) - f(v)| \leq K|u - v|\}$$

is compact. This follows from Ascoli's theorem.

Example 1.2.6. Let $(\mathcal{S}, \rho)$ denote the space of bounded continuous functions $f : [0, \infty) \to R^n$ with the supremum metric. Then $(\mathcal{S}, \rho)$ is a Banach space. The subset of $\mathcal{S}$,

$$L = \{f \in \mathcal{S} \mid \|f\| \leq 1, |f(u) - f(v)| \leq |u - v|\}$$

is not compact. But if we restrict L further by asking that there is a function $g : [0, \infty) \to [0, \infty)$ such that $g(t) \to 0$ as $t \to \infty$ and if $f \in L$ implies $|f(t)| \leq g(t)$, we can then use Theorem 1.2.2 to show that it is compact.

Further details, proofs, and more examples can be found in Burton [1985; pp. 164-170].

We will now state some useful fixed point theorems. The contraction mapping principle was given as Theorem 1.1.1, but it belongs here at the head of the group. It generally goes under the name of the Banach-Caccioppoli Theorem, or Banach's (1932) Contraction Mapping Principle. A proof can be found in many places such as Smart (1980; p. 2) or Burton (1985; p. 172). It gains more respect every day. The real power of the result lies in its application with cleverly chosen metrics. An interesting example may be found in a series of three recent papers: Lou (1999), De Pascale and De Pascale (2002), and Avramescu and Vladimirescu (2004). In those papers, the authors consider a type of mapping which does not appear to be a contraction. Other particular interesting forms and applications are found in Rus (2001) and in Rus, Petrusel and Petrusel (2002).

Theorem 1.2.3 The Contraction Mapping Principle. *Let $(\mathcal{S},\rho)$ be a complete metric space and let $P:\mathcal{S}\to\mathcal{S}$. If there is a constant $\alpha<1$ such that for each pair $\phi_1,\phi_2\in\mathcal{S}$ we have*

$$\rho(P\phi_1,P\phi_2)\le\alpha\rho(\phi_1,\phi_2),$$

then there is one and only one point $\phi\in\mathcal{S}$ with $P\phi=\phi$.

In the last section we considered an equation $x'=g(t,x)$ and introduced a linear term, $x'=Ax+[g(t,x)-Ax]$, where we needed $g(t,x)-Ax$ to be a contraction. As a trivial example, if we have the scalar equation $x'=-x^3=-x+(x-x^3)$, then the term $x-x^3$ is almost a local contraction. It fails near $x=0$. This is typical of difficulties encountered and we need a contraction theorem to take this into account. Such a result is found in Burton (1996).

Definition 1.2.6. *Let $(\mathcal{M},\rho)$ be a metric space and $B:\mathcal{M}\to\mathcal{M}$. B is said to be a* large contraction *if for each pair ϕ, $\psi\in\mathcal{M}$ with $\phi\ne\psi$ then $\rho(B\phi,B\psi)<\rho(\phi,\psi)$ and if for each $\varepsilon>0$ there exists $\delta<1$ such that $[\phi,\psi\in\mathcal{M},\ \rho(\phi,\psi)\ge\varepsilon]\Rightarrow\rho(B\phi,B\psi)\le\delta\rho(\phi,\psi)$.*

Theorem 1.2.4. *Let $(\mathcal{M},\rho)$ be a complete metric space and B be a large contraction. Suppose there is an $x\in\mathcal{M}$ and an $L>0$, such that $\rho(x,B^nx)\le L$ for all $n\ge 1$. Then B has a unique fixed point in $\mathcal{M}$.*

Proof. For that $x\in\mathcal{M}$, consider $\{B^nx\}$. If this is a Cauchy sequence, then the standard argument (cf. Smart (1980; pp. 1–3)) shows that the limit is a fixed point. By way of contradiction, if $\{B^nx\}$ is not a Cauchy sequence, then there exists $\varepsilon>0$, there exists $\{N_k\}\uparrow\infty$, $\exists\ n_k>N_k$, there exists $m_k>N_k$, $m_k>n_k$, with $\rho(B^{m_k}x,B^{n_k}x)\ge\varepsilon$. Thus,

$$\begin{aligned}\varepsilon&\le\rho(B^{m_k}x,B^{n_k}x)\le\rho(B^{m_k-1}x,B^{n_k-1}x)\\&\le\cdots\le\rho(x,B^{m_k-n_k}x);\end{aligned}$$

since B is a large contraction, for this $\varepsilon>0$ there is a $\delta<1$ with

$$\begin{aligned}\varepsilon&\le\rho(B^{m_k}x,B^{n_k}x)\le\delta\rho(B^{m_k-1}x,B^{n_k-1}x)\\&\le\cdots\le\delta^{n_k}\rho(x,B^{m_k-n_k}x)\\&\le\delta^{n_k}L,\end{aligned}$$

a contradiction for large n_k since $\delta < 1$ and $\varepsilon > 0$. This completes the proof.

Example 1.2.7. If $\|\cdot\|$ is the supremum metric, if

$$\mathcal{M} = \{\phi : [0, \infty) \to R \mid \phi \in \mathcal{C}, \|\phi\| \leq \sqrt{3}/3\},$$

and if $(H\phi)(t) = \phi(t) - \phi^3(t)$, then H is a large contraction of the set $\mathcal{M}$.

Proof. In the following computation, ϕ, ψ are evaluated at each t. We have $D := \left|H\phi - H\psi\right| = \left|\phi - \phi^3 - \psi + \psi^3\right| = \left|\phi - \psi\right| \left|1 - (\phi^2 + \phi\psi + \psi^2)\right|$. Then for

$$|\phi - \psi|^2 = \phi^2 - 2\phi\psi + \psi^2 \leq 2(\phi^2 + \psi^2)$$

and for $\phi^2 + \psi^2 < 1$ we have

$$\begin{aligned} D &\leq \left|\phi - \psi\right| \left[1 + |\phi\psi| - (\phi^2 + \psi^2)\right] \\ &\leq \left|\phi - \psi\right| \left[1 + \frac{\phi^2 + \psi^2}{2} - (\phi^2 + \psi^2)\right] \\ &= \left|\phi - \psi\right| \left[1 - \frac{\phi^2 + \psi^2}{2}\right]. \end{aligned}$$

What we have shown is that pointwise we have a large contraction. It is easy to see that this implies a large contraction in the supremum metric.

For a given $\varepsilon \in (0, 1)$, let $\phi, \psi \in \mathcal{M}$ with $\|\phi - \psi\| \geq \varepsilon$.

a) Suppose that for some t we have

$$\varepsilon/2 \leq |\phi(t) - \psi(t)|$$

so that

$$(\varepsilon/2)^2 \leq |\phi(t) - \psi(t)|^2 \leq 2(\phi^2(t) + \psi^2(t))$$

or

$$\phi^2(t) + \psi^2(t) \geq \varepsilon^2/8.$$

For all such t we have

$$|(H\phi)(t) - (H\psi)(t)| \leq |\phi(t) - \psi(t)|[1 - \frac{\varepsilon^2}{16}]$$

$$\leq \|\phi - \psi\| \left[1 - \frac{\varepsilon^2}{16} \right].$$

b) Suppose that for some t we have $|\phi(t) - \psi(t)| \leq \varepsilon/2$. Then

$$|(H\phi)(t) - (H\psi)(t)| \leq |\phi(t) - \psi(t)| \leq (1/2)\|\phi - \psi\|.$$

Thus, for all t we have

$$|(H\phi)(t) - (H\psi)(t)| \leq \min[1/2, 1 - \frac{\varepsilon^2}{16}]\|\phi - \psi\|.$$

We leave the contraction mappings for a while and turn to Schauder's theorems which require compactness instead of completeness and which yield a, possibly nonunique, fixed point. But we will soon return to contractions when we introduce several forms of Krasnoselskii's fixed point theorem which combines Banach's and Schauder's theorems. The next two results are found in Schauder (1930) and in Smart (1980).

Theorem 1.2.5 Schauder's first fixed point theorem. *Let $\mathcal{M}$ be a nonempty compact convex subset of a Banach space and let $P : \mathcal{M} \to \mathcal{M}$ be continuous. Then P has a fixed point in $\mathcal{M}$.*

Theorem 1.2.6 Schauder's second fixed point theorem. *Let $\mathcal{M}$ be a nonempty convex subset of a normed space and let $P : \mathcal{M} \to \mathcal{K}$ where $\mathcal{K}$ is a compact subset of $\mathcal{M}$. Then P has a fixed point in $\mathcal{K}$.*

Krasnoselskii (1958) (see Smart (1980; p. 31) studied a paper of Schauder (1932) and obtained the following working hypothesis: *The inversion of a perturbed differential operator yields the sum of a contraction and compact map.* Accordingly, he formulated the following fixed point theorem which is a combination of the contraction mapping principle and Schauder's fixed point theorems.

Theorem 1.2.7 Krasnoselskii. *Let $\mathcal{M}$ be a closed convex non-empty subset of a Banach space $(\mathcal{S}, \|\cdot\|)$. Suppose that A and B map $\mathcal{M}$ into $\mathcal{S}$ such that*

(i) $Ax + By \in \mathcal{M} (\forall x, y \in \mathcal{M})$,

(ii) *A is continuous and $A\mathcal{M}$ is contained in a compact set,*

(iii) *B is a contraction with constant $\alpha < 1$.*

Then there is a $y \in \mathcal{M}$ with $Ay + By = y$.

For our purposes, that result can be improved in two ways. In Burton (1996) we substituted large contraction for contraction. We have already mentioned that adding and subtracting a linear term can allow us to define a fixed point mapping, but may yield a non-contraction term. This theorem allows us to regain the contraction term.

Theorem 1.2.8. *Let $(\mathcal{S}, \|\cdot\|)$ be a Banach space, $\mathcal{M}$ a bounded convex nonempty subset of $\mathcal{S}$. Suppose that A, $B : \mathcal{M} \to \mathcal{M}$ and that*

(i) $x, y \in \mathcal{M} \Rightarrow Ax + By \in \mathcal{M}$,

(ii) *A is continuous and $A\mathcal{M}$ is contained in a compact subset of $\mathcal{M}$,*

(iii) *B is a large contraction.*

Then there exists $y \in \mathcal{M}$ with $Ay + By = y$.

Also, in Burton (1998) we note that the first part of Krasnoselskii's theorem is more restrictive than necessary in applications. It turns out that frequently we can not verify (i); but if we know that $x = Ax + By$ then it can become easy to show that the sum is in $\mathcal{M}$.

Theorem 1.2.9. *Let $\mathcal{M}$ be a closed, convex, and nonempty subset of a Banach space $(\mathcal{S}, \|\cdot\|)$. Suppose that $A : \mathcal{M} \to \mathcal{S}$ and $B : \mathcal{S} \to \mathcal{S}$ such that:*

(i) *B is a contraction with constant $\alpha < 1$,*

(ii) *A is continuous, $A\mathcal{M}$ resides in a compact subset of $\mathcal{S}$,*

(iii) $[x = Bx + Ay, y \in \mathcal{M}] \Rightarrow x \in \mathcal{M}$.

Then there is a $y \in \mathcal{M}$ with $Ay + By = y$.

We can substitute (iii) of Theorem 1.2.9 for (i) of Theorem 1.2.8.

Our next fixed point theorem is especially useful in proving the existence of periodic solutions. We will see it again in Chapter 5, Section 5.1. This formulation is found in Smart (1980; p. 29).

Theorem 1.2.10 Schaefer. *Let $(\mathcal{B}, \|\cdot\|)$ be a normed space, H a continuous mapping of $\mathcal{B}$ into $\mathcal{B}$ which is compact on each bounded subset X of $\mathcal{B}$. Then either*

(i) *the equation $x = \lambda Hx$ has a solution for $\lambda = 1$, or*

(ii) *the set of all such solutions x, for $0 < \lambda < 1$, is unbounded.*

These are the main definitions, theorems, and examples from analysis that will be needed in our stability work in this chapter.

1.3 Stability: Liapunov, Perron, and Krasovskii

Most of the problems we consider will have the form

$$x'(t) = g(t, x(t), x(t - r(t))) + \int_{t-h(t)}^{t} f(t, s, x(s))\, ds \tag{1.3.1}$$

where $r, h : [0, \infty) \to [0, \infty)$, $g : [0, \infty) \times R^n \times R^n \to R^n$, $f : R \times R \times R^n \to R^n$, and all the functions are at least continuous, $g(t, 0, 0) = f(t, s, 0) = 0$ for $t \geq 0$ and all s. Note that $x(t) \equiv 0$ is a solution. For a given $t_0 \geq 0$, let

$$\beta_0 = \min[\inf_{t \geq t_0} t - r(t), \inf_{t \geq t_0} t - h(t)]. \tag{1.3.2}$$

Then define

$$E_{t_0} = [\beta_0, t_0] \tag{1.3.3}$$

which we call the *initial interval*. Sometimes we let $t - h(t) = -\infty$ so that $E_{t_0} = (-\infty, t_0]$.

To specify a solution of (1.3.1) we need $t_0 \geq 0$ and a continuous function $\psi : E_{t_0} \to R^n$. We say that ψ is the *initial function* on the initial interval. Then a solution of (1.3.1) is a continuous function $x : [\beta_0, t_0 + \gamma) \to R^n$, $0 < \gamma \leq \infty$, where $x(t) = \psi(t)$ on E_{t_0} and satisfies (1.3.1) for $t_0 < t < \gamma$. Frequently, more than one solution will exist for a given set of initial conditions, but any such solution is denoted by

$$x(t, t_0, \psi).$$

Example 1.3.1. (a) The scalar equation

$$x'(t) = -x(t) + 2x(t - 5)$$

is a differential equation with constant delay. Here, $r(t) = 5$, $f(t, s, x) = 0$. For a given t_0 we have $\beta_0 = \inf_{t \geq t_0} t - 5 = t_0 - 5$, so the initial function must be defined on $[t_0 - 5, t_0]$;

(b) The scalar equation

$$x'(t) = x^3(t) + x(t - \sin^2 t)$$

is a differential equation with variable bounded delay. The initial interval has varying length.

(c) The vector equation

$$x'(t) = Ax(t) + \int_0^t C(t, s)x(s)ds,$$

where A and C are $n \times n$ matrices, is a Volterra integrodifferential equation with an unbounded delay, $h(t) = t$. If $t_0 = 0$, the initial interval is a point; but if $t_0 = 10$, the initial interval is $[0, 10]$.

(d) The vector equation

$$x'(t) = Ax(t) + \int_{-\infty}^{t} C(t,s)x(s)ds$$

is called a Volterra integrodifferential equation with infinite delay.

(e) The scalar equation

$$x'(t) = x^{1/3}(t)$$

has a delay which is identically zero. For $t_0 = 0$, $E_{t_0} = 0$, and if $\psi(0) = 0$ there is the solution $x(t) \equiv 0$; however, we may separate variables, integrate, and find another solution

$$x(t) = (\frac{2}{3}t)^{3/2}.$$

Notation. In these examples we have been careful to write $x(t)$ and $x(t-r)$. Generally, when a function is written without its argument, that argument will be t. If the argument is other than t, it will always be clearly specified.

We now look at the basic definitions of stability. Whenever we write $\psi \in C$ we mean that ψ is continuous on its interval of definition. Moreover, $\|\psi\|$ will denote the supremum metric of ψ on its interval of definition.

Definition 1.3.1. *The zero solution of (1.3.1) is said to be* stable *if for each $t_0 \geq 0$ and each $\varepsilon > 0$ there is a $\delta > 0$ such that*

$$[\psi : E_{t_0} \to R^n, \psi \in C, \|\psi\| < \delta, t \geq t_0]$$

implies that any solution of (1.3.1) satisfies $|x(t,t_0,\psi)| < \varepsilon$.

Definition 1.3.2. *The zero solution of (1.3.1) is said to be* uniformly stable *(US) if it is stable and if δ is independent of t_0.*

Definition 1.3.3. *The zero solution of (1.3.1) is said to be* asymptotically stable*(AS) if it is stable and if for each $t_0 \geq 0$ there is an $\eta > 0$ such that*

$$[\psi : E_{t_0} \to R^n, \psi \in C, \|\psi\| < \eta]$$

implies that any solution of (1.3.1) satisfies $x(t,t_0,\psi) \to 0$ as $t \to \infty$. If this is true for any t_0 and any η, then $x = 0$ is globally asymptotically stable *(GAS).*

Definition 1.3.4. *The zero solution of (1.3.1) is said to be uniformly asymptotically stable (UAS) if it is uniformly stable and if there is an $\eta > 0$ and for each $\mu > 0$ there exists $T > 0$ such that*

$$[t_0 \in [0, \infty),\ \psi : E_{t_0} \to R^n,\ \psi \in \mathcal{C},\ \|\psi\| < \eta,\ t \geq t_0 + T]$$

implies that any solution of (1.3.1) satisfies $|x(t, t_0, \phi)| < \mu$.

We will now look at some inter-related theory of Liapunov, Perron, and Krasovskii concerning some half-linear equations. It will bring us to a partial motivation of the study of stability by fixed point theory.

Let A be an $n \times n$ real constant matrix and consider the linear constant coefficient ordinary differential equation

$$x' = Ax. \tag{1.3.4}$$

If λ is a characteristic root of A, $\det(A - \lambda I) = 0$, and if x_0 is a characteristic vector belonging to λ, then $x(t) = x_0 e^{\lambda t}$ is a solution. Moreover, when the solution is complex, then the real part is a solution and the imaginary part is a solution. The following propositions can be found in any text on stability theory. See, for example, Burton (1985, pp. 31-41). In Example 1.3.2 we will show how to prove that (b) implies all the other parts of Proposition 1.3.2.

Proposition 1.3.1. *The following conditions are equivalent for (1.3.4).*

(a) The zero solution is stable.

(b) Every characteristic root λ of A satisfies $Re\,\lambda \leq 0$ and if $Re\,\lambda = 0$ and λ has multiplicity m, then there are m linearly independent characteristic vectors belonging to λ.

(c) Every solution is bounded for $t \geq 0$.

(d) The zero solution is uniformly stable.

Proposition 1.3.2. *The following conditions are equivalent for (1.3.4).*

(a) The zero solution is globally asymptotically stable.

(b) Every characteristic root λ of A satisfies $Re\,\lambda < 0$.

(c) Every solution tends to zero as $t \to \infty$.

(d) The zero solution is uniformly asymptotically stable.

Frequently investigators attempt to determine stability properties of a nonlinear equation by examining the linear part. Such a study will now be described.

Liapunov's direct method is the main classical technique for studying stability. Much can be found on the subject in Yoshizawa (1966) or Burton (1983, 1985). Suppose we are given a differential equation

$$x' = f(t, x) \tag{1.3.5}$$

where $f : [0, \infty) \times R^n \to R^n$ is continuous and $f(t, 0) = 0$. We look for a differentiable scalar function $V : [0, \infty) \times R^n \to [0, \infty)$ which is positive definite and is a type of measuring device showing properties of the distance from a solution of (1.3.5) to the origin. If $x(t)$ is an unknown solution of (1.3.5), then $V(t, x(t))$ is a scalar function of t and by the chain rule we can write

$$\frac{dV(t, x(t))}{dt} = \frac{\partial V}{\partial x_1}\frac{dx_1}{dt} + \cdots + \frac{\partial V}{\partial x_n}\frac{dx_n}{dt} + \frac{\partial V}{\partial t}. \tag{1.3.6}$$

While we do not know $x(t)$, we do know *directly from* (1.3.5) that

$$\frac{dx_i}{dt} = f_i(t, x)$$

so

$$\frac{dV(t, x(t))}{dt} = \text{grad}V \cdot f + \frac{\partial V}{\partial t}.$$

The derivative of V is determined *directly* from the differential equation itself; hence, we have the name Liapunov's direct method.

If we have chosen V so shrewdly that $\frac{dV}{dt} \le 0$, then $V(t, x(t))$ is not increasing. As V is assumed to be positive definite, we can then show that $|x_0|$ small implies $V(t, x(t))$ small; this, in turn, implies that $|x(t)|$ is small.

We now present the most prominent Liapunov function. The following result is proved in any text on Liapunov theory. See, for example, Burton (1985, p. 59). Many extensions to both ordinary and delay equations are found in Zhang (2001).

Theorem 1.3.1. *The matrix equation*

$$A^T B + BA = -I \tag{1.3.7}$$

can be solved for a unique positive definite symmetric matrix B if and only if all characteristic roots of A have negative real parts.

We now give a detailed example of construction of a Liapunov function and the stability consequences.

Example 1.3.2. Using (1.3.7) we will be able to show all parts of the process leading to the conclusion that the zero solution of (1.3.4) is uniformly asymptotically stable, as stated in Proposition 1.3.2 (d). If all characteristic roots of A have negative real parts, then we solve (1.3.7) for $B = B^T$ and form

$$V(x) = x^T Bx. \tag{1.3.8}$$

As B is positive definite we can find positive constants $\alpha^2 < \beta^2$ with

$$\alpha^2 x^T x \le x^T Bx \le \beta^2 x^T x. \tag{1.3.9}$$

Next, if we differentiate V along a solution of (1.3.4) by the product rule we have

$$\begin{aligned} V'(x(t)) &= (x^T)'Bx + x^T Bx' \\ &= x^T A^T Bx + x^T BAx \\ &= x^T[A^T B + BA]x \\ &= -x^T x \end{aligned}$$

or

$$V'(x(t)) = -x^T x \le -(1/\beta^2)V(x(t))$$

by (1.3.9). Hence, if $x(t) = x(t, t_0, x_0)$ then

$$\begin{aligned} \alpha^2 x^T(t)x(t) \le V(x(t)) &\le V(x(t_0))e^{-(1/\beta^2)(t-t_0)} \\ &\le \beta^2 x^T(t_0)x(t_0)e^{-(1/\beta^2)(t-t_0)}. \end{aligned} \tag{1.3.10}$$

Since $x^T x = |x|^2$, we can take square roots and obtain

$$|x(t)| \le (\beta/\alpha)|x(t_0)|e^{-(1/2\beta^2)(t-t_0)}. \tag{1.3.11}$$

From this we readily read off the stability relations. Given $\varepsilon > 0$ we take

$$(\beta/\alpha)\delta = \varepsilon \tag{1.3.12}$$

for the uniform stability. It is evident that we also have AS and UAS.

We will use (1.3.8) to prove the following result, obtained in 1892 by Liapunov (1992) and then by Perron (1930) in another way. It will be used to advance the Liapunov-Perron theorem to delay equations. Finally, we will use it to motivate the fixed point analysis.

Let $f : [0, \infty) \times R^n \to R^n$ be continuous and let A be an $n \times n$ constant real matrix. Consider the equation

$$x' = Ax + f(t, x). \tag{1.3.13}$$

The following result gives conditions under which f can simply be ignored for solutions starting near zero. It is likely the theorem most used by those who apply differential equations to their work.

In our computations, $|x|$ will denote the Euclidean norm on R^n and for an $n \times n$ matrix B, $|B|$ will denote any compatible norm so that $|Bx| \le |B||x|$.

Theorem 1.3.2. *If all characteristic roots of A have negative real parts and if*

$$\lim_{|x|\to 0} \frac{f(t,x)}{|x|} = 0 \tag{1.3.14}$$

uniformly for $0 \le t < \infty$, then the zero solution of (1.3.13) is uniformly asymptotically stable.

Proof. As before, we find B satisfying (1.3.7) so that B is positive definite and symmetric. Then define $V(x) = x^T Bx$. The derivative of V along a solution $x(t) = x(t, t_0, x_0)$ of (1.3.13) satisfies

$$\begin{aligned} V'(x(t)) &= [x^T A^T + f^T(t,x)]Bx + x^T B[Ax + f(t,x)] \\ &= x^T[A^T B + BA]x + f^T(t,x)Bx + x^T Bf(t,x) \\ &= -x^T x + 2x^T Bf(t,x) \\ &\le -|x|^2 + 2|x||B||f(t,x)| \\ &= |x|^2\left[-1 + \frac{2|B||f(t,x)|}{|x|}\right]. \end{aligned}$$

Using (1.3.14) we can find $\eta > 0$ such that for $|x| \le \eta$ then $2|B||f(t,x)| < |x|/2$. Then for $|x| \le \eta$ we have

$$V'(x(t)) \le -|x|^2/2.$$

We now have the same kinds of inequalities on V and V' as we did in the proof of Theorem 1.3.1, and these will yield uniform asymptotic stability.

Next, we consider the equation

$$x' = Ax + f(t, x) + g(t, x(t - r)) \tag{1.3.15}$$

in which A and f satisfy the conditions of Theorem 1.3.2, r is a positive constant, and $g : [0, \infty) \times R^n \to R^n$ is continuous. Condition (1.3.14) was a clear comparison between Ax and $f(t, x)$; Ax strongly dominates $f(t, x)$ near $x = 0$, uniformly in t. But there is no obvious sense in which Ax can dominate $g(t, x(t - r)$. Yet, Krasovskii (1963; p. 170) introduced a simple idea which enables us to do so. In line with (1.3.14) we ask that

$$\lim_{|x|\to 0} \frac{g(t + r, x)}{|x|} = 0 \tag{1.3.16}$$

uniformly for $0 \le t < \infty$. Our next theorem shows that f and g can simply be ignored in local stability questions. This is a major result.

Notation. In the proof of the next theorem we will begin using notation which is very common in functional differential equations. For a fixed positive number r and a continuous function $x : [-r, \infty) \to R^n$, by x_t for $t \ge 0$ we shall mean that element from the Banach space $(\mathcal{C}, \|\cdot\|)$ of continuous functions $\phi : [-r, 0] \to R^n$ with the supremum norm defined by $x_t(s) = x(t + s)$ for $-r \le s \le 0$. Also, for $\gamma > 0$, $\mathcal{C}_\gamma$ denotes the γ-ball in $\mathcal{C}$.

Theorem 1.3.3. *Let (1.3.16) and the conditions of Theorem 1.3.2 hold. Then the zero solution of (1.3.15) is uniformly asymptotically stable.*

Proof. Define B as in the proof of Theorem 1.3.2 and define a Liapunov functional by

$$V(t, x_t) = x^T Bx + k \int_{t-r}^{t} g^2(s + r, x(s))\, ds$$

where $k = 4|B|^2$ and $g^2 = g^T g$. If we take the derivative of V along a solution of (1.3.15) we have

$$\begin{aligned} V'(t, x_t) \le &-|x|^2 + 2|B||x||f(t, x)| + 2|B||x||g(t, x(t - r)| \\ &+ kg^2(t + r, x) - kg^2(t, x(t - r)). \end{aligned}$$

By (1.3.14) and (1.3.16) we can find $\gamma > 0$ such that

$[|x| \le \gamma, \quad t \in [0, \infty)]$

imply

$$2|B||f(t, x)| \le |x|/4 \quad \text{and} \quad 4|B|^2 g^2(t + r, x) \le |x|^2/4. \tag{1.3.17}$$

Now, for $|x| \leq \gamma$ we have

$$\begin{aligned} V'(t,x) \leq& -|x|^2 + (1/4)|x|^2 + \Big((1/4)|x|^2 + 4|B|^2 g^2(t, x(t-r))\Big) \\ &+ 4|B|^2 g^2(t+r,x) - 4|B|^2 g^2(t, x(t-r)) \\ \leq& -(1/4)|x|^2. \end{aligned}$$

We have arrived at three fundamental inequalities. For $\|x_t\| \leq \eta$ and $t \in [0, \infty)$ then

$$\begin{aligned} \alpha^2|x(t)|^2 \leq& V(t,x_t) \leq \beta^2|x(t)|^2 + (1/4)\int_{t-r}^{t} |x(s)|^2\, ds \\ \alpha^2|x(t)|^2 \leq& V(t,x_t) \leq (\beta^2 + (1/4)r)\|x_t\|^2 \\ V'(t,x_t) \leq& -(1/4)|x(t)|^2. \end{aligned} \tag{1.3.18}$$

To prove uniform stability, use $V' \leq 0$ and the second line of (1.3.18). For a given $\varepsilon \in (0, \gamma)$, choose $\delta > 0$ from the second line below

$$\begin{aligned} |x(t)|^2 &\leq [(\beta^2 + (r/4))/\alpha^2]\|\phi\|^2 \\ &\leq [(\beta^2 + (r/4))/\alpha^2]\delta^2 = \varepsilon^2. \end{aligned}$$

Next, review Definitions 1.3.2 and 1.3.4. For $\varepsilon = \gamma$ of (1.3.17), find δ of uniform stability and in Definition 1.3.4 take $\eta = \delta$. In Definition 1.3.4, let $\mu > 0$ be given, $\mu < \eta$. We will find $T > 0$ such that

$$[t_0 \geq 0, \quad \|\psi\| < \eta, \quad t \geq t_0 + T] \quad \text{implies} \quad |x(t, t_0, \psi)| < \mu.$$

Since $V' \leq 0$, looking at the first line of (1.3.18) we see that it will suffice to make $V(t, x_t) \leq \mu^2\alpha^2$. We also have

$$V(t_0, \phi) \leq (\beta^2 + (r/4))\eta^2 =: V_0.$$

Along the solution, we write $V(t, x_t) =: V(t)$.

Consider the intervals

$$I_n = [t_0 + (n-1)r, t_0 + nr]$$

for $n = 2, 3, \ldots$ From the first line of (1.3.18), if $V(t, x_t) \geq \mu^2\alpha^2$ on an I_n then at each $t \in I_n$ either

$$\beta^2|x(t)|^2 \geq \mu^2\alpha^2/2 \quad \text{or} \quad (1/4)\int_{t-r}^{t} |x(s)|^2\, ds \geq \mu^2\alpha^2/2. \tag{1.3.19}$$

If the first part of (1.3.19) holds for all $t \in I_n$, then integrate the last line in (1.3.18) and obtain

$$V(t_0+nr)-V(t_0+(n-1)r)\leq -(1/4)\int_{t_0+(n-1)r}^{t_0+nr}|x(s)|^2\,ds$$
$$\leq -\frac{r\mu^2\alpha^2}{8\beta^2}.$$

If the second relation in (1.3.19) holds at some $t\in I_n$ then

$$V(t)-V(t-r)\leq -\mu^2\alpha^2/2.$$

Notice that I_n may have been used twice! We must double n. Let

$$V_1:=\min\left[\frac{r\mu^2\alpha^2}{8\beta^2},\quad \mu^2\alpha^2/2\right].$$

Thus,

$$V(t_0+2nr)\leq V_0-nV_1$$

and the right-hand-side will be negative if $n>V(t_0)/V_1$. We then have our number $T\leq 2nr$.

Corollary. *If all characteristic roots of A have negative real parts and if there is a $\gamma>0$ for which (1.3.17) holds, then the zero solution of (1.3.15) is uniformly asymptotically stable.*

These simple and self-contained results provide the background for some general stability theorems which will be used throughout the book in comparing Liapunov theory with fixed point methods. Notice that (1.3.16) is a pointwise condition. Fixed point results contrast with these in that they involve averaged conditions.

For $\gamma>0$, consider the system

$$x'=F(t,x_t),\qquad F(t,0)=0 \tag{1.3.20}$$

where $F:[0,\infty)\times\mathcal{C}_\gamma\to R^n$ is continuous and takes bounded sets into bounded sets. Here, $\mathcal{C}_\gamma$ is the γ-ball in the space $(\mathcal{C},\|\cdot\|)$ of continuous functions $\phi:[-r,0]\to R^n$ with the supremum norm and $x_t(s)=x(t+s)$ for $-r\leq s\leq 0$ and r is a fixed positive constant. The prime example will be

$$x'=f\Big(t,x(t),x(t-h(t)),\int_{t-h(t)}^{t}g(s,x(s))\,ds\Big)$$

where $0\leq h(t)\leq r$ and h is continuous. Stability definitions for (1.3.20) are precisely those given in Definitions 1.3.2, 1.3.3, and 1.3.4.

We will employ functionals $V(t, x_t)$ for which we can define the derivative along an unknown solution of (1.3.20), usually using the chain rule. If V is continuous and only satisfies a local Lipschitz condition then we can employ the upper right-hand derivative. Details are found in Yoshizawa (1966, p. 3) or in Burton (1985; p. 260). For our purposes here we will always try to use the chain rule as we did in the motivating examples so that the derivative will be totally elementary. In any case, we denote the derivative of V along a solution of (1.3.20) by $V'_{(1.3.20)}$.

Definition 1.3.5. *A wedge is a continuous and strictly increasing function $W : [0, \infty) \to [0, \infty)$ with $W(0) = 0$.*

Wedges are denoted by W_i and such a symbol will always mean that there exists a wedge. Here is the basic result.

Theorem 1.3.4. *Suppose that for (1.3.20) and for some $\gamma > 0$ there is a scalar functional $V(t, x_t)$ defined, continuous in (t, x_t), and locally Lipschitz in x_t when $0 \le t < \infty$ and $x_t \in C_\gamma$. Suppose also that $V(t, 0) = 0$ and that $W_1(|x(t)|) \le V(t, x_t)$.*

(a) If $V'_{(1.3.20)}(t, x_t) \le 0$ for $0 \le t < \infty$ then the zero solution of (1.3.20) is stable.

(b) If, in addition to (a), $V(t, x_t) \le W_2(\|x_t\|)$, then $x = 0$ is uniformly stable.

(c) If there is an $M > 0$ with $|F(t, x_t)| \le M$ for $0 \le t < \infty$ and $x_t \in C_\gamma$, and if $V'_{(1.3.20)}(t, x_t) \le -W_3(|x(t)|)$, then $x = 0$ is asymptotically stable.

(d) If all of the above conditions hold, then the zero solution is uniformly asymptotically stable.

(e) If

$$W_1(|x(t)|) \le V(t, x_t) \le W_2(|x(t)|) + W_3\left(\int_{t-r}^{t} |x(s)|\, ds\right)$$

and if

$$V'_{(1.3.20)}(t, x_t) \le -W_4(|x(t)|)$$

then the zero solution is uniformly asymptotically stable.

Part (c) can be extended by asking that

(f) there is continuous function $\gamma : [0, \infty) \to [0, \infty)$ which is integrally positive with

$$V'_{(1.3.20)}(t, x_t) \leq -\gamma(t)W_3(|x(t)|).$$

Such a function, γ, is *integrally positive* in case for each sequence $\{t_n\} \uparrow +\infty$ and each $\delta > 0$ we have

$$\liminf_{n\to\infty} \int_{t_n}^{t_n+\delta} \gamma(t)\, dt > 0. \tag{1.3.21}$$

This concept arises in many problems typified by

$$x' = -\gamma(t)x(t) + b(t)x(t-r)$$

and we need to ask that γ be integrally positive when using Liapunov's direct method. Such requirements are unnecessary when we use fixed point theory. We find that $\gamma(t)$ can vanish on arbitrarily long intervals.

A proof of (a-d) is found in Yoshizawa (1966) or Burton (1985; pp. 262) and a proof of (e) is found in Burton (1978 or 1983). Several important generalizations are found in Hatvani (2002), Wang (1992, 1994a,b), and Zhang (1995).

There are very interesting parallels between Liapunov's direct method and fixed point methods for studying stability. Equation (1.3.6) tells us precisely how to find the derivative of a Liapunov function along the solutions of a differential equation; but finding the Liapunov function is an art. Often we rely on obtaining a form such as

$$x' = Ax + f(t, x) + g(t, x(t-r)), \tag{1.3.15}$$

possibly using differentiability, so that we can then have a start on the problem through the use of variants of

$$V(x) = x^T Bx \tag{1.3.8}$$

obtained from

$$A^T B + BA = -I, \tag{1.3.7}$$

as we saw in the proof of Theorem 1.3.2. In precisely the same way, if we can arrive at something like (1.3.15) then the use of the variation of parameters formula on (1.3.15) will give us an initial start on a mapping for the generation of a fixed point. Earlier we had mentioned simply adding and subtracting the linear term. While that is an option, virtually always we will change the problem into a neutral equation so that the linear term will appear as an integral part of the equation.

In so many of the examples to follow we will see scalar first order equations. This can be very deceptive. A linear n^{th}–order ordinary differential equation can be written as a scalar integral or integro-differential equation. The scalar equation

$$x'(t) = ax(t) + bx(t-r)$$

for r a positive constant and b nonzero is infinite dimensional; there is a sequence $\{\lambda_n\}$ of numbers with $Re\,\lambda_n \to -\infty$ such that any linear combination of terms $x_n(t) = e^{\lambda_n t}$ is a solution. The classical Lurie control problem

$$x' = Ax + bf(\sigma)$$
$$\sigma' = c^T b - rf(\sigma)$$

with A an $n \times n$ matrix can be written as a scalar integrodifferential equations.

1.4 The most elementary examples

We begin with an explanation of notation in equation numbering. This section, especially, is composed of separate examples. The main equation being considered will have a number, such as (1.4.1). Then all of the equations related to proving that (1.4.1) has a certain stability property will be numbered (1.4.1.1), (1.4.1.2), (1.4.1.3),... This will allow each problem to be rather self contained and it will be easy to recognize which conditions relate to a given equation. In effect, then, each main equation being studied represents the beginning of a subsection.

We will sometimes formally list exercises, but there will also be many implied exercises. In most of the proofs there will be a *Type I Implied Exercise.* We will give the proof of stability only for $t_0 = 0$; the interested reader should supply the details of the proof for an arbitrary t_0 and should consider whether or not the supplied proof yields uniform stability. There will also be a *Type II Implied Exercise* in which we are proving stability for an equation which is fully delayed, such as

$$x' = -ax(t-r),$$

which we will then write as a neutral equation

$$x' = -ax(t) + a\frac{d}{dt}\int_{t-r}^{t} x(s)\,ds$$

and the stability will be derived from the ode part, $x' = -ax(t)$. The reader may consider the parallel problem

$$x' = -bx(t) - ax(t-r)$$

written as

$$x' = -(a+b)x(t) + a\frac{d}{dt}\int_{t-r}^{t} x(s)\,ds.$$

In the last section we saw some very useful general stability criteria and we also stated some general stability results based on Liapunov functionals. Those first mentioned results depended on pointwise relations and on the delay being differentiable. In Theorem 1.3.4 c,d we noticed a strange requirement that $|f(t,x_t)|$ be bounded for x_t bounded. These are three, among many, difficulties which we hope to circumvent using fixed point theory.

The intent here is to build the problems slowly. We start with examples in which Liapunov functionals and contraction mappings give similar results and move on to problems where significant differences emerge. We refer to these problems as "elementary," but they have been the center of much investigation.

There are two striking differences between Liapunov and fixed point theory in the study of stability. In the former we rely on an existence theorem to tell us that a solution does exist and, hence, we can differentiate $V(t,x_t)$ along that solution. By contrast, the fixed point method proves existence of a solution. In addition, Liapunov theory establishes properties of any solution, while a fixed point theorem which does not assert the existence of a unique fixed point only gives information about one solution. Other considerations are then needed to ensure that all solutions behave in the same way.

Several of the examples of this section are found in Burton and Furumochi (2001a). We begin with the scalar equation

$$x'(t) = -ax(t) + bx(t-r) \tag{1.4.1}$$

where a, b, and r are constant, $r \geq 0$.

Three different proofs will be given and each one will be a pattern to which the reader may refer in other problems.

Theorem 1.4.1. *If $a > 0$ and $|b| < a$, then the zero solution of (1.4.1) is uniformly asymptotically stable for every $r \geq 0$.*

Proof. Proof I. Define the Liapunov functional following the Krasovskii idea in the proof of Theorem 1.3.3 by

$$V(x_t) = x^2(t) + |b| \int_{t-r}^{t} x^2(s)ds. \tag{1.4.1.1}$$

The derivative of V along an (unknown) solution of (1.4.1) satisfies

$$\begin{aligned} V'_{(1.4.1)}(x_t) &= 2x[-ax + bx(t-r)] + |b|x^2 - |b|x^2(t-r) \\ &\le -2ax^2 + |b|x^2 + |b|x^2(t-r) + |b|x^2 - |b|x^2(t-r) \\ &= [-2a + 2|b|]x^2 \\ &=: -\alpha x^2, \end{aligned}$$

where $\alpha > 0$. We have

$$|x(t)|^2 \le V(x_t) \le (1 + |b|r)\|x_t\|^2$$

and

$$V'(x_t) \le -2[a - |b|]|x|^2 = -\alpha x^2.$$

Referring to Theorem 1.3.4 either part (d) or (e) we can conclude uniform asymptotic stability.

Proof II. This is the first of two proofs using contraction mappings. Here, we will limit ourselves to taking $t_0 = 0$ and proving that for an arbitrary continuous initial function the unique solution will tend to zero. We will set up the problem to show an $\varepsilon - \delta$ proof of stability and leave the details to the interested reader.

Let $\psi : [-r, 0] \to R$ be a continuous initial function and let $(\mathcal{M}, \|\cdot\|)$ be the complete metric space of continuous function $\phi : [-r, \infty) \to R$ with $\phi_0 = \psi$ and $\phi(t) \to 0$ as $t \to \infty$. Recall that the notation $\phi_0 = \psi$ means that $\phi(t) = \psi(t)$ for $-r \le t \le 0$. Use the variation of parameters formula to write (1.4.1) as

$$x(t) = \psi(0)e^{-at} + \int_0^t e^{-a(t-s)} bx(s-r)\, ds$$

and use this to define a mapping $P : \mathcal{M} \to \mathcal{M}$ by $\phi \in \mathcal{M}$ and $-r \le t \le 0$ implies that $(P\phi)(t) = \psi(t)$, while $t > 0$ implies that

$$(P\phi)(t) = \psi(0)e^{-at} + \int_0^t e^{-a(t-s)} b\phi(s-r)\, ds. \tag{1.4.1.2}$$

Now, the convolution of an L^1–function (e^{-at}) with a function tending to zero ($\phi(s-r)$) tends to zero as $t \to \infty$ so it is clear that $P\phi \in \mathcal{M}$. To see that P is a contraction, let $\phi, \eta \in \mathcal{M}$. Then

$$\begin{aligned}|(P\phi)(t) - (P\eta)(t)| &\leq \int_0^t |b|e^{-a(t-s)}|\phi(s-r) - \eta(s-r)|\,ds \\ &\leq \|\phi - \eta\| \int_0^t |b|e^{-a(t-s)}\,ds \\ &\leq \|\phi - \eta\|(|b|/a).\end{aligned}$$

This is a contraction since $a > |b|$ and so P has a unique fixed point $\phi \in \mathcal{M}$. Since that fixed point ϕ resides in $\mathcal{M}$ it follows that $\phi(t) \to 0$. Of course, from (1.4.1.2) we see that the fixed point solves (1.4.1). Moreover, to prove stability, we apply norms to both sides of (1.4.1.2) at the fixed point and obtain

$$\|\phi\| \leq \|\psi\| + (|b|/a)\|\phi\|$$

from which the stability relation is easily derived.

Proof III. In this proof we will use a metric induced by a norm which admits only functions tending to zero exponentially. We arrive at the norm as follows. Determine $c > 0$ by $a = |b| + 2c$. Then find d with $0 < d < c$ so that

$$e^{dr}\frac{|b|}{a-c} < 1$$

and also

$$\frac{e^{dr}|b|}{a-d} =: \alpha < 1. \tag{1.4.1.3}$$

For a given continuous initial function $\psi : [-r, 0] \to R$ let $(\mathcal{M}, |\cdot|_d)$ be the complete metric space of continuous functions $\phi : [-r, \infty) \to R$ with $\phi_0 = \psi$ for which

$$|\phi|_d = \sup_{t\geq 0} e^{dt}|\phi(t)|$$

exists; thus, if $\phi, \eta \in \mathcal{M}$ then

$$|\phi - \eta|_d = \sup_{t\geq 0} e^{dt}|\phi(t) - \eta(t)|.$$

Define P by (1.4.1.2) once more. The continuity is clear. If $\phi \in \mathcal{M}$ then

$$\begin{aligned}
&|(P\phi)(t)|e^{dt} \\
&\leq |\psi(0)|e^{(-a+d)t} + \int_0^t e^{dt}e^{-d(t-s)+(-a+d)(t-s)}|b||\phi(s-r)|\,ds \\
&= |\phi(0)|e^{(-a+d)t} + \int_0^t |b|e^{(-a+d)(t-s)+dr+d(s-r)}|\phi(s-r)|\,ds \\
&\leq |\phi(0)|e^{(-a+d)t} + [|\phi|_d + \|\psi\|]\frac{|b|e^{dr}}{a-d}.
\end{aligned} \tag{1.4.1.4}$$

Hence, $P\phi \in \mathcal{M}$.

To see that P is a contraction, if $\phi, \eta \in \mathcal{M}$ then

$$\begin{aligned}
&|(P\phi)(t) - (P\eta)(t)|e^{dt} \\
&\qquad\leq \int_0^t e^{-a(t-s)+dt}|b||\phi(s-r) - \eta(s-r)|\,ds \\
&\qquad\leq |b|\int_0^t e^{(-a+d)(t-s)+dr+d(s-r)}|\phi(s-r) - \eta(s-r)|\,ds \\
&\qquad\leq \frac{|b|e^{dr}}{a-d}|\phi-\eta|_d = \alpha|\phi-\eta|_d.
\end{aligned} \tag{1.4.1.5}$$

Hence, P is a contraction with unique fixed point ϕ, a solution tending to zero of order e^{-dt}.

Recall that there is the Type I exercise in which the reader can carry out the proof for a general t_0. Moreover, it introduces a *Type III exercise.* In subsequent problems, the reader may profitably study the question of proving that asymptotic stability is exponential.

We motivate our next example by considering

$$x'(t) = -a(t)x(t) - b(t)x(t-r)$$

where a and b are bounded continuous functions,

$$a(t) \geq \delta > 0, \quad |b(t)| \leq \theta\delta, \quad \theta < 1.$$

Using the Liapunov functional

$$V(t, x_t) = (1/2)(x^2(t) + \delta\int_{t-r}^t x^2(s)\,ds)$$

with the triangle inequality we have

$$V' \leq (\delta/2)(\theta - 1)(x^2(t) + x^2(t-r)).$$

It can now be argued that this yields UAS. Note that this asks that a, b be bounded and that $|b(t)|$ is bounded by a all of the time.

The half-linear equation

$$x'(t) = -a(t)x(t) + b(t)g(x(t - r(t))) \tag{1.4.2}$$

presents severe challenges when we attempt to show that solutions tend to 0 using Liapunov functionals. We are interested in the case where a can be negative some of the time, a and b are related on average, while a, b, and r' can be unbounded.

Assume that a, b, and r are continuous, that

$$\int_0^t a(s)ds \to \infty \text{ as } t \to \infty, \tag{1.4.2.1}$$

there is an $\alpha < 1$ with

$$\int_0^t e^{-\int_s^t a(u)du} |b(s)|\, ds \leq \alpha, \quad t \geq 0, \tag{1.4.2.2}$$

$$0 \leq r(t), \quad t - r(t) \to \infty \text{ as } t \to \infty, \tag{1.4.2.3}$$

and there is an $L > 0$ so that if $|x|$, $|y| \leq L$ then

$$g(0) = 0 \text{ and } |g(x) - g(y)| \leq |x - y|. \tag{1.4.2.4}$$

Theorem 1.4.2. *If (1.4.2.1)-(1.4.2.4) hold, then every solution of (1.4.2) with small continuous initial function tends to 0 as $t \to \infty$. Moreover, the zero solution is stable.*

Proof. We will take $t_0 = 0$, leaving the general case as an exercise. For the α and L, find $\delta > 0$ with $\delta + \alpha L \leq L$. Let $\psi : (-\infty, 0] \to R$ be a given continuous function with $|\psi(t)| < \delta$ and let

$$\mathcal{S} = \{\phi : R \to R \,|\, \|\phi\| \leq L, \phi(t) = \psi(t) \text{ if } t \leq 0, \phi(t) \to 0 \text{ as } t \to \infty, \phi \in \mathcal{C}\}$$

where $\|\cdot\|$ is the supremum metric. Then $(\mathcal{S}, \|\cdot\|)$ is a complete metric space.

Define $P : \mathcal{S} \to \mathcal{S}$ by

$$(P\phi)(t) = \psi(t) \text{ if } t \leq 0$$

and for $t \geq 0$ let

$$(P\phi)(t) = e^{-\int_0^t a(s)ds}\psi(0) + \int_0^t e^{-\int_s^t a(u)du} b(s)g(\phi(s - r(s)))ds.$$

Clearly, $P\phi \in \mathcal{C}$. We now show that $(P\phi)(t) \to 0$ as $t \to \infty$. Let $\phi \in S$ and $\varepsilon > 0$ be given. Then $\|\phi\| \leq L$, there exists $t_1 > 0$ with $|\phi(t - r(t))| < \varepsilon$ if $t \geq t_1$, and there exists $t_2 > t_1$ such that $t > t_2$ implies that $e^{-\int_{t_1}^t a(u)\,du} < \varepsilon/(L\alpha)$.

Thus, $t > t_2$ implies that

$$\begin{aligned}
&\left|\int_0^t e^{-\int_s^t a(u)du} b(s)g(\phi(s - r(s)))\,ds\right| \\
&\qquad \leq \int_0^{t_1} e^{-\int_s^t a(u)du}|b(s)|Lds + \int_{t_1}^t e^{-\int_s^t a(u)\,du}|b(s)|\varepsilon\,ds \\
&\qquad \leq e^{-\int_{t_1}^t a(u)du} \int_0^{t_1} e^{-\int_s^{t_1} a(u)du}|b(s)|L\,ds + \alpha\varepsilon \\
&\qquad \leq \alpha L e^{-\int_{t_1}^t a(u)\,du} + \alpha\varepsilon \\
&\qquad \leq \varepsilon + \alpha\varepsilon.
\end{aligned}$$

To see that P is a contraction under the supremum metric, if $\phi, \eta \in \mathcal{S}$, then

$$|(P\phi)(t) - (P\eta)(t)| \leq \int_0^t e^{-\int_s^t a(u)du}|b(s)|\|\phi - \eta\|\,ds \leq \alpha\|\phi - \eta\|$$

with $\alpha < 1$ by (1.4.2.2).

Hence, for each such initial function, P has a unique fixed point in $\mathcal{S}$ which solves (1.4.2) and tends to 0.

To get stability for solutions starting at $t_0 = 0$, let $\varepsilon > 0$ be given and do the above work for $L = \varepsilon$.

We now begin looking at problems in which the size of the delay is important for reasons that are very clear. The classical Liapunov functional for

$$x' = -a(t)x + b(t)\int_{t-r}^t x(s)\,ds$$

is given by

$$V(t, x_t) = x^2 + \gamma \int_{-r}^{0} \int_{t+v}^{t} x^2(u)\, du\, dv,$$

where $|b(t)| \leq \gamma$, a constant. It is easy to see upon differentiation of V that problems arise if $a(t)$ becomes negative, if either $a(t)$ or $b(t)$ is not bounded, or if $r = r(t)$ has $r'(t)$ unbounded. These kinds of problems vanish with fixed point theory.

Consider the scalar equation

$$x' = -a(t)x + \int_{t-r(t)}^{t} b(t, s)g(x(s))\, ds \tag{1.4.3}$$

with a positive constant r_0 such that $0 \leq r(t) \leq r_0$. We suppose that there exists an $\alpha < 1$ with

$$\int_0^t e^{-\int_s^t a(u)du} \int_{s-r(s)}^{s} |b(s, u)|\, du\, ds \leq \alpha. \tag{1.4.3.1}$$

Also, for each $\varepsilon > 0$ there exist $t_1 > 0$ and $T > 0$ such that $t_2 \geq t_1$ and $t \geq t_2 + T$ imply that

$$e^{-\int_{t_2}^t a(s)\, ds} < \varepsilon \text{ and } e^{-\int_0^t a(s)\, ds} \to 0 \text{ as } t \to \infty. \tag{1.4.3.2}$$

Finally, there is an $L > 0$ such that $|x|$, $|y| \leq L$ imply that

$$|g(x) - g(y)| \leq |x - y| \text{ and } g(0) = 0. \tag{1.4.3.3}$$

Notice that stability is to come from the ordinary differential equation part and so in comparing a to b, the size of b must be related to the size of r_0. Also, $x(t)$ must be compared with the average of past values of x and that average depends on the size of r_0.

Theorem 1.4.3. *If (1.4.3.1)–(1.4.3.3) hold then the zero solution of (1.4.3) is asymptotically stable.*

Proof. As before, we take $t_0 = 0$. For the L and α find $\delta > 0$ so that $\delta + \alpha L \leq L$. Let $\psi : [-r_0, 0] \to R$ be a given continuous initial function with $|\psi(t)| < \delta$ and define

$$S = \{\phi : [-r_0, \infty) \to R \mid \|\phi\| \leq L, \phi \in C, \phi_0 = \psi, \phi(t) \to 0 \text{ as } t \to \infty\}$$

where $\|\cdot\|$ denotes the supremum metric.

Define $P : \mathcal{S} \to \mathcal{S}$ by

$$(P\phi)(t) = \psi(t) \text{ if } -r_0 \le t \le 0$$

and for $t \ge 0$ let

$$(P\phi)(t) = e^{-\int_0^t a(s)\,ds}\psi(0) + \int_0^t e^{-\int_s^t a(u)\,du} \int_{s-r(s)}^s b(s,u)g(\phi(u))\,du\,ds.$$

Now, let $\varepsilon > 0$ and $\phi \in \mathcal{S}$ be given. Then $\|\phi\| \le L$ and there exists $t_1 > 0$ such that $t \ge t_1 - r_0$ implies that $|\phi(t)| < \varepsilon$, while $t_2 \ge t_1$ and $t \ge t_2 + T$ imply that $e^{-\int_{t_2}^t a(s)ds} < \varepsilon$. Thus, by the second part of (1.4.3.2) we may suppose t_1 so large that $e^{-\int_0^{t_1} a(s)\,ds} < \varepsilon$. Hence, $t \ge t_2 + T$ implies that

$$\begin{aligned}|(P\phi)(t)| &\le \delta\varepsilon + \int_0^{t_2} e^{-\int_s^{t_2} a(u)\,du} e^{-\int_{t_2}^t a(u)\,du} \int_{s-r(s)}^s |b(s,u)|L\,du\,ds \\ &\quad + \int_{t_2}^t e^{-\int_s^t a(u)du} \int_{s-r(s)}^s |b(s,u)|\varepsilon\,ds \\ &\le \delta\varepsilon + \varepsilon\alpha L + \varepsilon\alpha.\end{aligned}$$

Thus, $(P\phi)(t) \to 0$ as $t \to \infty$. Also, $\|\phi\| \le L$ implies $\|P\phi\| \le L$ by choice of δ.

Finally,

$$\begin{aligned}|(P\phi)(t) - (P\eta)(t)| &\le \int_0^t e^{-\int_s^t a(u)\,du} \int_{s-r(s)}^s |b(s,u)|\,du\,ds\|\phi - \eta\| \\ &\le \alpha\|\phi - \eta\|.\end{aligned}$$

Hence, P has a unique fixed point in $\mathcal{S}$.

We now consider problems in which the ode part is not necessarily stable and the delay part will be used to stabilize it. Such work with Liapunov functions is found, for example, in Burton (1983; p. 139). The idea is to convert to a neutral equation. Here, the size of the delay will certainly play a role. The reader can confirm that the work we do here also carries through on a nonlinear equation

$$x'(t) = -a(t)x(t) + b(t)x(t-r) + h(t)g(x(t-r))$$

where g satisfies a local Lipschitz condition. But in order to not obscure the main part, we consider the scalar equation

$$x'(t) = -a(t)x(t) + b(t)x(t-r) \tag{1.4.4}$$

which we can write as

$$\begin{aligned} x' &= [-a(t) + b(t+r)]x(t) - (d/dt)\int_{t-r}^{t} b(s+r)x(s)\,ds \\ &=: A(t)x(t) - (d/dt)\int_{t-r}^{t} b(s+r)x(s)\,ds. \end{aligned} \tag{1.4.4.1}$$

It is assumed that

$$\int_0^t [-a(s) + b(s+r)]\,ds \to -\infty \text{ as } t \to \infty, \tag{1.4.4.2}$$

$$\int_{t-r}^{t} |b(u+r)|\,du + \int_0^t |A(s)e^{\int_s^t A(u)\,du}| \int_{s-r}^{s} |b(u+r)|\,du\,ds \le \alpha < 1, \tag{1.4.4.3}$$

and that whenever $\phi(t) \to 0$ as $t \to \infty$, then

$$\int_0^t A(s)e^{\int_s^t A(u)\,du} \int_{s-r}^{s} b(u+r)\phi(u)\,du\,ds \to 0 \text{ as } t \to \infty. \tag{1.4.4.4}$$

One may note that the inner integral in (1.4.4.4) tends to zero because of (1.4.4.3). If $A(t) \le 0$ then the classical proof that the convolution of an L^1-function with a function tending to zero does, itself, tend to zero can be used to prove that (1.4.4.4) holds. We will see such a proof in Section 1.6.

To fix ideas of how these conditions can be satisfied, one could let $a(t) = 0, b(t) = b < 0$ be constant. Then (1.4.4.2) and (1.4.4.4) are immediately satisfied, while (1.4.4.3) needs $-2br < 1$.

Theorem 1.4.4. *If (1.4.4.2)-(1.4.4.4) hold, then for every continuous $\psi : [-r, 0] \to R$, the solution of (1.4.4) satisfies $x(t, 0, \psi) \to 0$ as $t \to \infty$.*

Proof. Let

$$\mathcal{S} = \{\phi : [-r, \infty) \to R \mid \phi \in \mathcal{C}, \phi_0 = \psi, \phi(t) \to 0 \text{ as } t \to \infty\}.$$

We can integrate by parts and write the solution of (1.4.4.1) as

$$\begin{aligned}
x(t) &= e^{\int_0^t A(s)\,ds}\psi(0) - \int_0^t e^{\int_s^t A(u)\,du}(d/ds)\int_{s-r}^{s} b(u+r)x(u)\,du\,ds \\
&= e^{\int_0^t A(s)\,ds}\psi(0) - e^{\int_s^t A(u)\,du}\int_{s-r}^{s} b(u+r)x(u)\,du\Big|_0^t \\
&\quad - \int_0^t A(s)e^{\int_s^t A(u)\,du}\int_{s-r}^{s} b(u+r)x(u)\,du\,ds \\
&= e^{\int_0^t A(s)ds}\psi(0) \\
&\quad - \int_{t-r}^{t} b(u+r)x(u)du + e^{\int_0^t A(u)\,du}\int_{-r}^{0} b(u+r)\psi(u)\,du \\
&\quad - \int_0^t A(s)e^{\int_s^t A(u)\,du}\int_{s-r}^{s} b(u+r)x(u)\,du\,ds.
\end{aligned}$$

Now, define $P : \mathcal{S} \to \mathcal{S}$ by $(P\phi)(t) = \psi(t)$ if $-r \le t \le 0$ and for $t \ge 0$ then

$$\begin{aligned}
(P\phi)(t) &= e^{\int_0^t A(s)\,ds}\psi(0) - \int_{t-r}^{t} b(u+r)\phi(u)\,du \\
&\quad + e^{\int_0^t A(u)\,du}\int_{-r}^{0} b(u+r)\psi(u)\,du \\
&\quad - \int_0^t A(s)e^{\int_s^t A(u)\,du}\int_{s-r}^{s} b(u+r)\phi(u)\,du\,ds.
\end{aligned}$$

By (1.4.4.2) and (1.4.4.4), $(P\phi)(t) \to 0$ as $t \to \infty$. To see that P is a contraction, if $\phi, \eta \in S$ then

$$\begin{aligned}
|(P\phi)(t) - (P\eta)(t)| &\le \int_{t-r}^{t} |b(u+r)|\|\phi - \eta\|\,du \\
&\quad + \int_0^t |A(s)e^{\int_s^t A(u)\,du}|\int_{s-r}^{s} |b(u+r)|\,du\,ds\|\phi-\eta\| \\
&\le \alpha\|\phi - \eta\|
\end{aligned}$$

by (1.4.4.3). Hence, P is a contraction and there is a unique solution tending to zero.

To this point, all of our examples have been first order. We now give a very straight forward second order example in which the stability must come from the delay term. Here, the reader will see that problems will occur in the variation of parameters formula if we try to do this with variable coefficients in the part of the equation from which we hope to derive stability. But there are ways to choose the system from the second order equation so that the variable matrix $A(t)$ will commute with its integral. In that case, the work can proceed in the manner given below. The details are nontrivial and we will provide them later when we obtain stability using Schauder's fixed point theorem.

Consider the linear equation

$$x'' + ax' + bx(t-r) = 0 \tag{1.4.5}$$

and write it as the system

$$\begin{aligned} x' &= y \\ y' &= -ay - bx + (d/dt)\int_{t-r}^{t} bx(s)\,ds \end{aligned} \tag{1.4.5.1}$$

which is then expressed as the vector system

$$X' = AX + (d/dt)\int_{t-r}^{t} BX(s)\,ds \tag{1.4.5.2}$$

where A is the constant matrix and B is

$$\begin{pmatrix} 0 & 0 \\ b & 0 \end{pmatrix}.$$

By the variation of parameters formula

$$X(t) = e^{At}X_0 + \int_0^t e^{A(t-s)}(d/ds)\int_{s-r}^{s} BX(u)\,du\,ds \tag{1.4.5.3}$$

or upon integration by parts as in the last problem,

$$\begin{aligned} X(t) = e^{At}X_0 &+ \int_{t-r}^{t} BX(u)\,du - e^{At}\int_{-r}^{0} BX(u)\,du \\ &+ A\int_0^t e^{A(t-s)}\int_{s-r}^{s} BX(u)\,du\,ds. \end{aligned} \tag{1.4.5.4}$$

In order for $e^{At} \to 0$ as $t \to \infty$ we need

$$a > 0, \quad b > 0. \tag{1.4.5.5}$$

Let $\psi : [-r, 0] \to R^2$ be a given continuous initial function and define

$$\mathcal{S} = \{\phi : [-r, \infty] \to R^2 \mid \phi_0 = \psi, \phi \in \mathcal{C}, \phi(t) \to 0 \text{ as } t \to \infty\}.$$

Then define $P : \mathcal{S} \to \mathcal{S}$ by

$$(P\phi)(t) = e^{At}\psi(0) + \int_{t-r}^{t} B\phi(u)\,du - e^{At}\int_{-r}^{0} B\psi(u)\,du$$
$$+ A\int_{0}^{t} e^{A(t-s)}\int_{s-r}^{s} B\phi(u)\,du\,ds. \tag{1.4.5.6}$$

We denote by $|\cdot|$ a suitable matrix norm.

Theorem 1.4.5. *Let $a > 0, b > 0$. If, in addition,*

$$(br)(1 + \int_{0}^{t} |Ae^{A(t-s)}|ds) < 1 \tag{1.4.5.7}$$

then every solution of (1.4.5) tends to 0 as $t \to \infty$.

Proof. Since e^{At} is an L^1 function, if $\phi \in S$ then $(P\phi)(t) \to 0$ as $t \to \infty$. A calculation shows that P is a contraction just in case (1.4.5.7) holds.

This will acknowledge that much of the material in this section was first published in *Dynamic Systems and Applications*, as detailed in Burton and Furumochi (2001a), which is published by Dynamic Publishers.

1.5 Ultimate Boundedness and Periodicity

The reader who wants to pursue stability may skip this section. The next section will pick up the stability discussion where we left off in the previous section.

When we discussed

$$x'(t) = -a(t)x(t) + b(t)g(x(t - r(t))) \tag{1.4.2}$$

in the last section we began by talking of the challenges presented for Liapunov's direct method. Those challenges are multiplied many fold when we add a bounded perturbation, obtaining

$$x'(t) = -a(t)x(t) + b(t)g(x(t - r(t))) + p(t), \tag{1.5.1}$$

and hope to prove that solutions are bounded using Liapunov functionals. In a most pleasing way, those challenges vanish when we apply contraction mappings to the problem. Our mapping P in the proof of Theorem 1.4.2 now becomes

$$(P\phi)(t) = \psi(t) \text{ if } t \leq 0$$

and for $t \geq 0$ then

$$\begin{aligned}(P\phi)(t) &= e^{-\int_0^t a(s)\,ds}\psi(0) \\ &+ \int_0^t e^{-\int_s^t a(u)du} b(s)g(\phi(s - r(s)))\,ds + \int_0^t e^{-\int_s^t a(u)\,du} p(s)\,ds.\end{aligned}$$

It turns out that P will map a certain set of properly bounded functions into itself. But the really striking advantage is that when we seek to show that we have a contraction mapping, then the function p is totally eliminated from consideration. In marked contrast, when using Liapunov's direct method with Liapunov functional V we find a term in V' of the form $\frac{\partial V}{\partial x}p(t)$ which introduces endless difficulties. Moreover, even when we make severe assumptions and arrive at the classical pair

$$\begin{aligned}W_1(|x(t)|) &\leq V(t, x_t) \leq W_2(\|x_t\|) \\ V'(t, x_t) &\leq -W_3(|x(t)| + M\end{aligned}$$

for M a positive constant, we still face formidable challenges as may be seen in Yoshizawa (1966; pp. 206-207) and Burton (1985).

Stability is a form of boundedness and throughout the literature boundedness is linked to the existence of periodic solutions. Surprisingly late in history, Massera (1950) proved that if a scalar equation

$$x' = f(t, x)$$

enjoys unique solutions, possesses a bounded solution, and satisfies $f(t + L, x) = f(t, x)$ for all (t, x) and some $L > 0$, then it also has a periodic solution. He also proved a partial extension to second order equations. Burton (1985) traces a long line of results linking uniform boundedness and uniform ultimate boundedness to the existence of periodic solutions in general functional differential equations. More recently, Burton and Zhang (1990) showed that uniform ultimate boundedness alone suffices to yield a periodic solution under suitable continuity and periodic conditions.

Very early in the study of differential equations we see that boundedness of solutions of a perturbed equation is linked to stability of an unperturbed equation. If A is an $n \times n$ constant matrix all of whose characteristic roots have negative real parts and if $p : [0, \infty) \to R^n$ is continuous then any solution $x(t, t_0, x_0)$ of

$$x' = Ax \tag{1.5.2}$$

can be expressed as

$$x(t, t_0, x_0) = e^{A(t-t_0)}x_0 \tag{1.5.3}$$

so there are positive constants K and α with

$$|x(t, t_0, x_0)| \leq Ke^{-\alpha(t-t_0)}|x_0| \tag{1.5.4}$$

for $t \geq t_0$. We readily verify that the zero solution of (1.5.3) is uniformly asymptotically stable. By the variation of parameters formula any solution $x(t, t_0, x_0)$ of

$$x' = Ax + p(t) \tag{1.5.5}$$

can be written as

$$x(t, t_0, x_0) = e^{A(t-t_0)}x_0 + \int_{t_0}^{t} e^{A(t-s)}p(s)\, ds. \tag{1.5.6}$$

If there is a constant D with $|p(t)| \leq D$ then

$$|x(t, t_0, x_0)| \leq Ke^{-\alpha(t-t_0)}|x_0| + (KD/\alpha). \tag{1.5.7}$$

Relation (1.5.7) is the classical representation of uniform boundedness and uniform ultimate boundedness.

Definition 1.5.1. (i) *Solutions of (1.5.5) are said to be* uniformly bounded *if for each $B_1 > 0$ there exists $B_2 > 0$ such that*

$$[t_0 \geq 0, \quad |x_0| \leq B_1, \quad t \geq t_0] \Rightarrow |x(t, t_0, x_0)| < B_2. \tag{1.5.8}$$

(ii) *Solutions of (1.5.5) are said to be* uniformly ultimately bounded for bound B *if there is a $B > 0$ and for each $B_1 > 0$ there is a $T > 0$ such that*

$$[t_0 \geq 0, \quad |x_0| \leq B_1, \quad t \geq t_0 + T] \Rightarrow |x(t, t_0, x_0)| < B. \tag{1.5.9}$$

To translate these properties into functional differential equations terms, we simply replace x_0 by the initial function ψ and replace $|x_0|$ by $\|\psi\|$. If we take $B = 0$ we see the similarity to UAS. Kato (1980) has shown the independence of uniform ultimate boundedness and uniform boundedness for autonomous systems with a finite delay.

The ideas for this section are conveyed very well through the details of working with a perturbed form of (1.4.2). Thus, let

$$p : [0, \infty) \to R \quad \text{be continuous} \tag{1.5.10}$$

and suppose there is $Q > 0$ with

$$\int_0^t e^{-\int_s^t a(u)du}|p(s)|ds \leq Q. \tag{1.5.11}$$

We extend (1.4.2) and consider the half-linear equation

$$x'(t) = -a(t)x(t) + b(t)g(x(t - r(t))) + p(t) \tag{1.5.12}$$

where we are again interested in the case where a can be negative some of the time, a and b are related on average, while a, b, and r' can be unbounded. In particular, we assume that a, b, and r are continuous, that

$$\int_0^t a(s)ds \to \infty \text{ as } t \to \infty, \tag{1.5.12.1}$$

there is an $\alpha < 1$ with

$$\int_0^t e^{-\int_s^t a(u)du}|b(s)|\, ds \leq \alpha, \quad t \geq 0, \tag{1.5.12.2}$$

$$0 \leq r(t), \quad t - r(t) \to \infty \text{ as } t \to \infty, \tag{1.5.12.3}$$

and if $x, y \in R$ then

$$g(0) = 0 \text{ and } |g(x) - g(y)| \leq |x - y|. \tag{1.5.12.4}$$

In a corollary, that last requirement will become local.

One of our main advantages with fixed point theory is allowing $a(t)$ to change sign. By (1.5.12.1) for each t_0 we can find $c = c(t_0) > 1$ with $e^{-\int_{t_0}^t a(s)ds} \leq c(t_0)$ for $t \geq t_0$. We are going to give two proofs in which we state that we have a uniform boundedness property at $t_0 = 0$. That is sufficient in so many problems of interest. But if we want the same result for any t_0 then we must ask for a uniform $c \neq c(t_0)$. In this form, the results we prove are usually referred to in the literature as "equi" instead of "uniformly".

Theorem 1.5.1. *If (1.5.10), (1.5.11), and (1.5.12.1)-(1.5.12.4) hold, then solutions of (1.5.12) are uniformly bounded at $t_0 = 0$.*

Proof. By (1.5.12.1) we can find $c \geq 1$ with

$$e^{-\int_0^t a(s)\,ds} \leq c.$$

Let $B_1 > 0$ be given, $\psi : (-\infty, 0] \to R$ be continuous with $\|\psi\| \leq B_1$ and determine $B_2 > 0$ with

$$cB_1 + \alpha B_2 + Q = B_2.$$

We will show that $|x(t, 0, \psi)| \leq B_2$ for $t \geq 0$. Let

$$\mathcal{S} = \{\phi : (-\infty, \infty) \to R \mid \phi \in C, \phi_0 = \psi, \|\phi\| \leq B_2\}$$

and define $P : \mathcal{S} \to \mathcal{S}$ by

$$(P\phi)(t) = \psi(t) \text{ if } t \leq 0$$

and for $t \geq 0$ let

$$(P\phi)(t) = e^{-\int_0^t a(s)\,ds}\psi(0) + \int_0^t e^{-\int_s^t a(u)du} b(s) g(\phi(s - r(s)))\,ds + \int_0^t e^{-\int_s^t a(u)du} p(s)\,ds. \tag{1.5.12.5}$$

Notice that

$$|(P\phi)(t)| \leq cB_1 + \alpha B_2 + Q = B_2 \tag{1.5.12.6}$$

so $P : S \to S$. Clearly, P is a contraction as we saw in the proof of Theorem 1.4.2. Hence, there is a unique fixed point which is a solution residing in $\mathcal{S}$ and that completes the proof.

We would like to have essentially the same kind of result for use in proving the existence of periodic solutions without asking for a global Lipschitz condition on g. That can be done quite simply by focusing on (1.5.12.4). There are many periodic results in which it is stated that there is a periodic solution if the forcing function is not too large. Thus, if α is fixed less than 1 and if c is nearly 1 (so that $a(t)$ is not too negative) then we can make B_1 and Q small enough to satisfy the relation (1.5.12.4) keeping B_2 small.

Corollary. *Let all the conditions of Theorem 1.5.1 hold except (1.5.12.4). Suppose there are $B_1 > 0, B_2 > 0$ with $cB_1 + \alpha B_2 + Q = B_2$ and when $|x|, |y| \leq B_2$ then*

$$g(0) = 0 \quad \text{and} \quad |g(x) - g(y)| \leq |x - y|. \tag{1.5.12.7}$$

If ψ is an initial function with $\|\psi\| \leq B_1$ then $|x(t, 0, \psi)| \leq B_2$ for $t \geq 0$.

The proof of the next result contains an unexpected turn of events. We define a mapping set which should lead us to ultimate boundedness which is not uniform. But a simple observation shows the stronger conclusion. It is an idea which has other applications. For example, the reader can now go back to Theorem 1.4.2 and prove UAS by strengthening (1.4.2.1) through following the discussion before Theorem 1.5.1 concerning the properties of c.

Theorem 1.5.2. *If the conditions of Theorem 1.5.1 hold then solutions of (1.5.12) are uniformly ultimately bounded for bound B at $t_0 = 0$.*

Proof. Let $B_1 > 0$ be given and find the numbers c and B_2 in the proof of Theorem 1.5.1 so that $\|\psi\| \leq B_1$ implies that $|x(t,0,\psi)| \leq B_2$ for $t \geq 0$. We will first show that for

$$B := 2Q/(1-\alpha)$$

and for B^* the B-ball, then the distance from $x(t,0,\psi)$ to B^*, denoted by $\rho(x(t,0,\psi),B^*)$, tends to 0 as $t \to \infty$. Then, we will notice that the solutions enter and remain in a ball of radius $B+\varepsilon$ for $t > T_3$, where T_3 depends only on B_1.

Define

$$\mathcal{S} = \{\phi : R \to R \mid \phi \in \mathcal{C}, \phi_0 = \psi, \|\phi\| \leq B_2, \rho(\phi(t),B^*) \to 0 \quad \text{as} \quad t \to \infty\}.$$

Notice that $(\mathcal{S}, \|\cdot\|)$ is a complete metric space. It is contained in the Banach space of bounded continuous functions with the supremum norm. Any Cauchy sequence has a limit and that limit is in $\mathcal{S}$. To see this, let $\{\phi_n\}$ be a Cauchy sequence in $\mathcal{S}$ with limit ϕ which is a bounded continuous function. We will show that $\rho(\phi(t),B^*) \to 0$. Let $\varepsilon > 0$ be given and find N such that $n,m \geq N$ imply that $\|\phi_n - \phi_m\| < \varepsilon$ and $\|\phi_n - \phi\| < 2\varepsilon$. Thus, there is a $T(N) > 0$ such that $\rho(\phi_N(t),B^*) < \varepsilon$ if $t \geq T(N)$ so

$$\rho(\phi(t),B^*) \leq \rho(\phi(t),\phi_N(t)) + \rho(\phi_N(t),B^*) < 3\varepsilon.$$

Define $P : \mathcal{S} \to \mathcal{S}$ by (1.5.12.5) again:

$$(P\phi)(t) = \psi(t) \text{ if } t \leq 0$$

and for $t \geq 0$ let

$$(P\phi)(t) = e^{-\int_0^t a(s)\,ds}\psi(0) + \int_0^t e^{-\int_s^t a(u)du}b(s)g(\phi(s-r(s)))\,ds$$
$$+ \int_0^t e^{-\int_s^t a(u)\,du}p(s)\,ds. \qquad (1.5.12.5)$$

Fix $\phi \in \mathcal{S}$, let $\varepsilon > 0$ be given, and then find $T > 0$ so that $t \geq T$ implies that $\rho(\phi(t - r(t)), B^*) < \varepsilon/2$. For the fixed T and the fixed B_1, first find $T_1 > T$ with

$$B_1 e^{-\int_0^t a(s)\,ds} < \varepsilon/4 \quad \text{if} \quad t \geq T_1.$$

Next, find $T_2 > T_1$ with

$$B_2 e^{-\int_T^t a(u)\,du} \alpha < \varepsilon/4 \quad \text{if} \quad t \geq T_2.$$

If $t \geq T_2$ then

$$\begin{aligned}
|(P\phi)(t)| &\leq e^{-\int_0^t a(s)\,ds}|\psi(0)| + \int_0^T e^{-\int_s^t a(u)\,du}|b(s)||g(\phi(s - r(s))|\,ds \\
&+ \int_T^t e^{-\int_s^t (u)du}|b(s)|(B + \frac{\varepsilon}{2})\,ds + \int_0^t e^{-\int_s^t a(u)du}|p(s)|\,ds \\
&\leq (\varepsilon/4) + B_2 e^{-\int_T^t a(u)\,du} \int_0^T e^{-\int_s^T a(u)du}|b(s)|\,ds \\
&+ (B + \frac{\varepsilon}{2}) \int_T^t e^{-\int_s^t a(u)du}|b(s)|\,ds + Q \\
&\leq (\varepsilon/4) + B_2 e^{-\int_T^t a(u)\,du}\alpha + (B + \frac{\varepsilon}{2})\alpha + Q \\
&\leq (\varepsilon/2) + (B + \frac{\varepsilon}{2})\alpha + \frac{(1-\alpha)B}{2} \\
&= \frac{\varepsilon}{2} + B\alpha + \frac{\varepsilon\alpha}{2} + \frac{B}{2} - \frac{\alpha B}{2} \\
&= \frac{\varepsilon}{2} + \frac{B\alpha}{2} + \frac{\varepsilon\alpha}{2} + \frac{B}{2} < \varepsilon + B.
\end{aligned}$$

Thus, $P : \mathcal{S} \to \mathcal{S}$ and P is a contraction with unique fixed point ϕ. Moreover, we see that for that fixed point ϕ we have $|(P\phi)(t)| \leq B + \varepsilon$ if $t \geq T_2$. Hence, $|\phi(t)| = |(P\phi)(t)| \leq B + \varepsilon$ if $t \geq T_3$. This is the required uniform ultimate boundedness at $t_0 = 0$.

These results on boundedness are proved for the forced equation without strengthening the conditions listed in the previous section for the unforced equation. This is a typical advantage of contraction mappings over Liapunov's direct method. If $p(t)$, $a(t)$, $b(t)$, $r(t)$ are all periodic with the same period, say L, then the uniform boundedness and uniform ultimate boundedness will imply the existence of a periodic solution. See, for example, Burton (1985; p. 249). The result is proved using Horn's (1970) fixed point theorem and the Poincaré map defined by a translation of t by an amount L. Much of Burton (1985) is devoted to that type of mapping. Moreover, when g does not satisfy a Lipschitz condition, then we still have the mappings defined and we can use Horn's fixed point theorem to obtain periodic solutions.

But when the equation is half linear then there is a much simpler proof of periodicity using contraction mappings with a considerably different mapping equation. Recall that we began looking at fixed point theory by integrating the differential equation to get a mapping equation. That method worked well for simple existence of solutions, but we ran into difficulty with that method when we tried to study stability. Half linear equations utilized the variation of parameters formula and produced a mapping which was highly effective in the study of stability, circumventing many difficulties seen in Liapunov's direct method. The real difficulty with both of these mappings in the search for periodic solutions is that we have no idea what the initial condition for the periodic solution should be. Thus, we need to avoid initial conditions entirely.

Let us consider again (1.5.12) which we write as

$$x'(t) = -a(t)x(t) + b(t)g(x(t - r(t))) + p(t) \tag{1.5.13}$$

with $a(t)$, $b(t)$, $r(t)$, $p(t)$ all continuous and periodic of period $L > 0$. Again we ask that

$$\int_0^t a(s)\, ds \to \infty \text{ as } t \to \infty, \tag{1.5.12.1}$$

$$\int_0^t e^{-\int_s^t a(u)du}|b(s)|\, ds \leq \alpha < 1, \quad t \geq 0, \tag{1.5.12.2}$$

and if $x, y \in R$ then

$$g(0) = 0 \text{ and } |g(x) - g(y)| \leq |x - y|. \tag{1.5.12.4}$$

Theorem 1.5.3. *Let (1.5.12.1), (1.5.12.2), and (1.5.12.4) hold, together with the periodic assumptions. Then (1.5.13) has a unique L–periodic solution.*

Proof. Because the equation is half linear we can use an integrating factor to write

$$(x(t)e^{\int_0^t a(u)du})' = e^{\int_0^t a(u)du}[b(t)g(x(t - r(t)) + p(t)]$$

and under the assumption (later justified) that there is a solution bounded on $(-\infty, t)$, integrate both sides of that expression over the indicated interval and obtain, after rearrangement,

$$x(t) = \int_{-\infty}^t e^{-\int_s^t a(u)\, du}[b(s)g(x(s - r(s))) + p(s)]\, ds.$$

Let $(\mathcal{X}, \|\cdot\|)$ be the Banach space of continuous L-periodic functions with the supremum norm and define a mapping $P : \mathcal{X} \to \mathcal{X}$ by the equation $\phi \in \mathcal{X}$ implies that

$$(P\phi)(t) = \int_{-\infty}^{t} e^{-\int_v^t a(s)\,ds}[b(v)g(\phi(v - r(v))) + p(v)]\,dv.$$

By (1.5.12.2) the mapping is a contraction with unique fixed point $\phi \in \mathcal{X}$. Differentiate the equation for the fixed point and obtain a solution of (1.5.12).

We will return to periodic theory later when we deal with equations which are not half linear.

1.6 Fixed Points or Liapunov Functions

In this section we will study certain very standard delay equations and integrodifferential equations. We will show that under one set of assumptions we obtain better results using contraction mappings, while better results are obtained using Liapunov theory under a different set of assumptions. There are several numerical examples which are worked using crude manual methods with results only close enough to demonstrate that one method is better than the other. That is our only goal here. These examples provide interesting exercises for computations using *Maple* or *Mathmatica.* Most of the material in this section is found in Burton (2003a)

Let $a : [0, \infty) \to R$ be bounded and continuous, let r be a positive constant, and let

$$x' = -a(t)x(t - r). \tag{1.6.1}$$

Although we can treat solutions with any initial time, we will always study a solution $x(t) := x(t, 0, \psi)$ where $\psi : [-r, 0] \to R$ is a given continuous initial function and $x(t, 0, \psi) = \psi(t)$ on $[-r, 0]$. It is then known that there is a unique continuous solution $x(t)$ satisfying (1.6.1) for $t > 0$ and with $x(t) = \psi(t)$ on $[-r, 0]$.

With such ψ in mind, we can write (1.6.1) as

$$x' = -a(t + r)x(t) + (d/dt)\int_{t-r}^{t} a(s + r)x(s)\,ds \tag{1.6.2}$$

so that by the variation of parameters formula, followed by integration by parts, we have

$$\begin{aligned}x(t) &= x(0)e^{-\int_0^t a(s+r)\,ds} + \int_{t-r}^{t} a(u+r)x(u)\,du \\ &\quad - e^{-\int_0^t a(u+r)\,du} \int_{-r}^{0} a(u+r)x(u)\,du \\ &\quad - \int_0^t a(s+r)e^{-\int_s^t a(u+r)\,du} \int_{s-r}^{s} a(u+r)x(u)\,du\,ds. \end{aligned} \tag{1.6.3}$$

In a space to be defined and with a mapping defined from (1.6.3) we will find that we have a contraction mapping just in case there is a constant $\alpha < 1$ with

$$\int_{t-r}^{t} |a(u+r)|\,du + \int_0^t |a(s+r)|e^{-\int_s^t a(u+r)\,du} \int_{s-r}^{s} |a(u+r)|\,du\,ds \leq \alpha. \tag{1.6.4}$$

As we are interested in asymptotic stability we will need

$$\int_0^t a(s+r)\,ds \to \infty \text{ as } t \to \infty. \tag{1.6.5}$$

This section compares results from a certain application of fixed point theory with a certain common Liapunov functional. In theory, there is no comparison at all. It is known that if we have a strong type of stability, then there exists a Liapunov functional of a certain type. The fact that we can not find that Liapunov functional gives validity to this type of comparison. With that in mind, from (1.6.4) it is easy to see one of the advantages of fixed point theory over Liapunov theory. The latter usually requires $a(t+r) > 0$. If $a(t+r) \geq 0$, then a very good bound is obtained in (1.6.4) with little effort (see Example 1.6.2). If $a(t+r)$ changes sign then (1.6.4) can still hold, although a good bound on the second integral is more difficult (see Example 1.6.3).

Theorem 1.6.1. *Let (1.6.4) and (1.6.5) hold. Then for every continuous initial function $\psi : [-r, 0] \to R$ the solution $x(t, 0, \psi)$ of (1.6.1) is bounded and tends to zero as $t \to \infty$.*

Proof. Let $(\mathcal{B}, \|\cdot\|)$ be the Banach space of bounded and continuous functions $\phi : [-r, \infty) \to R$ with the supremum norm. Let $(\mathcal{S}, \|\cdot\|)$ be the complete metric space with supremum metric consisting of functions $\phi \in \mathcal{B}$ such that $\phi(t) = \psi(t)$ on $[-r, 0]$ and $\phi(t) \to 0$ as $t \to \infty$.

Define $P : \mathcal{S} \to \mathcal{S}$ by

$$(P\phi)(t) = \psi(t) \text{ on } [-r, 0]$$

and

$$\begin{aligned}(P\phi)(t) &= \psi(0)e^{-\int_0^t a(s+r)\,ds} + \int_{t-r}^t a(u+r)\phi(u)\,du \\ &\quad - e^{-\int_0^t a(u+r)\,du} \int_{-r}^0 a(u+r)\psi(u)\,du \\ &\quad - \int_0^t a(s+r)e^{-\int_s^t a(u+r)\,du} \int_{s-r}^s a(u+r)\phi(u)\,du\,ds.\end{aligned}$$

Clearly, $P\phi$ is continuous, $(P\phi)(0) = \psi(0)$, and from (1.6.4) it follows that $P\phi$ is bounded. Also, P is a contraction by (1.6.4).

We can show that the last term tends to zero by using the classical proof that the convolution of an L^1–function with a function tending to zero, does also tend to zero. Here are the details. Let $\phi \in \mathcal{S}$ be fixed and let $0 < T < t$. Denote the supremum of $|\phi|$ on $[0, \infty)$ by $\|\phi\|$ and the supremum of $|\phi|$ on $[T, \infty)$ by $\|\phi\|_{[T,\infty)}$. Consider (1.6.4) and (1.6.5). We have

$$\begin{aligned}&\int_0^t |a(s+r)|e^{-\int_s^t a(u+r)du} \int_{s-r}^s |a(u+r)\phi(u)|\,du\,ds \\ &\le \int_0^T |a(s+r)|e^{-\int_s^T a(u+r)du} \int_{s-r}^s |a(u+r)|\,du\,ds\|\phi\|e^{-\int_T^t a(u+r)\,du} \\ &+ \int_T^t |a(s+r)|e^{-\int_s^t a(u+r)du} \int_{s-r}^s |a(u+r)|\,du\,ds\|\phi\|_{[T-r,\infty)} \\ &\le \alpha\|\phi\|e^{-\int_T^t a(u+r)du} + \alpha\|\phi\|_{[T-r,\infty)}.\end{aligned}$$

For a given $\varepsilon > 0$ take T so large that $\alpha\|\phi\|_{[T-r,\infty)} < \varepsilon/2$. For that fixed T, take t^* so large that $\alpha\|\phi\|e^{-\int_T^t a(u+r)du} < \varepsilon/2$ for all $t > t^*$. We then have that last term smaller than ε for all $t > t^*$. Thus, $P : \mathcal{S} \to \mathcal{S}$ is a contraction with unique fixed point in $\mathcal{S}$.

The following example is designed to be compared with Example 6.1.4 using Liapunov's direct method.

Example 1.6.1. In (1.6.1), let

$$a(t) = 1.1 + \sin t.$$

The conditions of Theorem 1.6.1 are satisfied if

$$2(1.1r + 2\sin(r/2)) < 1.$$

This is approximated by $0 < r < .2$.

Proof. We first estimate

$$\int_{t-r}^{t} |a(u+r)|\, du = \int_{t}^{t+r} (1.1 + \sin u)\, du.$$

It is easy to see that for $r < 1$ this integral is dominated by

$$\int_{(\pi-r)/2}^{(\pi+r)/2} (1.1 + \sin u)\, du = 1.1r + 2\sin(r/2).$$

The second integral in (1.6.4) is seen to be bounded by

$$1.1r + 2\sin(r/2)$$

as well. Hence, to satisfy (1.6.4) we need

$$2(1.1r + 2\sin(r/2)) < 1.$$

A very rough estimate (taking $\sin(r/2) = r/2$) yields the requirement

$$0 < r < \frac{1}{4.2}.$$

We will see that this can be compared to a result using a Liapunov functional and, in this case, the Liapunov functional yields a significantly better result. But when $a(t)$ is allowed to vanish, or even change sign, then the situation reverses and the fixed point method is better. We will see this in two steps.

First, if $a(t) \geq 0$, then the Liapunov functional fails to address the problem, while the fixed point theorem yields a result fully consistent with that of Example 1.6.1.

Next, if $a(t)$ becomes negative, then the Liapunov functional fails, while the fixed point theorem yields a stability result which is significantly poorer than in the first two cases because of inherent difficulties in estimating the integrals in (1.6.4). Application of *Maple* or *Mathematica* yields interesting results.

Example 1.6.2. In (1.6.1), let

$$a(t) = 1 + \sin t.$$

The conditions of Theorem 1.6.1 are satisfied if

$$2(r + 2\sin(r/2)) < 1.$$

Proof. The first integral in (1.6.4) is again estimated in the same way as in Example 1.6.1 by

$$r + 2\sin(r/2).$$

The second integral in (1.6.4) is estimated by the same amount. We need

$$2(r + 2\sin(r/2)) < 1.$$

A crude estimate with $\sin(r/2) = r/2$ asks that $r < 1/4$.

Intuitively, Example 1.6.1 should be more strongly stable than Example 1.6.2. Yet, a look at (1.6.4) readily reveals why our results state the opposite. We conjecture that a different fixed point mapping might reverse the relation.

Example 1.6.3. Consider the scalar equation

$$x'(t) = -(1 + 2\sin t)x(t - r) \tag{1.6.6}$$

where $0 < r < 1$ and r is to be restricted further later. The stability must not only be derived from $x(t - r)$, but the coefficient function

$$a(t) = 1 + 2\sin t \tag{1.6.7}$$

changes sign. We need to establish some bounds toward fulfilling the conditions of Theorem 1.6.1. The calculations are crude, but establish that we can still have stability when the coefficient changes sign.

We will conclude asymptotic stability when

$$(r + 4\sin(r/2))(2 + 2e^2) < 1.$$

This is approximately $0 \leq r < .02$ which will be established using the following two lemmas.

Lemma 1. *For $a(t) = 1 + 2\sin t$ and $0 < r < 1$ then*

$$\int_{t-r}^{t} |a(u + r)|\, du \leq 2((r/2) + 2\sin(r/2)). \tag{1.6.8}$$

Proof. We have

$$\begin{aligned}\int_{t-r}^{t} |a(u+r)|\, du &= \int_{t}^{t+r} |1+2\sin u|\, du \\ &\leq \int_{(\pi-r)/2}^{(\pi+r)/2} (1+2\sin u)\, du \\ &= 2((r/2)+2\sin(r/2)).\end{aligned}$$

Lemma 2. *Under the same conditions as in Lemma 1,*

$$\begin{aligned}&J := \\ &\int_0^t |1+2\sin(s+r)| e^{-\int_s^t (1+2\sin(u+r))\, du} \int_{s-r}^{s} |(1+2\sin(u+r))|\, du\, ds \\ &\leq (r+4\sin(r/2))(1+2e^2). \qquad (1.6.9)\end{aligned}$$

Proof. Note that

$$\int_{s-r}^{s} |(1+2\sin(u+r))| du \leq r+4\sin(r/2)$$

by the mean value theorem. Thus,

$$\begin{aligned}J &\leq (r+4\sin(r/2)) \int_0^t (1+2\sin(s+r)+2) e^{-\int_s^t (1+2\sin(u+r))\, du}\, ds \\ &\leq (r+4\sin(r/2))\left(1+2\int_0^t e^{-\int_s^t (1+2\sin(u+r))\, du}\, ds\right) \\ &\leq (r+4\sin(r/2))\left(1+2e^2\int_0^t e^{-\int_s^t du} ds\right) \\ &\leq (r+4\sin(r/2))(1+2e^2).\end{aligned}$$

Putting the results of the two lemmas together and referring to (1.6.4) we require that

$$(r+4\sin(r/2))(2+2e^2) < 1. \qquad (1.6.10)$$

We will now see what might be done with Liapunov's direct method as a comparison to fixed point methods. Return to

$$x' = -a(t)x(t-r) \tag{1.6.1}$$

and

$$x' = -a(t+r)x + (d/dt)\int_{t-r}^{t} a(s+r)x(s)\,ds. \tag{1.6.2}$$

Theorem 1.6.2. *If there is a $\delta > 0$ with*

$$a(t+r) \geq \delta, \text{ for all } t \geq 0, \tag{1.6.11}$$

an $\varepsilon > 0$ with

$$a(t+r)\int_{t-r}^{t} a(s+r)\,ds - 2 + r \leq -\varepsilon \text{ for all } t \geq 0, \tag{1.6.12}$$

and if there is a $\gamma > 0$ with

$$\gamma[a(t) + a(t+r)] \leq (\varepsilon/2)a(t+r) \text{ for all } t \geq 0 \tag{1.6.13}$$

then the zero solution of (1.6.1) is uniformly asymptotically stable.

Proof. From (1.6.2) we have

$$(x - \int_{t-r}^{t} a(s+r)x(s)\,ds)' = -a(t+r)x$$

and we select a first Liapunov functional as

$$V_1(t, x_t) = \left(x - \int_{t-r}^{t} a(s+r)x(s)\,ds\right)^2 + \int_{-r}^{0}\int_{t+s}^{t} a(u+r)x^2(u)\,du\,ds$$

so that the derivative along a solution of (1.6.2) is

$$\begin{aligned}
V_1'(t,x_t) &= 2\left(x - \int_{t-r}^{t} a(s+r)x(s)\,ds\right)(-a(t+r)x) \\
&\quad + \int_{-r}^{0} a(t+r)x^2(t)\,du - \int_{-r}^{0} a(t+s+r)x^2(t+s)\,ds \\
&\le -2a(t+r)x^2 \\
&\quad + a^2(t+r)\int_{t-r}^{t} a(s+r)\,dsx^2 + \int_{t-r}^{t} a(s+r)x^2(s)\,ds \\
&\quad + a(t+r)x^2(t)r - \int_{t-r}^{t} a(s+r)x^2(s)\,ds \\
&= [-2a(t+r) + a^2(t+r)\int_{t-r}^{t} a(s+r)ds + ra(t+r)]x^2 \\
&= a(t+r)[-2 + a(t+r)\int_{t-r}^{t} a(s+r)\,ds + r]x^2 \\
&\le -\varepsilon a(t+r)x^2 \\
&\quad \text{(In Remark 1.6.1 below, use this and (1.6.13).)} \\
&\le -\varepsilon\delta x^2.
\end{aligned}$$

Now, we need to define a second Liapunov functional and add them together to make a positive definite Liapunov functional. Define

$$V_2(t,x_t) = \gamma[x^2 + \int_{t-r}^{t} a(s+r)x^2(s)\,ds]$$

so that the derivative along a solution of (1.6.1) is

$$\begin{aligned}
V_2'(t,x_t) &\le \gamma[-2a(t)xx(t-r) + a(t+r)x^2 - a(t)x^2(t-r)] \\
&\le \gamma[a(t)x^2 + a(t)x^2(t-r) + a(t+r)x^2 - a(t)x^2(t-r)] \\
&= \gamma[a(t) + a(t+r)]x^2 \\
&\le (\varepsilon/2)a(t+r)x^2.
\end{aligned}$$

If we define

$$V(t,x_t) = V_1(t,x_t) + V_2(t,x_t)$$

then we have

$$V'(t,x_t) \le -(\varepsilon/2)\delta x^2.$$

We can now find wedges with

$$W_1(|x(t)|) \leq V(t, x_t) \leq W_2(\|x_t\|)$$

and, since (1.6.11) and (1.6.12) imply that $x(t)$ is bounded, conclude that the zero solution is uniformly asymptotically stable.

Theorem 1.6.2 requires (1.6.11) so it can not be compared to Example 1.6.2 or 1.6.3. But it does compare very favorably with Example 1.6.1.

Example 1.6.4. Let

$$a(t) = 1.1 + \sin t.$$

Theorem 1.6.2 holds if there is an $\varepsilon > 0$ with

$$2.1(1.1r + 2\sin(r/2)) - 2 + r < -\varepsilon.$$

Proof. Note that (1.6.11) is satisfied with $\delta = .1$. To satisfy (12) we have

$$\begin{aligned}(1.1 + \sin(t+r)) \int_{t-r}^{t} (1.1 + \sin(s+r))\, ds - 2 + r \\ \leq (1.1 + \sin(t+r))(1.1r + 2\sin(r/2)) - 2 + r \\ \leq 2.1(1.1r + 2\sin(r/2)) - 2 + r < -\varepsilon.\end{aligned}$$

We make another very rough estimate by taking $\sin(r/2) = r/2$ and say that we need

$$5.41r - 2 < -\varepsilon$$

or

$$r < 2/5.41.$$

Remark 1.6.1. *We have remarked earlier that, once a Liapunov functional is found which is positive definite and whose derivative is not positive, then there are many* ad hoc *techniques for getting much out of the Liapunov functional. Here is a typical example. If the conditions of Theorem 1.6.2 hold, except that $a(t+r) \geq \delta > 0$ is replaced by $a(t+r) \geq 0$ and $\int_0^\infty a(t)dt = \infty$, then asymptotic stability can be concluded. In fact, by*

$V_1'(t,x_t) \le -\varepsilon a(t+r)x^2$, we have $\int_0^\infty a(t+r)x^2(t)\,dt < \infty$. This implies that $\liminf_{t\to\infty} x^2(t) = 0$. Notice that

$$\left|\frac{d}{dt}x^2(t)\right| = 2a(t)|x(t)||x(t-r)|$$
$$\le a(t)|x(t)|^2 + a(t)|x(t-r)|^2.$$

Since $a(t)|x(t-r)|^2$ and $a(t)|x(t)|^2$ are in $L^1[0,\infty)$, we have $\frac{d}{dt}x^2(t) \in L^1[0,\infty)$. Thus, $\lim_{t\to\infty} x(t) = 0$.

Remark 1.6.2. *We do not see a clear way to interpret a relation between the fixed point condition (1.6.4) and the Liapunov condition (1.6.12). Sometimes one is better than the other. Sometimes the fixed point condition gives results when the Liapunov condition can not. But the next problem will conclude with a much clearer relation.*

We now consider a Volterra equation. The Liapunov functional just used was motivated by the one derived by Burton and Mahfoud (1983) for

$$x'(t) = A(t)x + \int_0^t C(t,s)x(s)\,ds \tag{1.6.14}$$

where we suppose there is a function $G(t,s)$ with $(\partial G(t,s)/\partial t) = C(t,s)$; for example, $G(t,s) = -\int_t^\infty C(u,s)du$. Then we can write (1.6.14) as

$$x' = A(t)x - G(t,t)x + (d/dt)\int_0^t G(t,s)x(s)\,ds$$

or

$$x' = Q(t)x + (d/dt)\int_0^t G(t,s)x(s)ds. \tag{1.6.15}$$

In Burton and Mahfoud (1983) there is the following basic result for such equations.

Theorem 1.6.3. *Suppose there are constants Q_1, Q_2, J, R with $R < 2$ such that*

$$0 < Q_1 \le |Q(t)| \le Q_2, \tag{1.6.16}$$

$$\int_0^t |G(t,s)|\,ds \le J < 1, \tag{1.6.17}$$

and

$$\int_0^t |G(t,s)|\,ds + \int_t^\infty |G(u,t)|\,du \le RQ_1/Q_2. \tag{1.6.18}$$

Suppose also that there is a continuous function $h : [0, \infty) \to [0, \infty)$ with $|G(t, s)| \leq h(t - s)$ and $h(u) \to 0$ as $t \to \infty$. Then the zero solution of (1.6.14) is stable if and only if $Q(t) < 0$.

The proof is based on a Liapunov functional

$$V(t, x_t) = (x - \int_0^t G(t, s)x(s)\, ds)^2 + Q_2 \int_0^t \int_t^\infty |G(u, s)|\, du x^2(s)\, ds.$$

Slight additional assumptions (not stated here) yield asymptotic stability.

We may obtain a corollary stating that if Q is constant and G is of convolution type then (1.6.16)-(1.6.18) can be replaced by $\int_0^t |G(t-s)|ds \leq J < 1$. And this demands exactly half as much as the fixed point theorem requires.

The next example shows asymptotic stability using this theorem when $n > 24$. After Theorem 1.6.6 we will consider the same equation using fixed point techniques and obtain asymptotic stability for $n > 4.2$. This, again, shows that sometimes the fixed point theory yields better results and sometimes the Liapunov function does.

Example 1.6.5. Let

$$x' = (\sin t)x - (1.1)(n - 1) \int_0^t (1 + t - s)^{-n} x(s)\, ds$$

where $n > 25.1$. Then the conditions of Theorem 1.6.3 hold and the zero solution is stable (actually it is asymptotically stable).

Proof. The equation can be written as

$$x' = (\sin t - 1.1)x + (d/dt)(1.1) \int_0^t (1 + t - s)^{1-n} x(s)\, ds$$

with

$$G(t - s) = (1.1)(1 + t - s)^{1-n}.$$

We have

$$Q_1 = .1, \quad Q_2 = 2.1.$$

To satisfy (1.6.18) we calculate

$$\int_0^t (1.1)(1 + t - s)^{1-n}\, ds + \int_t^\infty (1.1)(1 + u - t)^{1-n} du \leq 2.2/(n - 2).$$

Then

$$RQ_1/Q_2 \leq 2(.1)/2.1 = 2/21.$$

A sufficient condition for (1.6.18) is $2.2/(n - 2) < 2/21$ or $n > 25.1$.

We are interested in comparing this result with those obtained by fixed point theorems. Two cases will be considered. First, it is interesting to see what happens in the dividing case when $Q(t) = 0$. It turns out to be stable and periodic perturbations generate periodic response in the limiting equations. Next, we replace the conditions by averaging conditions and obtain asymptotic stability results.

The case

$$Q(t) = 0 \tag{1.6.19}$$

is interesting in its own right. In addition, the method of proof in the above theorem completely fails in that case. We examine the problem using fixed point theory.

Theorem 1.6.4. *Let Q be defined in (1.6.15) and let (1.6.17) and (1.6.19) hold. Then for each $x(0) \neq 0$ there is a unique bounded and continuous solution $x(t, 0, x(0))$ of (1.6.14) on $[0, \infty)$. Moreover, that solution does not converge to zero.*

Proof. Let $(\mathcal{B}, \|\cdot\|)$ be the Banach space of bounded continuous $\phi : [0, \infty) \to R$ with the supremum norm and define $P : \mathcal{B} \to \mathcal{B}$ by $\phi \in \mathcal{B}$ implies

$$(P\phi)(t) = x(0) + \int_0^t G(t, s)\phi(s)\, ds. \tag{1.6.20}$$

Clearly, if $P\phi = \phi$ then $\phi(0) = x(0)$, ϕ is continuous, and $\phi'(t) = (d/dt)\int_0^t G(t, s)\phi(s)\, ds$ and so ϕ satisfies (1.6.15) and (1.6.14). Also, P does map $\mathcal{B}$ into itself and we have

$$|(P\phi)(t) - (P\psi)(t)| \leq \int_0^t |G(t, s)||\phi(s) - \psi(s)|\, ds \leq J\|\phi - \psi\|$$

so the unique fixed point is established. If $\phi(t) \to 0$ then we readily prove that

$$\int_0^t G(t, s)\phi(s)\, ds \to 0$$

leaving $\phi(t) \to x(0) \neq 0$, a contradiction.

To see that the zero solution is stable, every solution $x(t, 0, x(0))$ of (1.6.14) can be expressed as $x(t) = x(0)x(t, 0, 1)$. If $|x(t, 0, 1)| \leq M$, then for a given $\varepsilon > 0$ we take $\delta = \varepsilon/M$. Then $|x(0)| \leq \delta$ implies $|x(t)| \leq \delta M = \varepsilon$. This completes the proof.

It is interesting to determine just how strongly stable the zero solution of (1.6.14) is when (1.6.17) and (1.6.19) hold. One measure of strength is to perturb (1.6.14) with a continuous function having a bounded integral. Then the same proof will yield a bounded continuous solution. Specialize that to ask that the perturbation be periodic. Then develop the limiting equation. We find that the limiting equation will have a periodic solution. Here are the details.

Consider

$$x' = A(t)x + \int_0^t C(t,s)x(s)\,ds + p(t)$$

and suppose there is a positive constant T with

$$p(t+T) = p(t), \quad \int_0^T p(s)ds = 0,$$
$$C(t+T, s+T) = C(t,s), \quad G(t+T, s+T) = G(t,s). \tag{1.6.21}$$

The limiting equation is obtained by defining $y(t) = x(t+nT)$ for $n = 1, 2, 3, ...$ and then letting $n \to \infty$. We write the result as

$$z' = A(t)z + \int_{-\infty}^t C(t,s)z(s)\,ds + p(t) \tag{1.6.22}$$

under the assumption that

$$\int_{-\infty}^t |C(t,s)|\,ds \tag{1.6.23}$$

is bounded and continuous. See Burton (1985; p. 89).

Write (1.6.22) as in (1.6.15) in the form

$$z' = Q(t)z + (d/dt)\int_{-\infty}^t G(t,s)z(s)\,ds + p(t). \tag{1.6.24}$$

Theorem 1.6.5. *In (1.6.24) we suppose that (1.6.21) holds and that*

$$Q(t) = 0 \text{ and } \int_{-\infty}^t |G(t,s)|\,ds \le J < 1. \tag{1.6.25}$$

Then (1.6.22) has a T-periodic solution.

Proof. Let $(\mathcal{B}_T, \|\cdot\|)$ be the Banach space of continuous T-periodic functions with the supremum norm. Define $P : \mathcal{B}_T \to \mathcal{B}_T$ by $\phi \in \mathcal{B}_T$ implies that

$$(P\phi)(t) = \int_{-\infty}^t G(t,s)\phi(s)\,ds + \int_0^t p(s)\,ds.$$

If $P\phi = \phi$ then ϕ satisfies (1.6.24) and (1.6.22). By (1.6.25) we see that P is a contraction.

Remark 1.6.3. *If $G(t,s) = G(t-s)$ and if p is almost periodic with bounded integral, then $p : AP \to AP$ and (1.6.22) has an AP solution.*

We could also show that if (1.6.14) is given an $L^1[0,\infty)$ perturbation then a companion of Theorem 1.6.3 would give a bounded solution. It would be interesting to show that if (1.6.14) were given a perturbation $p(t)$ for which $\int_{t-1}^{t} |p(s)|ds \to 0$ as $t \to \infty$, then there is still a bounded solution. All of these results offer a measure of the strength of the stability in case $Q(t) = 0$.

We now consider the case in which $\int_0^t Q(s)ds \to -\infty$ as $t \to \infty$. Return to

$$x' = A(t)x + \int_0^t C(t,s)x(s)\,ds \tag{1.6.14}$$

and

$$x' = Q(t)x + (d/dt)\int_0^t G(t,s)x(s)\,ds \tag{1.6.15}$$

where we suppose that for any fixed $T > 0$ then

$$\int_0^T |G(t,v)|dv \to 0 \text{ and } \int_0^t Q(s)ds \to -\infty \text{ as } t \to \infty. \tag{1.6.26}$$

Write (1.6.15) as

$$\begin{aligned} x(t) &= x_0 e^{\int_0^t Q(s)\,ds} + \int_0^t e^{\int_s^t Q(u)du}(d/ds)\int_0^s G(s,v)x(v)\,dv\,ds \\ &= x_0 e^{\int_0^t Q(s)\,ds} + e^{\int_s^t Q(u)du}\int_0^s G(s,v)x(v)dv\Big|_0^t \\ &\quad - \int_0^t e^{\int_s^t Q(u)du}Q(s)\int_0^s G(s,v)x(v)\,dv\,ds \end{aligned}$$

or

$$\begin{aligned} x(t) &= x_0 e^{\int_0^t Q(s)\,ds} + \int_0^t G(t,v)x(v)\,dv \\ &\quad - \int_0^t Q(s)e^{\int_s^t Q(u)\,du}\int_0^s G(s,v)x(v)\,dv\,ds. \end{aligned} \tag{1.6.27}$$

Theorem 1.6.6. *Suppose that (1.6.26) holds and that there is an $\alpha < 1$ with*

$$\int_0^t [|G(t,v)| + |Q(v)|e^{\int_v^t Q(u)\,du}\int_0^v |G(v,u)|\,du]\,dv \le \alpha. \tag{1.6.28}$$

Then for each x_0 there is a unique bounded solution $x(t,0,x_0)$ of (1.6.14) which tends to 0 as $t \to \infty$.

We define a mapping from (1.6.27) and the proof proceeds just as before. Recall that in Theorem 1.6.5 we needed $Q(t) < 0$.

Example 1.6.6. Let

$$x' = (\sin t)x - (1.1)(n-1)\int_0^t (1+t-s)^{-n}x(s)\,ds$$

where $n > 4.2$. Then (1.6.26) and (1.6.28) are satisfied and the conclusion of Theorem 1.6.6 holds.

Proof. The equation can be written as

$$x' = (\sin t - 1.1)x + (d/dt)(1.1)\int_0^t (1+t-s)^{1-n}x(s)\,ds$$

with

$$G(t-s) = (1.1)(1+t-s)^{1-n}.$$

Then

$$\int_0^v |G(v,u)|\,du = \int_0^v (1.1)(1+v-u)^{1-n}\,du \le 1.1/(n-2).$$

A calculation then shows that (1.6.28) holds when

$$2(1.1)/(n-2) < 1$$

as required.

Compare this to the result using a Liapunov functional in Example 1.6.5. In fact, the fixed point condition of Theorem 1.6.6 yields a stability condition at least 5 times as good as that of the Liapunov condition (1.6.18) in Theorem 1.6.3.

Example 1.6.7. If $n > 4$, then the zero solution of

$$x'(t) = (\sin t)x - (n-1)\int_0^t (1+t-s)^{-n}x(s)\,ds$$

is asymptotically stable.

Proof. To prove this, we mainly have to show that (1.6.28) is satisfied. We can write the equation as

$$x' = (\sin t - 1)x + (d/dt)\int_0^t (1+t-s)^{1-n}x(s)\,ds$$

so that

$$Q(t) = \sin t - 1 \text{ and } G(t-s) = (1+t-s)^{1-n}.$$

Substituting these values into (1.6.28) yields

$$\begin{aligned}
&\int_0^t (1+t-v)^{1-n}\,dv \\
&+ \int_0^t \left[(1-\sin v)e^{\int_v^t (\sin u - 1)du}\left|\int_0^v (1+v-u)^{1-n}du\right|\right]dv \\
&\le [1/(n-2)]\left[1 + \int_0^t (1-\sin v)e^{-[t+\cos t]+[v+\cos v]}dv\right] \\
&\le ([1/(n-2)][1+1] < 1
\end{aligned}$$

provided that $n > 4$.

Liapunov theory is now more than one hundred years old and it has been a very fruitful area. The fixed point theory used in stability seems to be in very early stages. The investigator will get better results by using both methods than by using only one of them.

We will see several other comparisons between Liapunov's direct method and fixed point techniques. Section 2.2 is devoted to four classical problems for which the fixed point technique is better in several ways than the Liapunov technique. But Section 2.3 is devoted to several forms of the delayed Liénard equation for which Liapunov techniques seem much better.

This will acknowledge that much of the material of this section was first published in *Fixed Point Theory*, as described in Burton (2003a), which is published by the House of the Book of Science.

1.7 Necessary and sufficient conditions

In this section we present recent work of Bo Zhang (2004, 2005) on linear and half-linear problems in which he obtains some sufficient conditions for stability, as well as some necessary and sufficient conditions.

Let $R = (-\infty, \infty), R^+ = [0, \infty)$, and $R^- = (-\infty, 0]$ respectively. Consider the scalar equation

$$x'(t) = -a(t)x(t) + g(t, x_t) \tag{1.7.1}$$

in which $a : R^+ \to R$ and $g : R^+ \times \mathcal{C} \to R$ are continuous with $\mathcal{C}$ being the Banach space of bounded continuous functions $\phi : R^- \to R$ with the supremum norm $\|\cdot\|$. If $x : R \to R$ is a bounded continuous function and if $t \geq 0$ is a fixed number, then x_t denotes the restriction of x to the interval $(-\infty, t]$ so that x_t is an element of $\mathcal{C}$ defined by $x_t(s) = x(t+s)$ for $s \in R^-$.

Consider first the case in which (1.7.1) is in a simple form, say

$$x'(t) = -a(t)x(t) + b(t)x(t - r(t)) \tag{1.7.2}$$

where b, $r : R^+ \to R$ are continuous functions with

$$r(t) \geq 0, \ \ t - r(t) \to \infty \quad \text{as } t \to \infty \tag{1.7.3}$$

and

$$a(t) \geq \alpha, \ \ J|b(t)| \leq a(t), \ \ t \geq 0 \tag{1.7.4}$$

for some constants $\alpha > 0$, $J > 1$. One can apply a Razumikhin technique to solve the problem. Taking the derivative of the Liapunov function $V(x) = x^2$ along a solution $x = x(t)$ of (1.7.2), we obtain

$$V'(x) \leq -\mu x^2(t), \ \ \mu > 0$$

whenever $x^2(s) < Jx^2(t)$ for all $s \in [t - r(t), t]$. It can now be argued that the zero solution of (1.7.2) is asymptotically stable.

It seems severe to ask that $a(t)$ be bounded away from zero and that $|b(t)|$ is "bounded" by $a(t)$ all the time. When $g(t, \phi)$ is not a linear functional, we may find many fundamental difficulties in the process of constructing a Liapunov function or functional.

For each $\gamma > 0$, define $\mathcal{C}(\gamma) = \{\phi \in \mathcal{C} : \|\phi\| \leq \gamma\}$. Given a function $\psi : R \to R$, we define $\|\psi\|^{[s,t]} = \sup\{|\psi(u)| : s \leq u \leq t\}$. Also, for $A > 0$ a continuous function $x : (-\infty, A) \to R$ is called a solution of (1.7.1) through $(t_0, \phi) \in R^+ \times \mathcal{C}$ if $x_{t_0} = \phi$ and x satisfies (1.7.1) on $[t_0, A)$.

Theorem 1.7.1. *Suppose that there exist positive constants α, L, and a continuous function $b : R^+ \to R^+$ such that the following conditions hold:*

(*i*) $\liminf_{t\to\infty} \int_0^t a(s)\, ds > -\infty$.

(ii) $\int_0^t e^{-\int_s^t a(u)du} b(s)\,ds \le \alpha < 1$ *for all* $t \ge 0$.

(iii) $|g(t,\phi) - g(t,\psi)| \le b(t)\|\phi - \psi\|$ *for all* $\phi, \psi \in C(L)$, $g(t,0) = 0$.

(iv) For each $\varepsilon > 0$ *and* $t_1 \ge 0$, *there exists a* $t_2 > t_1$ *such that* $[t \ge t_2,\ x_t \in C(L)]$ *imply*

$$|g(t,x_t)| \le b(t)\left(\varepsilon + \|x\|^{[t_1,t]}\right).$$

Then the zero solution of (1.7.1) is asymptotically stable if and only if

(v) $\int_0^t a(s)ds \to \infty$ as $t \to \infty$

Proof. First, suppose that (v) holds. Let $t_0 \ge 0$ and find $\delta_0 > 0$ such that $\delta_0 K + \alpha L \le L$, where

$$K = \sup_{t \ge t_0}\{e^{-\int_{t_0}^t a(s)\,ds}\}. \tag{1.7.5}$$

Let $\phi \in C(\delta_0)$ be fixed and set

$$\mathcal{S} = \{x : R \to R : x_{t_0} = \phi,\ x_t \in C(L) \text{ for } t \ge t_0,\ x(t) \to 0 \text{ as } t \to \infty\}.$$

Then $\mathcal{S}$ is a complete metric space with metric $\rho(x,y) = \sup_{t \ge t_0}\{|x(t) - y(t)|\}$. Define $P : \mathcal{S} \to \mathcal{S}$ by

$$(Px)(t) = \phi(t) \quad \text{for } t \le t_0$$

and

$$(Px)(t) = \phi(t_0)e^{-\int_{t_0}^t a(s)\,ds} + \int_{t_0}^t e^{-\int_s^t a(u)\,du} g(s,x_s)\,ds \quad \text{for } t \ge t_0.$$

Clearly, $(Px) : R \to R$ is continuous with $(Px)_{t_0} = \phi$ and

$$\begin{aligned}|(Px)(t)| &\le |\phi(t_0)|e^{-\int_{t_0}^t a(s)\,ds} + \int_{t_0}^t e^{-\int_s^t a(u)\,du} b(s)\|x_s\|\,ds \\ &\le \delta_0 K + \alpha L \le L \quad \text{for } t \ge t_0.\end{aligned}$$

Thus, $(Px)_t \in C(L)$ for $t \ge t_0$. We now show that $(Px)(t) \to 0$ as $t \to \infty$. Let $x \in \mathcal{S}$ and $\varepsilon > 0$ be given. Since $x(t) \to 0$ as $t \to \infty$, there exists

$t_1 > t_0$ such that $|x(t)| < \varepsilon$ for all $t \geq t_1$. Since $|x(t)| \leq L$ for all $t \in R$, by (iv) there is $t_2 > t_1$ such that $t \geq t_2$ implies

$$|g(t, x_t)| \leq b(t) \left(\varepsilon + \|x\|^{[t_1,t]} \right).$$

For $t \geq t_2$, we have

$$\begin{aligned}
&\left| \int_{t_0}^{t} e^{-\int_s^t a(u)\,du} g(s, x_s)\, ds \right| \\
&\leq \int_{t_0}^{t_2} e^{-\int_s^t a(u)du} |g(s, x_s)|\, ds + \int_{t_2}^{t} e^{-\int_s^t a(u)du} |g(s, x_s)| ds \\
&\leq \int_{t_0}^{t_2} e^{-\int_s^t a(u)du} b(s) \|x_s\|\, ds + \int_{t_2}^{t} e^{-\int_s^t a(u)\,du} b(s) \left(\varepsilon + \|x\|^{[t_1,s]} \right) ds \\
&\leq \alpha L e^{-\int_{t_2}^t a(s)\,ds} + 2\alpha\varepsilon.
\end{aligned}$$

By (v), there exists $t_3 > t_2$ such that

$$\delta_0 e^{-\int_{t_0}^t a(s)\,ds} + L e^{-\int_{t_2}^t a(s)\,ds} < \varepsilon.$$

Thus, for $t \geq t_3$, we have

$$|(Px)(t)| \leq \delta_0 e^{-\int_{t_0}^t a(s)\,ds} + \alpha L e^{-\int_{t_2}^t a(s)\,ds} + 2\alpha\varepsilon < 3\varepsilon.$$

This proves $(Px)(t) \to 0$ as $t \to \infty$, and hence $(Px) \in S$.

To see that P is a contraction mapping, observe that for $t \geq t_0$

$$\begin{aligned}
|(Px)(t) - (Py)(t)| &\leq \int_{t_0}^{t} e^{-\int_s^t a(u)\,du} |g(s, x_s) - g(s, y_s)|\, ds \\
&\leq \int_{t_0}^{t} e^{-\int_s^t a(u)\,du} b(s) \|x_s - y_s\|\, ds \leq \alpha\rho(x, y)
\end{aligned}$$

or

$$\rho(Px, Py) \leq \alpha\rho(x, y).$$

By the contraction mapping principle, P has a unique fixed point x in S which is a solution of (1.7.1) with $\phi \in \mathcal{C}(\delta_0)$ and $x(t) = x(t, t_0, \phi) \to 0$ as $t \to \infty$.

To prove asymptotic stability, we need to show that the zero solution of (1.7.1) is stable. Let $\varepsilon > 0$ ($\varepsilon < L$) be given. Choose $\delta > 0$ ($\delta < \varepsilon$) with $\delta K + \alpha\varepsilon < \varepsilon$. If $x(t) = x(t, t_0, \phi)$ is a solution of (1.7.1) with $\|\phi\| < \delta$, then

$$x(t) = \phi(t_0)e^{-\int_{t_0}^t a(s)\,ds} + \int_{t_0}^t e^{-\int_s^t a(u)\,du} g(s, x_s)\,ds.$$

We claim that $|x(t)| < \varepsilon$ for all $t \geq t_0$. Notice that $|x(t_0)| < \varepsilon$. If there exists $t^* > t_0$ such that $|x(t^*)| = \varepsilon$ and $|x(s)| < \varepsilon$ for $t_0 \leq s < t^*$, then

$$\begin{aligned} |x(t^*)| &\leq \delta e^{-\int_{t_0}^{t^*} a(s)\,ds} + \int_{t_0}^{t^*} e^{-\int_s^{t^*} a(u)\,du} b(s)\|x_s\|\,ds \\ &\leq \delta K + \alpha\varepsilon < \varepsilon \end{aligned}$$

which contradicts the definition of t^*. Thus, $|x(t)| < \varepsilon$ for all $t \geq t_0$ and the zero solution of (1.7.1) is stable. This shows that the zero solution of (1.7.1) is asymptotically stable if (v) holds.

Conversely, suppose (v) fails. Then by (i) there exists a sequence $\{t_n\}$, $t_n \to \infty$ as $n \to \infty$ such that

$$\lim_{n\to\infty} \int_0^{t_n} a(s)\,ds = \ell$$

for some $\ell \in R$. We may also choose a positive constant Q satisfying

$$-Q \leq \int_0^{t_n} a(s)\,ds \leq Q$$

for all $n = 1, 2, \cdots$. By (ii), we have

$$\int_0^{t_n} e^{-\int_s^{t_n} a(u)du} b(s)\,ds \leq \alpha.$$

This yields

$$\int_0^{t_n} e^{\int_0^s a(u)du} b(s)\,ds \leq \alpha e^{\int_0^{t_n} a(u)\,du} \leq e^Q.$$

The sequence $\{\int_0^{t_n} e^{\int_0^s a(u)du} b(s)\,ds\}$ is bounded so there exists a convergent subsequence. For brevity in notation, we may assume

$$\lim_{n\to\infty} \int_0^{t_n} e^{\int_0^s a(u)\,du} b(s)\,ds = \gamma$$

for some $\gamma \in R^+$ and choose a positive integer $\bar{k}$ so large that

$$\int_{t_{\bar{k}}}^{t_n} e^{\int_0^s a(u)du} b(s)\, ds < (1-\alpha)/(2K^2)$$

for all $n \geq \bar{k}$.

By (i), K in (1.7.5) is well defined. We now consider the solution $x(t) = x(t, t_{\bar{k}}, \phi)$ with $\phi(s) \equiv \delta_0$ for $s \leq t_{\bar{k}}$. Then $|x(t)| \leq L$ for all $t \geq t_{\bar{k}}$ and

$$\begin{aligned} |x(t)| &\leq \delta_0 e^{-\int_{t_{\bar{k}}}^t a(s)\, ds} + \int_{t_{\bar{k}}}^t e^{-\int_s^t a(u)\, du} b(s) \|x_s\|\, ds \\ &\leq \delta_0 K + \alpha \|x_t\|. \end{aligned}$$

This implies that

$$|x(t)| \leq \frac{\delta_0 K}{1-\alpha} =: \beta$$

for all $t \geq t_k$. On the other hand, for $n \geq \bar{k}$, we also have

$$\begin{aligned} |x(t_n)| &\geq \delta_0 e^{-\int_{t_{\bar{k}}}^{t_n} a(s)\, ds} - \int_{t_{\bar{k}}}^{t_n} e^{-\int_s^{t_n} a(u)\, du} b(s) \|x_s\|\, ds \\ &\geq \delta_0 e^{-\int_{t_{\bar{k}}}^{t_n} a(s)\, ds} - \beta e^{-\int_0^{t_n} a(u)\, du} \int_{t_{\bar{k}}}^{t_n} e^{\int_0^s a(u)du} b(s)\, ds \\ &\geq e^{-\int_{t_{\bar{k}}}^{t_n} a(s)\, ds} \left[\delta_0 - \beta K \int_{t_{\bar{k}}}^{t_n} e^{\int_0^s a(u)\, du} b(s) ds \right] \\ &\geq \frac{1}{2} \delta_0 e^{-\int_{t_{\bar{k}}}^{t_n} a(s)\, ds} \geq \frac{1}{2} \delta_0 e^{-2Q}. \end{aligned}$$

This implies $x(t) \not\to 0$ as $t \to \infty$. Thus, condition (v) is necessary for the asymptotic stability of the zero solution of (1.7.1). The proof is complete.

Example 1.7.1. Consider the half-linear equation

$$x'(t) = -a(t)x(t) + b(t)q(x(t - r(t))) \tag{1.7.6}$$

where b, $r : R^+ \to R$ and $q : R \to R$ are continuous with

(i*) $\liminf_{t \to \infty} \int_0^t a(s)\, ds > -\infty$,

(ii*) $\sup_{t\geq 0} \int_0^t e^{-\int_s^t a(u)du}|b(s)|\, ds < 1,$

(iii*) there exists an $L > 0$ so that if $|x|, |y| \leq L$, then

$$|q(x) - q(y)| \leq |x - y| \quad \text{and } q(0) = 0,$$

(iv*) $r(t) \geq 0, \ t - r(t) \to \infty \quad \text{as } t \to \infty.$

Then the zero solution of (1.7.6) is asymptotically stable if and only if

(v*) $\int_0^t a(s)\, ds \to \infty \quad \text{as} \quad t \to \infty.$

To prove the statement, apply Theorem 1.7.1 by replacing $b(t)$ by $|b(t)|$ in (ii)-(iv) and using the fading memory condition (iv*) in (iv).

Remark 1.7.1. *If $q(x) \equiv x$, then (1.7.6) reduces to (1.7.2). It is clear that (1.7.4) implies (ii*) and (v*), and conditions here are significantly improved. We can see from (ii*) that $a(t)$ can be negative some of the time, $a(t)$ and $b(t)$ are related on average, and all functions involved can be unbounded.*

Remark 1.7.2. *It was noted in Seifert (1973) that a fading memory condition such as (iv) or (iv*) is required for the asymptotic stability of a general delay equation. This means that the equation representing a physical system should remember its past, but the memory should fade with time.*

Example 1.7.2. Consider the Volterra equation

$$x'(t) = -a(t)x(t) + \int_{-\infty}^t E(t, s, x(s))\, ds \tag{1.7.7}$$

where $a : R^+ \to R$ and $E : \Omega \times R \to R$, $\Omega = \{(t,s) \in R^2 : t \geq s\}$ are continuous. Suppose there exist a constant $L > 0$ and a continuous function $q : \Omega \to R^+$ such that

($\tilde{\text{i}}$) $\liminf_{t\to\infty} \int_0^t a(s)ds > -\infty,$

($\tilde{\text{ii}}$) $\sup_{t\geq 0} \int_0^t e^{-\int_s^t a(u)\, du} \int_{-\infty}^s q(s,\tau)d\tau\, ds < 1,$

($\widetilde{\text{iii}}$) if $|x|, |y| \le L$, then

$|E(t,s,x) - E(t,s,y)| \le q(t,s)|x-y|$ and $E(t,s,0) = 0$ for all $(t,s) \in \Omega$,

($\widetilde{\text{iv}}$) for each $\varepsilon > 0$ and $t_1 \ge 0$, there exists a $t_2 > t_1$ such that $t \ge t_2$ implies

$$\int_{-\infty}^{t_1} q(t,s)\,ds \le \varepsilon \int_{-\infty}^{t} q(t,s)\,ds. \tag{1.7.8}$$

Then the zero solution of (1.7.7) is asymptotically stable if and only if

($\tilde{\text{v}}$) $\int_0^t a(s)ds \to \infty$ as $t \to \infty$.

Proof. We only need to verify that (iii) and (iv) hold. Indeed, letting $g(t,\phi) = \int_{-\infty}^{0} E(t,t+s,\phi(s))\,ds$ and $b(t) = \int_{-\infty}^{t} q(t,s)\,ds$, we have

$$\begin{aligned}|g(t,\phi) - g(t,\psi)| &= \left|\int_{-\infty}^{0} E(t,t+s,\phi(s)) - \int_{-\infty}^{0} E(t,t+s,\psi(s))\,ds\right| \\ &\le \int_{-\infty}^{0} q(t,t+s)\,ds\|\phi-\psi\| = b(t)\|\phi-\psi\|\end{aligned}$$

for all $\phi, \psi \in C(L)$. This shows (iii) holds. Next, let $\varepsilon > 0$ and $t_1 \ge 0$ be given. By ($\widetilde{\text{iv}}$), there exists a $t_2 > t_1$ such that

$$L\int_{-\infty}^{t_1} q(t,s)\,ds < \varepsilon \int_{-\infty}^{t} q(t,s)\,ds$$

for all $t \ge t_2$. Let $x : R \to R$ be continuous with $x_t \in C(L)$. If $t \ge t_2$, then

$$\begin{aligned}|g(t,x_t)| &\le \int_{-\infty}^{t_1} |E(t,s,x(s))|\,ds + \int_{t_1}^{t} |E(t,s,x(s))|\,ds \\ &\le \int_{-\infty}^{t_1} q(t,s)\,dsL + \int_{t_1}^{t} q(t,s)|x(s)|\,ds \\ &\le \varepsilon \int_{-\infty}^{t} q(t,s)\,ds + \int_{t_1}^{t} q(t,s)\,ds\|x\|^{[t_1,t]} \\ &\le b(t)\left(\varepsilon + \|x\|^{[t_1,t]}\right).\end{aligned}$$

This implies that (iv) is satisfied, and by Theorem 1.7.1, the zero solution of (1.7.7) is asymptotically stable if and only if ($\tilde{\text{v}}$) holds.

This will acknowledge that the above work was first published in *Dynamic Systems and Applications*, as detailed in Zhang (2004a), which is published by Dynamic Publishers.

Next, we consider the scalar delay equation

$$x'(t) = -b(t)x(t-\tau(t)) \tag{1.7.9}$$

and its generalization

$$x'(t) = -\sum_{j=1}^{N} b_j(t)x(t-\tau_j(t)) \tag{1.7.10}$$

where $b,\ b_j \in C(R^+, R)$ and $\tau,\ \tau_j \in C(R^+, R^+)$ with $t-\tau(t) \to \infty$ and $t-\tau_j(t) \to \infty$ as $t \to \infty$. Note that (1.7.10) becomes (1.7.9) for $N = 1$.

Let $R = (-\infty, \infty), R^+ = [0, +\infty)$, and $R^- = (-\infty, 0]$, respectively. Also, $C(S_1, S_2)$ denotes the set of all continuous functions $\phi : S_1 \to S_2$. For each t_0, define

$$m_j(t_0) = \inf\{s - \tau_j(s) : s \geq t_0\}, \quad m(t_0) = \min\{m_j(t_0) : 1 \leq j \leq N\},$$

and $C(t_0) = C([m(t_0), t_0], R)$ with the supremum norm $\|\cdot\|$. Define the inverse of $t - \tau_i(t)$ by $g_i(t)$ if it exists and set

$$Q(t) = \sum_{j=1}^{N} b_j(g_j(t))$$

$$\theta(t) = \sum_{j=1}^{N} \int_0^t e^{-\int_s^t Q(u)du} |b_j(s)||\tau_j'(s)|\, ds \tag{1.7.11}$$

if $\tau_j(t)$ is differentiable. For each $(t_0, \phi) \in R^+ \times C(t_0)$, a solution of (1.7.10) through (t_0, ϕ) is a continuous function $x : [m(t_0), t_0 + \alpha) \to R^n$ for some positive constant $\alpha > 0$ such that $x(t)$ satisfies (1.7.10) on $[t_0, t_0 + \alpha)$ and $x(s) = \phi(s)$ for $s \in [m(t_0), t_0]$. We denote such a solution by $x(t) = x(t, t_0, \phi)$. For each $(t_0, \phi) \in R^+ \times C(t_0)$, there exists a unique solution $x(t) = x(t, t_0, \phi)$ of (1.7.10) defined on $[t_0, +\infty)$. For fixed t_0, we define

$$\|\phi\| = \max\{|\phi(s)| : m(t_0) \leq s \leq t_0\}.$$

Theorem 1.7.2. *Suppose that τ_j is differentiable, the inverse function $g_j(t)$ of $t - \tau_j(t)$ exists, and there exists a constant $\alpha \in (0, 1)$ such that for $t \geq 0$*

(i) $\liminf_{t\to\infty} \int_0^t Q(s)\,ds > -\infty$

(ii)

$$\sum_{j=1}^{N} \left[\int_{t-\tau_j(t)}^{t} |b_j(g_j(s))|\,ds + \int_0^t e^{-\int_s^t Q(u)\,du} |Q(s)| \right.$$
$$\left. \times \int_{s-\tau_j(s)}^{s} |b_j(g_j(\nu))| d\nu\,ds \right] + \theta(t) \le \alpha \tag{1.7.12}$$

Then the zero solution of (1.7.10) is asymptotically stable if and only if

(iii) $\int_0^t Q(s)\,ds \to \infty$ as $t \to \infty$.

Proof. First, suppose that (iii) holds. For each $t_0 \ge 0$, we set

$$K = \sup_{t\ge 0}\{e^{-\int_0^t Q(s)\,ds}\}. \tag{1.7.13}$$

Let $\phi \in C(t_0)$ be fixed and define

$$\mathcal{S} = \{\, x \in C([m(t_0),\infty),\, R) : x(t) \to 0 \quad \text{as } t \to \infty,$$
$$x(s) = \phi(s) \quad \text{for} \quad s \in [m(t_0), t_0]\,\} \tag{1.7.14}$$

Then $\mathcal{S}$ is a complete metric space with metric

$$\rho(x,y) = \sup_{t\ge t_0}\{|x(t) - y(t)|\}.$$

Define $P : \mathcal{S} \to \mathcal{S}$ by $(Px)(t) = \phi(t)$ for $t \in [m(t_0), t_0]$ and

$$(Px)(t) = \left(\phi(t_0) - \sum_{j=1}^{N} \int_{t_0-\tau_j(t_0)}^{t_0} b_j(g_j(s))\phi(s)\,ds \right) e^{-\int_{t_0}^t Q(u)du}$$

$$+ \sum_{j=1}^{N} \int_{t-\tau_j(t)}^{t} b_j(g_j(s))x(s)\,ds$$

$$- \int_{t_0}^{t} e^{-\int_s^t Q(u)du} \sum_{j=1}^{N} b_j(s)\tau_j'(s)x(s-\tau_j(s))\,ds$$

$$- \int_{t_0}^{t} e^{-\int_s^t Q(u)\,du} Q(s) \left(\sum_{j=1}^{N} \int_{s-\tau_j(s)}^{s} b_j(g_j(\nu))x(\nu)d\nu \right) ds \tag{1.7.15}$$

for $t \geq t_0$. It is clear that $(Px) \in C([m(t_0), \infty), R)$. We now show that $(Px)(t) \to 0$ as $t \to \infty$. Since $x(t) \to 0$ and $t - \tau_j(t) \to \infty$ as $t \to \infty$, for each $\varepsilon > 0$, there exists a $T_1 > t_0$ such that $s \geq T_1$ implies $|x(s - \tau_j(s)| < \varepsilon$ for $j = 1, 2, \cdots, N$. Thus, for $t \geq T_1$, the last term I_4 in (1.7.15) satisfies

$$
\begin{aligned}
&|I_4| \\
&= \left| \int_{t_0}^{t} e^{-\int_s^t Q(u)\,du} Q(s) \left(\sum_{j=1}^{N} \int_{s-\tau_j(s)}^{s} b_j(g_j(\nu)) x(\nu)\, d\nu \right) ds \right| \\
&\leq \int_{t_0}^{T_1} e^{-\int_s^t Q(u)du} |Q(s)| \left(\sum_{j=1}^{N} \int_{s-\tau_j(s)}^{s} |b_j(g_j(\nu))| |x(\nu)|\, d\nu \right) ds \\
&+ \int_{T_1}^{t} e^{-\int_s^t Q(u)\,du} |Q(s)| \left(\sum_{j=1}^{N} \int_{s-\tau_j(s)}^{s} |b_j(g_j(\nu))| |x(\nu)|\, d\nu \right) ds \\
&\leq \sup_{\sigma \geq m(t_0)} |x(\sigma)| \int_{t_0}^{T_1} e^{-\int_s^t Q(u)du} |Q(s)| \\
&\times \left(\sum_{j=1}^{N} \int_{s-\tau_j(s)}^{s} |b_j(g_j(\nu))|\, d\nu \right) ds \\
&+ \varepsilon \int_{T_1}^{t} e^{-\int_s^t Q(u)du} |Q(s)| \left(\sum_{j=1}^{N} \int_{s-\tau_j(s)}^{s} |b_j(g_j(\nu))| d\nu \right) ds.
\end{aligned}
$$

By (iii), there exists $T_2 > T_1$ such that $t \geq T_2$ implies

$$
\sup_{\sigma \geq m(t_0)} |x(\sigma)| \int_{t_0}^{T_1} e^{-\int_s^t Q(u)du} |Q(s)| \left(\sum_{j=1}^{N} \int_{s-\tau_j(s)}^{s} |b_j(g_j(\nu))|\, d\nu \right) ds < \varepsilon.
$$

Apply (ii) to obtain $|I_4| \leq \varepsilon + \varepsilon\alpha < 2\varepsilon$. Thus, $I_4 \to 0$ as $t \to \infty$. Similarly, we can show that the rest of the terms in (1.7.15) approach zero as $t \to \infty$. This yields $(Px)(t) \to 0$ as $t \to \infty$, and hence $Px \in S$. Also, by (ii), P is a contraction mapping with contraction constant α. By the contraction mapping principle, P has a unique fixed point x in S which is a solution of (1.7.10) with $x(s) = \phi(s)$ on $[m(t_0), t_0]$ and $x(t) = x(t, t_0, \phi) \to 0$ as $t \to \infty$.

To prove asymptotic stability, we need to show that the zero solution of (1.7.10) is stable. Let $\varepsilon > 0$ be given and choose $\delta > 0$ ($\delta < \varepsilon$) satisfying $2\delta K e^{\int_0^{t_0} Q(u)du} + \alpha\varepsilon < \varepsilon$. If $x(t) = x(t, t_0, \phi)$ is a solution of (1.7.10) with

$\|\phi\| < \delta$, then $x(t) = (Px)(t)$ which was defined in (1.7.15). We claim that $|x(t)| < \varepsilon$ for all $t \geq t_0$. Notice that $|x(s)| < \varepsilon$ on $[m(t_0), t_0]$. If there exists $t^* > t_0$ such that $|x(t^*)| = \varepsilon$ and $|x(s)| < \varepsilon$ for $m(t_0) \leq s < t^*$, then it follows from (1.7.15) that

$$\begin{aligned}
|x(t^*)| \leq \|\phi\| &\left(1 + \sum_{j=1}^{N} \int_{t_0-\tau_j(t_0)}^{t_0} |b_j(g_j(s))|\, ds\right) e^{-\int_{t_0}^{t^*} Q(u)du} \\
&+ \varepsilon \sum_{j=1}^{N} \int_{t^*-\tau_j(t^*)}^{t^*} |b_j(g_j(s))| ds \\
&+ \varepsilon \int_{t_0}^{t^*} e^{-\int_s^{t^*} Q(u)\, du} \sum_{j=1}^{N} |b_j(s)||\tau_j'(s)|\, ds \\
&+ \varepsilon \int_{t_0}^{t^*} e^{-\int_s^{t^*} Q(u)\, du} |Q(s)| \left(\sum_{j=1}^{N} \int_{s-\tau_j(s)}^{s} |b_j(g_j(\nu))|\, d\nu\right) ds \\
&\leq 2\delta K e^{\int_0^{t_0} Q(u)\, du} + \alpha\varepsilon < \varepsilon
\end{aligned} \tag{1.7.16}$$

which contradicts the definition of t^*. Thus, $|x(t)| < \varepsilon$ for all $t \geq t_0$, and the zero solution of (1.7.10) is stable. This shows that the zero solution of (1.7.10) is asymptotically stable if (iii) holds.

Conversely, suppose (iii) fails. Then by (i) there exists a sequence $\{t_n\}$, $t_n \to \infty$ as $n \to \infty$ such that $\lim_{n\to\infty} \int_0^{t_n} Q(u)du = \ell$ for some $\ell \in R$. We may also choose a positive constant J satisfying

$$-J \leq \int_0^{t_n} Q(s)\, ds \leq J$$

for all $n \geq 1$. To simplify expressions, we define

$$\omega(s) = \sum_{j=1}^{N} \left[|b_j(s)||\tau_j'(s)| + |Q(s)| \int_{s-\tau_j(s)}^{s} |b_j(g_j(\nu))|\, d\nu\right]$$

for all $s \geq 0$. By (ii), we have

$$\int_0^{t_n} e^{-\int_s^{t_n} Q(u)\, du} \omega(s)\, ds \leq \alpha.$$

This yields

$$\int_0^{t_n} e^{\int_0^s Q(u)\, du} \omega(s) ds \leq \alpha e^{\int_0^{t_n} Q(u)\, du} \leq e^J.$$

The sequence $\{\int_0^{t_n} e^{\int_0^s Q(u)\,du}\omega(s)\,ds\}$ is bounded, so there exists a convergent subsequence. For brevity in notation, we may assume

$$\lim_{n\to\infty}\int_0^{t_n} e^{\int_0^s Q(u)\,du}\omega(s)\,ds = \gamma$$

for some $\gamma \in R^+$ and choose a positive integer $\bar{k}$ so large that

$$\int_{t_{\bar{k}}}^{t_n} e^{\int_0^s Q(u)du}\omega(s)\,ds < \delta_0/4K$$

for all $n \geq \bar{k}$, where $\delta_0 > 0$ satisfies $4\delta_0 K e^J + \alpha < 1$.

By (i), K in (1.7.13) is well defined. We now consider the solution $x(t) = x(t, t_{\bar{k}}, \phi)$ of (1.7.10) with $\phi(t_{\bar{k}}) = \delta_0$ and $|\phi(s)| \leq \delta_0$ for $s \leq t_{\bar{k}}$. An argument similar to that in (1.7.16) shows $|x(t)| \leq 1$ for $t \geq t_{\bar{k}}$. We may choose ϕ so that

$$\phi(t_{\bar{k}}) - \sum_{j=1}^{N}\int_{t_{\bar{k}}-\tau_j(t_{\bar{k}})}^{t_{\bar{k}}} b_j(g_j(s))\phi(s)\,ds \geq \frac{1}{2}\delta_0. \tag{1.7.17}$$

It follows from (1.7.15) with $x(t) = (Px)(t)$ that for $n \geq t_{\bar{k}}$,

$$\begin{aligned}
&\left| x(t_n) - \sum_{j=1}^{N}\int_{t_n-\tau_j(t_n)}^{t_n} b_j(g_j(s))x(s)\,ds \right| \\
&\geq \frac{1}{2}\delta_0 e^{-\int_{t_{\bar{k}}}^{t_n} Q(u)\,du} - \int_{t_{\bar{k}}}^{t_n} e^{-\int_s^{t_n} Q(u)du}\omega(s)\,ds \\
&= \frac{1}{2}\delta_0 e^{-\int_{t_{\bar{k}}}^{t_n} Q(u)\,du} - e^{-\int_0^{t_n} Q(u)\,du}\int_{t_{\bar{k}}}^{t_n} e^{\int_0^s Q(u)\,du}\omega(s)ds \\
&= e^{-\int_{t_{\bar{k}}}^{t_n} Q(u)\,du}\left[\frac{1}{2}\delta_0 - e^{-\int_0^{t_{\bar{k}}} Q(u)\,du}\int_{t_{\bar{k}}}^{t_n} e^{\int_0^s Q(u)\,du}\omega(s)\,ds\right] \\
&\geq e^{-\int_{t_{\bar{k}}}^{t_n} Q(u)du}\left[\frac{1}{2}\delta_0 - K\int_{t_{\bar{k}}}^{t_n} e^{\int_0^s Q(u)du}\omega(s)\,ds\right] \\
&\geq \frac{1}{4}\delta_0 e^{-\int_{t_{\bar{k}}}^{t_n} Q(u)du} \geq \frac{1}{4}\delta_0 e^{-2J} > 0.
\end{aligned} \tag{1.7.18}$$

On the other hand, if the zero solution of (1.7.10) is asymptotically stable, then $x(t) = x(t, t_{\bar{k}}, \phi) \to 0$ as $t \to \infty$. Since $t_n - \tau_j(t_n) \to \infty$ as $n \to \infty$ and (ii) holds, we have

$$x(t_n) - \sum_{j=1}^{N}\int_{t_n-\tau_j(t_n)}^{t_n} b_j(g_j(s))x(s)ds \to 0 \quad \text{as } t \to \infty$$

which contradicts (1.7.18). Hence, condition (iii) is necessary for the asymptotic stability of the zero solution of (1.7.10). The proof is complete.

Remark 1.7.3. *It follows from the first part of the proof of Theorem 1.7.2 that the zero solution of (1.7.10) is stable under (i) and (ii). Moreover, Theorem 1.7.2 still holds if (ii) is satisfied for $t \geq t_\sigma$ for some $t_\sigma \in R^+$.*

Remark 1.7.4. *If $Q(t) \geq 0$ and $\theta(t) \leq \ell$ (see (1.7.11)) for $\ell \in [0,1)$, then (i) is satisfied and (ii) can be replaced by*

$$\sup_{t\geq 0} \sum_{j=1}^{N} \int_{t-\tau_j(t)}^{t} |b_j(g_j(s))|\, ds < (1-\ell)/2.$$

Corollary. *Let $N = 1$ and suppose that τ is differentiable, the inverse function $g(t)$ of $t - \tau(t)$ exists, and there exists a constant $\alpha \in (0,1)$ such that for $t \geq 0$*

(i)* $\liminf_{t\to\infty} \int_0^t b(g(s))\, ds > -\infty$

(ii)*

$$\int_{t-\tau(t)}^{t} |b(g(s))|\, ds + \int_0^t e^{-\int_s^t b(g(u))\, du} |b(g(s))| \times \int_{s-\tau(s)}^{s} |b(g(\nu))|\, d\nu\, ds + \theta(t) \leq \alpha$$

where $\theta(t) = \int_0^t e^{-\int_s^t b(g(u))\, du} |b(s)||\tau'(s)|\, ds$. Then the zero solution of (1.7.9) is asymptotically stable if and only if

(iii)* $\int_0^t b(g(s))ds \to \infty, \quad \text{as } t \to \infty.$

The constants in the examples below can be improved, often with a program such as *Maple* or *Mathematica.*

Example 1.7.3. Consider the scalar equation

$$x'(t) = -b_1(t)x(t-\tau_1(t)) - b_2(t)x(t-\tau_2(t)) \tag{1.7.19}$$

where $\tau_1(t) = \mu(2t - \sin t)$, $\tau_2(t) = 2\mu t$, and

$$b_1(t) = \frac{\gamma}{1+t}, \quad b_2(t) = \frac{2\gamma \sin t}{1+t}.$$

To be specific, let $\mu = 0.0025$ and $\gamma = 0.4$. Then the zero solution of (1.7.19) is asymptotically stable.

Proof. The functions $t - \tau_j(t)$, $j = 1, 2$, are increasing with $t - \tau_j(t) \to \infty$ as $t \to \infty$. We denote their inverses by $g_j(t)$, respectively. It is clear that $g_2(u) = u/(1 - 2\mu)$. Since

$$f_1(t) =: (1 - 2\mu)t - 1 \le t - \tau_1(t) \le (1 - \mu)t =: f_2(t),$$

we have $f_2^{-1}(u) \le g_1(u) \le f_1^{-1}(u)$; that is, $u/(1-\mu) \le g_1(u) \le (u + 1)/(1 - 2\mu)$ for all $u \ge 0$. Observe that

$$\int_{t-\tau_1(t)}^{t} |b_1(g_1(s))|\, ds \le \int_{(1-2\mu)t-\mu}^{t} \frac{\gamma}{1 + \frac{s}{1-\mu}}\, ds \le \gamma(1 - \mu)|\ln(1 - 2\mu)|$$

and

$$\int_{t-\tau_2(t)}^{t} |b_2(g_2(s))|\, ds \le \int_{(1-2\mu)t}^{t} \frac{2\gamma}{1 + \frac{s}{1-2\mu}} ds \le 2\gamma(1 - 2\mu)|\ln(1 - 2\mu)|.$$

Thus,

$$\sum_{j=1}^{2} \int_{t-\tau_j(t)}^{t} |b_j(g_j(s))|\, ds \le \gamma(3 - 5\mu)|\ln(1 - 2\mu)| \le 0.006. \qquad (1.7.20)$$

Use the fact that $|\int_a^b \sin\nu/(1 + \nu)\, d\nu| \le 3$ for $b \ge a \ge 0$ to obtain

$$\begin{aligned} -\int_s^t Q(u)du &= -\int_s^t \frac{\gamma}{1 + g_1(u)} du - \int_s^t \frac{2\gamma \sin g_2(u)}{1 + g_2(u)} du \\ &\le -\int_s^t \frac{\gamma}{1 + g_1(u)}\, du + 6\gamma(1 - 2\mu). \end{aligned} \qquad (1.7.21)$$

Notice that

$$|Q(t)| \le \frac{\gamma}{1 + g_1(t)} + \frac{2\gamma|\sin(g_2(t))|}{1 + g_2(t)} \le \frac{\gamma(3 - 5\mu)}{(1 - 2\mu) + t}.$$

Since $(a + 1 + t)/(a + t)$ is decreasing in $t \in R^+$ for $a \ge 0$, we have

$$\begin{aligned} |Q(t)| &\le \frac{(1 - 2\mu) + 1 + t}{(1 - 2\mu) + t} \cdot \frac{\gamma(3 - 5\mu)}{(1 - 2\mu) + 1 + t} \\ &\le \frac{(1 - 2\mu) + 1}{(1 - 2\mu)} \cdot \frac{\gamma(3 - 5\mu)}{(1 - 2\mu) + 1 + t} \\ &\le \frac{2(1 - \mu)}{(1 - 2\mu)^2} \cdot \frac{\gamma(3 - 5\mu)}{1 + g_1(t)} \end{aligned} \qquad (1.7.22)$$

Using (1.7.20),(1.7.21), and (1.7.22), we get

$$\int_0^t e^{-\int_s^t Q(u)\,du}|Q(s)|\left(\sum_{j=1}^{2}\int_{s-\tau_j(s)}^{s}|b_j(g_j(\nu))|d\nu\right)ds$$

$$\leq \gamma(3-5\mu)|\ln(1-2\mu)|\int_0^t e^{-\int_s^t Q(u)du}|Q(s)|\,ds$$

$$\leq \gamma(3-5\mu)|\ln(1-2\mu)|e^{6\gamma}\,\frac{2(1-\mu)(3-5\mu)}{(1-2\mu)^2}$$

$$\times\int_0^t e^{-\int_s^t \frac{\gamma}{1+g_1(u)}\,du}\frac{\gamma}{1+g_1(s)}\,ds$$

$$\leq \gamma(3-5\mu)|\ln(1-2\mu)|e^{6\gamma}\,\frac{2(1-\mu)(3-5\mu)}{(1-2\mu)^2}<0.398. \tag{1.7.23}$$

Notice that

$$\sum_{j=1}^{2}|b_j(s)| \;\leq\; \frac{3\gamma}{1+s}=\frac{3\gamma}{(1-2\mu)+1+s}\cdot\frac{(1-2\mu)+1+s}{1+s}$$

$$\leq\; \frac{3\gamma[2(1-\mu)]}{(1-2\mu)+1+s}\leq\frac{6(1-\mu)}{1-2\mu}\cdot\frac{\gamma}{1+g_1(s)}$$

since $\frac{(1-2\mu)+1+s}{1+s}$ is decreasing in s for $s\geq 0$. As $|\tau_j'(t)|\leq 3\mu$, we arrive at

$$\int_0^t e^{-\int_s^t Q(u)du}\sum_{j=1}^{2}|b_j(s)||\tau_j'(s)|ds$$

$$\leq 3\mu e^{6\gamma}\frac{6(1-\mu)}{1-2\mu}\int_0^t e^{-\int_s^t \frac{\gamma}{1+g_1(u)}du}\frac{\gamma}{1+g_1(s)}ds$$

$$\leq 3\mu e^{6\gamma}\frac{6(1-\mu)}{1-2\mu}<0.498 \tag{1.7.24}$$

Combining (1.7.20), (1.7.23), and (1.7.24), we see that condition (ii) of Theorem 1.7.2 holds with $\alpha=0.902$. One can show that $\int_0^\infty Q(s)\,ds=\infty$. Thus, the asymptotic stability of the zero solution of (1.7.19) follows from Theorem 1.7.2.

This will acknowledge that the material from this section was first published in Dynamic Systems and Applications, as detailed in Zhang (2004), published by Dynamic Publishers and in Proceedings of the World Congress on Nonlinear Analysis, published in Nonlinear Analysis, as detailed in Zhang (2005), which is published by Elsevier Science.

Chapter 2

Classical Problems, Harmless Perturbations

2.1 The Cooke-Yorke Hypothesis

We will study some equations in which it is true that the difference of a pair of terms, say $A - B$, seems to have little or no effect on the behavior of the solutions of the system. But the real advantage of identification of such a pair is that we sometimes encounter an equation, say

$$x' = F(t, x_t) + A,$$

which is most intractable but

$$x' = F(t, x_t) + B$$

would be very easy to analyze. The idea, then, is to use the time-honored technique of adding and subtracting the same thing with the hope that

$$x' = F(t, x_t) + A - B + B$$

will have essentially the behavior of the second mentioned equation. The results in this section are taken from Burton (2004d).

We come upon the idea by studying a paper by Cooke and Yorke (1973) which continues to attract enormous attention. Their paper develops a theory of biological growth and epidemics based on several equations which are basically of the form $x'(t) = g(x(t)) - g(x(t - L))$ where L is a positive constant. The fundamental *hypothesis* is that g is an arbitrary Lipschitz function. That equation has the property that every constant function is a solution and every solution (under certain regularity conditions) approaches a constant; if the regularity conditions fail then a solution may tend to $\pm\infty$ on its maximal right-interval of definition. As we look at the problem today we are compelled to conjecture that if $x' = F(t, x_t)$ enjoys stability properties, those properties should be shared with

$$x' = F(t, x_t) + g(x(t)) - g(x(t - L))$$

under mild conditions on g and L; $g(x(t)) - g(x(t-L))$ should be a harmless perturbation. That is, indeed, the case and it is the topic of this chapter. Cooke and Yorke consider several important questions and several variations of the equation. It was the thesis of Burton (2004d) that every one of those questions could be answered by simple contraction mapping arguments. Each of those forms yields another harmless perturbation.

Our *goal* is to construct a set of harmless perturbations based on the problems posed in the Cooke-Yorke paper and to apply those to different problems in later sections. The Cooke-Yorke paper inspired a great many other papers considering a host of equations in which every constant is a solution and every solution approaches a constant.

It is our *conjecture* that each of those equations represent new harmless perturbations which will be of enormous value in the study of stability.

Cooke and Yorke modeled a population as follows. If $x(t)$ denotes the number of individuals in a population at time t, the number of births $B(t)$ is some function of $x(t)$, say $B(t) = g(x(t))$. Let us assume at first that every individual has life span L, a constant. Inasmuch as every individual dies at age L, the number of deaths per unit time at time t is $g(x(t - L))$. Since the difference $g(x(t)) - g(x(t - L))$ is net change in population per unit time, the growth of the population is governed by the equation

$$x'(t) = g(x(t)) - g(x(t - L)). \tag{2.1.1}$$

Here, they emphasize that g is allowed to be any Lipschitz function. This put their investigations far ahead of so many similar investigations, both before and after publication of that paper. The freedom to take g as an arbitrary Lipschitz function is critical in real-world problems because of uncertainties. In many population problems there is endless speculation on the form of the functions generating growth and decline. The classic monograph by Maynard-Smith (1974) devotes Chapters 2 and 3 to describing several choices for growth functions with less than strong reason for choosing one over the other. Pielou (1969; p. 20) suggests a mechanical choice by picking the simplest part of a Taylor series which will generate an S-shaped curve; this yields $g(x) = ax - bx^2$ with a and b positive constants. It is to be noted that in so many of the papers motivated by the Cooke-Yorke paper, monotonicity and sign conditions on g are assumed. Thus, those papers seem to have missed a main aspect of the theory.

It should be mentioned early on that if $\int_0^x g(s)ds \to -\infty$ as $|x| \to \infty$ then the trivial Liapunov function will show that all solutions of (2.1.1) are bounded. We pointedly do not make that assumption, but for completeness note that an appropriate Liapunov functional is

$$V(x(\cdot)) = -2\int_0^x g(s)\,ds + \int_{t-L}^t g^2(x(s))\,ds$$

and its derivative along solutions of (2.1.1) satisfies

$$\begin{aligned}V'_{(2.1.1)} &= -2g^2(x) + 2g(x)g(x(t-L)) + g^2(x) - g^2(x(t-L))\\ &= -g^2(x) + 2g(x)g(x(t-L) - g^2(x(t-L))\\ &= -(g(x) - g(x(t-L))^2.\end{aligned}$$

Thus, $V(x) \to \infty$ as $|x| \to \infty$ and $V' \leq 0$ yields all solutions bounded and the maximal interval of definition is $[0, \infty)$. When solutions depend continuously on initial conditions, then a classical result by Krasovskii (1963; p. 153) shows that every bounded solution approaches the set where $V' = 0$. That set consists of those functions where $g(x(t)) = g(x(t-L))$ and so $x'(t) = 0$ and $x(t)$ is constant. The particular proof given by Krasovskii bears more study. It seems likely that the required continuous dependence of solutions on initial conditions might be reduced.

We now investigate appropriate assumptions. Certainly, when g is arbitrary there is danger of finite escape time and we will constantly be wary of that. To specify a solution of (2.1.1) we require an initial function on an initial interval. Typically, we need a continuous function $\psi : [-L, 0] \to R$ and obtain a continuous function $x(t, 0, \psi)$ with $x(t, 0, \psi) = \psi(t)$ on $[-L, 0]$, while x satisfies (2.1.1) for $t > 0$. Although x is continuous, the derivative may fail at $t = 0$. Equation (2.1.1) is autonomous so we lose nothing by starting the solution at $t_0 = 0$. At times we will consider non-autonomous equations and may still start at zero for simple convenience, keeping in mind the type I and II exercises.

In addition to the fact that each constant function is a solution of (2.1.1), it is also critical to note that (2.1.1) has a first integral

$$x(t) = \int_{t-L}^t g(x(s))\,ds + c \tag{2.1.1.1}$$

where c is a constant of integration.

We come now to a surprising alternate use of an apparently irrelevant condition. Cooke and Yorke claim that to have a correct biological interpretation we must have $c = 0$. In fact, for a given initial function ψ, then

$$c = \psi(0) - \int_{-L}^{0} g(\psi(s))\, ds. \tag{2.1.1.2}$$

As our interest is not particularly centered on biological problems, we ignore that condition and focus on (2.1.1.1) with c defined by (2.1.1.2) and we denote (2.1.1.1) by

$$x(t) = \psi(0) - \int_{-L}^{0} g(\psi(s))\, ds + \int_{t-L}^{t} g(x(s))\, ds. \tag{2.1.1.3}$$

While we have no intention of asking $c = 0$, we note that if we were to ask that $c = 0$, that $g(x) = Kx$, and that $x = k$ is a nonzero solution of (2.1.1.3), then (2.1.1.2) would become

$$0 = k - KkL$$

so that

$$KL = 1 \tag{2.1.1.4}$$

since we ask for a solution $x = k \neq 0$. It turns out that if we ask $KL < 1$ then the Cooke-Yorke problems are easily solved by contractions. Not only does (2.1.1.4) give us guidance on admissible assumptions, but it has meaningful interpretation in terms of the mechanics of the problems posed. Imagine that we are guiding a moving object having position $x(t)$ at time t. We apply a force $g(x(t - L))$ to the object at time $t - L$ and then the same force in the opposite direction at time t. Solving the Cooke-Yorke problem by making $KL < 1$ is to say that we can make the object approach a straight line path if the applied force is not too large and the time for the opposite correction is not too long. Later we will show exactly this situation in a problem of Minorsky in the automatic guidance of a large ship.

To solve our problem we assume that there is a fixed positive constant K such that $x, y \in R$ implies that

$$|g(x) - g(y)| \leq K|x - y|, \tag{2.1.1.5}$$

a global Lipschitz condition. Next, we need a slightly tighter control on L than (2.1.1.4) and we suppose there is a positive constant $\alpha < 1$ with

$$KL \leq \alpha. \tag{2.1.1.6}$$

Theorem 2.1.1. *Let (2.1.1.5) and (2.1.1.6) hold and let $\psi : [-L, 0] \to R$ be a given continuous function. Then there is a unique constant k satisfying*

$$k = \psi(0) + g(k)L - \int_{-L}^{0} g(\psi(s))\,ds \tag{2.1.1.7}$$

such that the unique solution of (2.1.1) with initial function ψ satisfies $x(t, 0, \psi) \to k$ as $t \to \infty$.

Proof. Let us first show that (2.1.1.7) has a unique solution. Define a mapping $Q : R \to R$ by $k \in R$ implies that

$$Qk = \psi(0) + g(k)L - \int_{-L}^{0} g(\psi(s))\,ds.$$

Then for $k, d \in R$ and $|\cdot|$ denoting absolute value we have

$$|Qk - Qd| \le L|g(k) - g(d)| \le LK|k - d| \le \alpha|k - d|,$$

so Q is a contraction on the complete metric space $(R, |\cdot|)$. Thus Q has a unique fixed point k.

Next, let $(\mathcal{M}, \|\cdot\|)$ be the complete metric space of bounded continuous functions $\phi : [-L, \infty) \to R$ with $\phi(t) = \psi(t)$ on $[-L, 0]$, $\phi(t) \to k$ as $t \to \infty$, and $\|\cdot\|$ is the supremum metric.

Define $P : \mathcal{M} \to \mathcal{M}$ by $\phi \in \mathcal{M}$ implies that $(P\phi)(t) = \psi(t)$ on $[-L, 0]$ and for $t > 0$ let

$$(P\phi)(t) = \psi(0) - \int_{-L}^{0} g(\psi(s))\,ds + \int_{t-L}^{t} g(\phi(s))\,ds; \tag{2.1.1.8}$$

we have used (2.1.1.3) to define P and a fixed point will solve (2.1.1) and (2.1.1.3). Notice that since $\phi(t) \to k$ we have $\int_{t-L}^{t} g(\phi(s))ds \to g(k)L$ as $t \to \infty$. Using this, (2.1.1.8), and then (2.1.1.7) we see that $(P\phi)(t) \to k$ as $t \to \infty$. Thus, $P : \mathcal{M} \to \mathcal{M}$. To see that P is a contraction we note that for $\phi, \eta \in \mathcal{M}$ then

$$\begin{aligned} |(P\phi)(t) - (P\eta)(t)| &\le \int_{t-L}^{t} |g(\phi(s)) - g(\eta(s))|\,ds \\ &\le LK\|\phi - \eta\| \end{aligned}$$

so P is a contraction with unique fixed point $\phi \in \mathcal{M}$. By the way (2.1.1.8) was constructed, ϕ satisfies (2.1.1.3).

Remark 2.1.1. *This theorem tells us precisely what the limit of each solution will be, based on its initial function. We note that in Cooke-Yorke (1973; p. 84) it is shown that if g of a solution stays above a certain value for L time units, then g of it will never go below that value. They also show that if g of the solution stays below a certain value for L time units, then g of it will stay below that value forever. These results can require care in interpretation unless $g(x)$ has the sign of x, a condition which is studiously avoided here. The contraction mapping argument gives a sharper result; given an initial function, we know exactly the limiting value of the corresponding solution. There is no need at all for the solution to stay above a certain value over an interval of length L; it merely needs to do so on some well-defined type of average.*

Remark 2.1.2. *Examine Equation (2.1.1.7) which determines the unique constant k to which the solution $x(t,0,\psi)$ converges. Notice that ψ enters as*

$$\psi(0) - \int_{-L}^{0} g(\psi(s))\,ds.$$

If $\eta : [-L,0] \to R$ is continuous and satisfies

$$\eta(0) - \int_{-L}^{0} g(\eta(s))\,ds = \psi(0) - \int_{-L}^{0} g(\psi(s))\,ds$$

then Theorem 2.1.1 will also show that $x(t,0,\eta) \to k$. For a given ψ, there is an infinite set of functions η which qualify and $\|\eta\|$ is unbounded. We can think of k as being a "selective global attractor". The same observation can be made in all of the subsequent problems.

Remark 2.1.3. *Equation (2.1.1) is also an extremely stable control problem. Given a desired target k, solve (2.1.1.7) for*

$$\int_{-L}^{0} g(\psi(s))\,ds - \psi(0) = g(k)L - k.$$

The right-hand-side is a fixed constant. Pick any ψ satisfying that equation. Use the chosen ψ as the initial function. The resulting solution will approach the desired target, k. This will have significant application in a second order control problem to be discussed later.

Theorem 2.1.1 is in the way of a stability result. Continual dependence of solutions on initial functions tells us that solutions which start close will remain close on finite intervals. But under conditions of Theorem 2.1.1 they remain close forever and their asymptotic constants are close.

Theorem 2.1.2. *Under the conditions of Theorem 2.1.1, every continuous initial function is stable: for each $\varepsilon > 0$ there is a $\delta > 0$ such that $\|\psi_1 - \psi_2\| < \delta$ implies that $|x(t,0,\psi_1) - x(t,0,\psi_2)| < \varepsilon$ for $t \geq 0$. In particular, if $x(t,0,\psi_1) \to k_1$ and $x(t,0,\psi_2) \to k_2$ then $|k_1 - k_2| < \varepsilon$.*

Proof. We will use the notation of the proof of Theorem 2.1.1 and we have also denoted the supremum of initial functions ψ on $[-L,0]$ by $\|\psi\|$, even though we also use that as the supremum metric on $[-L,\infty)$. Let ψ_1 be fixed and let ψ_2 be any other continuous initial function. Then by Theorem 2.1.1 there are unique ϕ_1, ϕ_2, k_1, k_2 such that if P_i is the mapping defined with ψ_i then

$$P_1\phi_1 = \phi_1, \quad P_2\phi_2 = \phi_2, \quad \phi_1(t) \to k_1, \quad \phi_2(t) \to k_2.$$

Notice that since $\phi_i = \psi_i$ on $[-L,0]$ we have

$$\begin{aligned}
|\phi_1(t) - \phi_2(t)| &= |(P_1\phi_1)(t) - (P_2\phi_2)(t)| \\
&\leq |\psi_1(0) - \psi_2(0)| + \int_{-L}^{0} |g(\psi_1(s)) - g(\psi_2(s))|\, ds \\
&\quad + \int_{t-L}^{t} |g(\phi_1(s)) - g(\phi_2(s))|\, ds \\
&\leq |\psi_1(0) - \psi_2(0)| + KL\|\psi_1 - \psi_2\| + KL\|\phi_1 - \phi_2\|
\end{aligned}$$

or

$$\|\phi_1 - \phi_2\| \leq \frac{KL+1}{1-KL}\|\psi_1 - \psi_2\| < \varepsilon$$

provided that

$$\|\psi_1 - \psi_2\| < \frac{\varepsilon(1-KL)}{1+KL} =: \delta.$$

This proves the first part.

For the second part, we have $|\phi_i(t) - k_i| \to 0$ as $t \to \infty$ and so

$$\begin{aligned}
|k_1 - k_2| &= |k_1 - \phi_1(t) + \phi_1(t) - \phi_2(t) + \phi_2(t) - k_2| \\
&\leq |k_1 - \phi_1(t)| + \|\phi_1 - \phi_2\| + |\phi_2(t) - k_2|
\end{aligned}$$

and the last term tends to $\|\phi_1 - \phi_2\| < \varepsilon$. This completes the proof.

When we come to apply these ideas in harmless perturbation problems, we will inquire how things may change if, instead of $g(x)$ we have $g(t,x)$.

First, it is natural to conjecture that if $g(t, x)$ is periodic in t then instead of solutions approaching a constant, they might approach a periodic solution. That was a conjecture of Cooke and Yorke, but it is not true. We can show by simple inspection that the only solutions of period L are constant functions. As in Theorem 2.1.1, for a given ψ there is a constant to which the solution converges. But the idea of periodicity is exactly what makes the process work of adding and subtracting a quantity as we discussed in

$$x' = F(t, x) + A - B + B$$

at the start of this section.

Consider the equation

$$x'(t) = f(t, x(t)) - f(t, x(t - L)) \tag{2.1.2}$$

where f is continuous and

$$f(t + L, x) = f(t, x) \tag{2.1.2.1}$$

for all x. By (2.1.2.1) we can write (2.1.2) as

$$\begin{aligned} x'(t) &= f(t, x(t)) - f(t - L, x(t - L)) \\ &= \frac{d}{dt} \int_{t-L}^{t} f(s, x(s))\, ds. \end{aligned} \tag{2.1.2.2}$$

Proposition 2.1.1. *If (2.1.2.1) is satisfied and if (2.1.2) has a periodic solution of period L, then that solution is constant.*

Proof. If $x(t)$ is a solution of (2.1.2) with period L, then we can integrate (2.1.2.2) and write

$$x(t) = x(0) + \int_{t-L}^{t} f(s, x(s))\, ds - \int_{-L}^{0} f(s, x(s))\, ds. \tag{2.1.2.3}$$

As $x(t + L) = x(t)$, it follows from (2.1.2.1) that

$$F(t) := f(t, x(t))$$

satisfies $F(t+L) = F(t)$. Thus, the integral of F over any period of length L has the same constant value. It then follows from (2.1.2.3) that $x(t) = x(0)$ for all t. This completes the proof.

The application of this idea in dealing with harmless perturbations runs as follows. We suppose that we have an equation $x' = F(t, x_t) + g(t, x(t-L))$ and we would like to replace g by a term without a delay. Obviously, we would not expect g to be $L-$periodic in t. We write

$$x' = F(t, x_t) + g(t, x(t-L)) - g(t+L, x(t)) + g(t+L, x(t))$$

with the idea that the two middle terms constitute a harmless perturbation. Even though g is not periodic in t we still write this as

$$x' = F(t, x_t) + g(t+L, x) - \frac{d}{dt}\int_{t-L}^{t} g(s+L, x(s))\, ds$$

with the last term being a harmless perturbation.

We have established that if (2.1.2) has a periodic solution, then it is constant. Now we continue to show that all solutions approach predetermined constants just as in (2.1.1); thus, $f(t,x) - f(t, x(t-L))$ should be another harmless perturbation to add to our list.

We will need a counterpart of (2.1.1.5). Suppose there is a constant K such that $t, x, y \in R$ implies that

$$|f(t,x) - f(t,y)| \leq K|x-y|. \tag{2.1.2.4}$$

Theorem 2.1.3. *Let (2.1.1.6), (2.1.2.1), and (2.1.2.4) hold for (2.1.2) and let $\psi : [-L, 0] \to R$ be a given continuous function. Then there is a unique constant k satisfying*

$$k = \psi(0) + \int_{t-L}^{t} f(s,k)\, ds - \int_{-L}^{0} f(s, \psi(s))\, ds \tag{2.1.2.5}$$

and the unique solution of (2.1.2) with this initial function satisfies $x(t, 0, \psi) \to k$ as $t \to \infty$.

Here is a brief sketch. The proof proceeds exactly as that of Theorem 2.1.1 when we notice in (2.1.2.5) that $f(s,k)$ has period L in s so that $\int_{t-L}^{t} f(s,k)ds$ is constant; thus, Q has a fixed point. Defining P from (2.1.2.3) and defining $\mathcal{M}$ as before, it readily follows that $\phi(t) \to k$ implies that $(P\phi)(t) \to k$. We will see more detail of this type in later, more difficult, theorems.

Cooke and Yorke (1973; p. 87) continue the study and propose a model in which they postulate a time lag L_1 between conception and birth. Then the number of births at time t is $g(x(t-L_1))$ and (2.1.1) is replaced by

$$x'(t) = g(x(t-L_1)) - g(x(t-L_1-L_2))$$

or in integrated form

$$x(t) = c + \int_{t-L_1-L_2}^{t-L_1} g(x(s))\,ds.$$

They state that this integral is the number of individuals born in the past generation $[t-L_2, t]$ (conceived in $[t-L_1-L_2, t-L_1]$).

In fact, it is easy to show that the behavior of solutions in this equation is unchanged from that in (2.1.1). We select a constant in advance from the initial condition and show that the solution converges to that constant.

We turn now to problems with a distributed delay and see that we again have the same kind of behavior. Most of this leaves the Cooke-Yorke paper and focuses on a formulation of Haddock and Terjéki (1990), although we do not make the monotonicity assumptions found in that paper. Thus, we consider

$$x'(t) = g(x(t)) - \int_{t-L}^{t} p(s-t)g(x(s))\,ds \tag{2.1.3}$$

with p continuous,

$$\int_{-L}^{0} p(s)ds = 1, \tag{2.1.3.1}$$

and assume that there is a constant K such that for all real x, y we have

$$|g(x) - g(y)| \le K|x-y|. \tag{2.1.3.2}$$

For the contraction condition we will need $\alpha < 1$ such that

$$K\int_{-L}^{0} |p(s)|(-s)\,ds \le \alpha. \tag{2.1.3.3}$$

Remark 2.1.4. *If $p(t) = 1/L$ then (2.1.3.1) is satisfied, while (2.1.3.3) yields $KL/2 \le \alpha$, a weaker requirement than in the constant delay case.*

We can write (2.1.3) as

$$x'(t) = \frac{d}{dt}\int_{-L}^{0} p(s)\int_{t+s}^{t} g(x(u))\,du\,ds. \tag{2.1.3.4}$$

Then to specify a solution we need a continuous function $\psi : [-L, 0] \to R$ so that we can write (2.1.3) as

$$x(t) = \psi(0) + \int_{-L}^{0} p(s)\int_{t+s}^{t} g(x(u))\,du\,ds \\ - \int_{-L}^{0} p(s)\int_{s}^{0} g(\psi(u))\,du\,ds. \tag{2.1.3.5}$$

The following theorem is actually a corollary to Theorem 2.1.5 but the proof is so simple it seems wrong to embed it in that framework with a much more complicated proof.

Theorem 2.1.4. *Suppose that (2.1.3.1), (2.1.3.2), and (2.1.3.3) hold. For the given initial function ψ there is a unique constant k satisfying*

$$k = \psi(0) + g(k)\int_{-L}^{0} p(s)(-s)\,ds - \int_{-L}^{0} p(s)\int_{s}^{0} g(\psi(u))\,du\,ds \tag{2.1.3.6}$$

and the unique solution $x(t, 0, \psi)$ of (2.1.3.5) converges to k as $t \to \infty$.

Proof. Use (2.1.3.6) to define a mapping Q as we did in the proof of Theorem 2.1.1. The mapping will be a contraction because of (2.1.3.3). We then define

$$\mathcal{M} = \{\phi : [-L, \infty) \to R \mid \phi \in \mathcal{C}, \\ \phi(t) = \psi(t) \quad \text{on} \quad [-L, 0], \ \lim_{t\to\infty} \phi(t) = k\}. \tag{2.1.3.7}$$

With the supremum metric, $\mathcal{M}$ is a complete metric space. Use (2.1.3.5) to define a mapping P of that space into itself. In particular, we note that $\phi \in \mathcal{M}$ implies $(P\phi)(t) \to k$ as $t \to \infty$. Also, P will be a contraction because of (2.1.3.3) with unique fixed point. This completes the proof.

With distributed delay there seems to be a possibility of a periodic solution. Consider the scalar equation

$$x'(t) = g(t, x(t)) - \int_{t-L}^{t} p(s-t)g(s, x(s))\,ds \tag{2.1.4}$$

where g and p are continuous, there is a $K > 0$ such that $x, y \in R$ implies

$$|g(t, x) - g(t, y)| \le K|x - y|, \tag{2.1.4.1}$$

$$g(t + L, x) = g(t, x), \tag{2.1.4.2}$$

and p satisfies (2.1.3.1).

We can write (2.1.4) as

$$x'(t) = \frac{d}{dt}\int_{-L}^{0} p(s)\int_{t+s}^{t} g(u, x(u))\,du\,ds. \tag{2.1.4.3}$$

Then for a continuous initial function ψ we can write (2.1.4.3) as

$$x(t) = \int_{-L}^{0} p(s)\int_{t+s}^{t} g(u, x(u))\,du\,ds + \psi(0) - \int_{-L}^{0} p(s)\int_{s}^{0} g(u, x(u))\,du\,ds.$$

In the next result we could also prove that the solutions are stable following the ideas in the proof of Theorem 2.1.2.

Theorem 2.1.5. *If (2.1.3.1), (2.1.3.3), (2.1.4.1), and (2.1.4.2) hold, then for a given continuous function $\psi : [-L, 0] \to R$ there is a unique periodic solution γ of (2.1.4) having period L and the unique solution $x(t, 0, \psi)$ of (2.1.4) tends to γ as $t \to \infty$. Also, γ is constant only if there is a $k \in R$ with*

$$g(t, k) = \int_{-L}^{0} p(s) g(t + s, k)\,ds.$$

Proof. Let $c = \psi(0) - \int_{-L}^{0} p(s) \int_{s}^{0} g(u, \psi(u))\,du\,ds$ and let $(\mathcal{X}, \|\cdot\|)$ be the Banach space of continuous L-periodic functions with the supremum norm. Define $Q : \mathcal{X} \to \mathcal{X}$ by $\phi \in \mathcal{X}$ implies that

$$(Q\phi)(t) = c + \int_{-L}^{0} p(s)\int_{t+s}^{t} g(u, \phi(u))\,du\,ds.$$

Note that

$$\begin{aligned}(Q\phi)(t+L) &= c + \int_{-L}^{0} p(s)\int_{t+L+s}^{t+L} g(u, \phi(u))\,du\,ds \\ &= c + \int_{-L}^{0} p(s)\int_{t+s}^{t} g(u+L, \phi(u+L))\,du\,ds \\ &= (Q\phi)(t).\end{aligned}$$

To see that Q is a contraction, if $\phi, \eta \in \mathcal{X}$ then

$$|(Q\phi)(t) - (Q\eta)(t)| \le \int_{-L}^{0} |p(s)| \int_{t+s}^{t} K\,du\,ds\|\phi - \eta\| \le \alpha\|\phi - \eta\|.$$

Hence, Q has a unique fixed point $\gamma \in \mathcal{X}$, an L-periodic solution of (2.1.4).

Now, let $(\mathcal{M}, \|\cdot\|)$ be the complete metric space of continuous functions ϕ with $\phi(t) = \psi(t)$ on $[-L, 0]$ and $\phi(t) \to \gamma(t)$ as $t \to \infty$. We take the supremum metric.

Define a mapping $P : \mathcal{M} \to \mathcal{M}$ by $\phi \in \mathcal{M}$ implies that $(P\phi)(t) = \psi(t)$ on $[-L, 0]$, and for $t > 0$ define

$$\begin{aligned}(P\phi)(t) &= \psi(0) - \int_{-L}^{0} p(s) \int_{s}^{0} g(u, \psi(u))\, du\, ds \\ &\quad + \int_{-L}^{0} p(s) \int_{t+s}^{t} g(u, \phi(u))\, du\, ds \\ &= c + \int_{-L}^{0} p(s) \int_{t+s}^{t} g(u, \phi(u))\, du\, ds.\end{aligned}$$

We must show that $(P\phi)(t) \to \gamma(t)$. From the definition of γ and P for $t > L$ we have

$$\begin{aligned}|(P\phi)(t) - \gamma(t)| &= |\int_{-L}^{0} p(s) \int_{t+s}^{t} g(u, \phi(u)) du \\ &\quad - \int_{-L}^{0} p(s) \int_{t+s}^{t} g(u, \gamma(u)) du\, ds| \\ &\leq \int_{-L}^{0} |p(s)| \int_{t+s}^{t} K|\phi(u) - \gamma(u)|\, du\, ds \\ &\leq \int_{-L}^{0} |p(s)| \int_{t-L}^{t} K|\phi(u) - \gamma(u)|\, du\, ds\end{aligned}$$

and the last term tends to zero as $t \to \infty$ since $|\phi(u) - \gamma(u)| \to 0$ as $u \to \infty$ and p is continuous.

The details for showing that P is a contraction are identical to those of showing that Q is a contraction. Hence, P has a unique fixed point in $\mathcal{M}$ and that fixed point does converge to γ.

If (2.1.4) has a constant solution k, then $x'(t) = k' = 0$ so in (2.1.4.3)

$$\int_{-L}^{0} p(s)[g(t, k) - g(t + s, k)]\, ds = 0$$

or

$$g(t, k) \int_{-L}^{0} p(s) ds = \int_{-L}^{0} p(s) g(t + s, k)\, ds$$

and since $\int_{-L}^{0} p(s)\, ds = 1$ we have

$$g(t, k) = \int_{-L}^{0} p(s) g(t + s, k)\, ds.$$

This completes the proof.

Remark 2.1.5. *It is worth thinking about the proof for a moment. One is tempted to say that P and Q are the same, but they are not because P is defined separately on $[-L, 0]$. The solution ϕ will be periodic only if one is so fortunate as having guessed an initial function ψ whose periodic extension to all of R is a fixed point of Q.*

The right-hand-sides of the equations numbered as (2.1.n) will turn out to be harmless perturbations under the proper conditions. We now illustrate how two different harmless perturbations can be put together. Consider the equation

$$x'(t) = \int_{t-L}^{t} p(s-t)g(x(s))\,ds - \int_{-\infty}^{t} q(s-t)g(x(s))\,ds \tag{2.1.5}$$

where we suppose that

$$\int_{-L}^{0} p(s)\,ds = 1 \tag{2.1.5.1}$$

while

$$\int_{-\infty}^{0} q(s)\,ds = 1 \quad \text{and} \quad \int_{-\infty}^{0}\int_{-\infty}^{v} |q(u)|\,du\,dv \quad \text{exists}\,. \tag{2.1.5.2}$$

Although the equation does not have a ready first integral, there is a simple fixed point solution. There is also a price to pay. Write (2.1.5) as

$$\begin{aligned} x'(t) = &\int_{t-L}^{t} p(s-t)g(x(s))\,ds \\ &- g(x(t)) + g(x(t)) - \int_{-\infty}^{t} q(s-t)g(x(s))\,ds \end{aligned} \tag{2.1.5.3}$$

and then write

$$\begin{aligned} x'(t) = &-\frac{d}{dt}\int_{-L}^{0} p(s)\int_{t+s}^{t} g(x(u))\,du\,ds \\ &+ \frac{d}{dt}\int_{-\infty}^{t}\int_{-\infty}^{s-t} q(u)du g(x(s))\,ds. \end{aligned} \tag{2.1.5.4}$$

If c is any constant then

$$\begin{aligned} x(t) = c &- \int_{-L}^{0} p(s)\int_{t+s}^{t} g(x(u))\,du\,ds \\ &+ \int_{-\infty}^{t}\int_{-\infty}^{s-t} q(u)\,du g(x(s))\,ds \end{aligned} \tag{2.1.5.5}$$

is a solution of (2.1.5.4).

Let $\psi : (-\infty, 0] \to R$ be a bounded continuous function and write (2.1.5.4) as

$$\begin{aligned} x(t) = &-\int_{-L}^{0} p(s) \int_{t+s}^{t} g(x(u))\,du\,ds + \int_{-\infty}^{t} \int_{-\infty}^{s-t} q(u) du g(x(s))\,ds \\ &+ \psi(0) + \int_{-L}^{0} p(s) \int_{s}^{0} g(\psi(u))\,du\,ds \\ &- \int_{-\infty}^{0} \int_{-\infty}^{s} q(u)\,du g(\psi(s))\,ds \end{aligned} \tag{2.1.5.6}$$

since

$$\begin{aligned} c := \psi(0) &+ \int_{-L}^{0} p(s) \int_{s}^{0} g(\psi(u))\,du\,ds \\ &- \int_{-\infty}^{0} \int_{-\infty}^{s} q(u)\,du g(\psi(s))\,ds. \end{aligned} \tag{2.1.5.7}$$

We suppose there is a $K > 0$ such that $x, y \in R$ implies that

$$|g(x) - g(y)| \leq K|x - y| \tag{2.1.5.8}$$

and there is an $\alpha < 1$ such that

$$\int_{-L}^{0} |p(s)|(-s)\,ds + \int_{-\infty}^{0} \int_{-\infty}^{v} |q(u)|\,du\,dv \leq \alpha/K. \tag{2.1.5.9}$$

Theorem 2.1.6. *Let (2.1.5.1), (2.1.5.2), (2.1.5.8), and (2.1.5.9) hold. Then there is a unique constant k satisfying*

$$k = c - g(k) \int_{-L}^{0} p(s)(-s) ds + g(k) \int_{-\infty}^{0} \int_{-\infty}^{v} q(u)\,du\,dv \tag{2.1.5.10}$$

where c is defined in (2.1.5.7). The unique solution $x(t, 0, \psi)$ of (2.1.5.6) approaches k as $t \to \infty$.

Proof. Use (2.1.5.10) to define a mapping $Q : R \to R$ by

$$Qk = c - g(k) \int_{-L}^{0} p(s)(-s) ds + g(k) \int_{-\infty}^{0} \int_{-\infty}^{v} q(u)\,du\,dv$$

so that for $k, d \in R$ we have

$$\begin{aligned} |Qk - Qd| &\leq K|k - d| \int_{-L}^{0} |p(s)s|\,ds + K|k - d| \int_{-\infty}^{0} \int_{-\infty}^{v} |q(u)|\,du\,dv \\ &\leq \alpha|k - d| \end{aligned}$$

by (2.1.5.9). This yields the unique k.

For the given ψ and the fixed point k, define $\mathcal{M}$ as the set of continuous functions $\phi : (-\infty, \infty) \to R$ with $\phi(t) = \psi(t)$ on $(-\infty, 0]$ and $\phi(t) \to k$ as $t \to \infty$. Use the supremum metric. Define $P : \mathcal{M} \to \mathcal{M}$ by $\phi \in \mathcal{M}$ implies that $(P\phi)(t) = \psi(t)$ on $(-\infty, 0]$ and for $t \geq 0$ then use (2.1.5.6) and (2.1.5.7) to define

$$(P\phi)(t) = c - \int_{-L}^{0} p(s) \int_{t+s}^{t} g(\phi(u))\, du\, ds + \int_{-\infty}^{t} \int_{-\infty}^{s-t} q(u) du g(\phi(s))\, ds.$$

We must show that $(P\phi)(t) \to k$ as $t \to \infty$.

For a fixed $\phi \in \mathcal{M}$, let $\varepsilon > 0$ be given and find positive numbers J and T such that $|g(\phi(t)) - g(k)| \leq J$ for all t and $|\phi(t) - k| < \varepsilon$ if $T \leq t < \infty$.

First, it is clear that

$$\int_{-L}^{0} p(s) \int_{t+s}^{t} g(\phi(u))\, du\, ds \to g(k) \int_{-L}^{0} p(s)(-s)\, ds$$

as $t \to \infty$. Next,

$$\begin{aligned}
&\left| \int_{-\infty}^{t} \int_{-\infty}^{s-t} q(u)\, du (g(\phi(s)) - g(k))\, ds \right| \\
&\leq J \int_{-\infty}^{T} \int_{-\infty}^{s-t} |q(u)| du ds + \int_{T}^{t} \int_{-\infty}^{s-t} |q(u)|\, du K\varepsilon\, ds \\
&= J \int_{-\infty}^{T-t} \int_{-\infty}^{s} |q(u)|\, du\, ds + K\varepsilon \int_{-\infty}^{t} \int_{-\infty}^{s-t} |q(u)|\, du\, ds.
\end{aligned}$$

In the last line, the first term tends to zero as $t \to \infty$ because of the assumed convergence in (2.1.5.9). The second term is bounded by $K\varepsilon\alpha/K = \alpha\varepsilon$ by (2.1.5.9). Hence,

$$\int_{-\infty}^{t} \int_{-\infty}^{s-t} q(u)\, du g(\phi(s))\, ds \to g(k) \int_{-\infty}^{t} \int_{-\infty}^{s-t} q(u)\, du\, ds$$

as $t \to \infty$. Comparing these results with (2.1.5.10) shows that $(P\phi)(t) \to k$.

To see that P is a contraction, if $\phi, \eta \in \mathcal{M}$ then

$$\begin{aligned}
&|(P\phi)(t) - (P\eta)(t)| \\
&\leq K\|\phi - \eta\| \int_{-L}^{0} |p(s)s| ds + K\|\phi - \eta\| \int_{-\infty}^{t} \int_{-\infty}^{s-t} |q(u)|\, du\, ds \\
&\leq \alpha\|\phi - \eta\|.
\end{aligned}$$

There is a fixed point in $\mathcal{M}$ which meets the conditions in the theorem.

Remark 2.1.6. *This is the only infinite delay problem we will do here, but the interested reader will see that all of our delays could be changed to infinite delays and the analysis would be parallel to that just given. We could also follow the proof of Theorem 2.1.5 and introduce periodicity in t in $g(t,x)$. Solutions are stable and each limit constant is a selective global attractor. In problems to come we will see that (2.1.5.1) and the first part of (2.1.5.2) are, indeed, remarkable. They are asking only that p and q be positive on average.*

Considerable effort has gone into extending the theory of equations where each constant is a solution to neutral equations. And there is good reason for doing so. The problem has its roots in mathematical biology, a subject in which neutral-type behavior is ubiquitous. Every parent and every gardener has observed a living organism displaying ordinary or subordinary growth. Suddenly, growth accelerates and acceleration gives birth to more acceleration until the observer may claim to actually see the growth taking place. This is typical of neutral growth. Present growth rate depends not only on the past state, but on the past growth rate. A typical example in our context is

$$\frac{d}{dt}(x(t) - h(x(t - L_1))) = g(x(t)) - g(x(t - L_2)). \tag{2.1.6}$$

Clearly, any constant function is a solution. We suppose that $g, h : R \to R$ and that there are positive constants K_1, K_2 so that for all $x, y \in R$ we have

$$|h(x) - h(y)| \leq K_1|x - y| \tag{2.1.6.1}$$

and

$$|g(x) - g(y)| \leq K_2|x - y| \tag{2.1.6.2}$$

with an $\alpha < 1$ such that

$$K_1 + L_2K_2 \leq \alpha. \tag{2.1.6.3}$$

If $L = \max(L_1, L_2)$ and if $\psi : [-L, 0] \to R$ is continuous, then we can write our equation with that initial function as

$$\begin{aligned} x(t) = h(x(t - L_1)) + \int_{t-L_2}^{t} g(x(s))\,ds + \psi(0) - h(\psi(-L_1)) \\ - \int_{-L_2}^{0} g(\psi(s))\,ds. \end{aligned} \tag{2.1.6.4}$$

Theorem 2.1.7. *Let (2.1.6.1)–(2.1.6.3) hold and let ψ be a given continuous function. There is a unique constant k satisfying*

$$k = h(k) + g(k)L_2 + \psi(0) - h(\psi(-L_1)) - \int_{-L_2}^{0} g(\psi(s))\,ds \tag{2.1.6.5}$$

and the unique solution $x(t, 0, \psi)$ of (2.1.6.4) tends to k as $t \to \infty$.

Proof. Use (2.1.6.5) to define a mapping Q as before. By (2.1.6.3) it will be a contraction with unique fixed point. For the given ψ, let $(\mathcal{M}, \|\cdot\|)$ be the complete metric space with supremum metric of continuous functions $\phi : [-L, \infty) \to R$ with $\phi(t) = \psi(t)$ on $[-L, 0]$ and $\phi(t) \to k$ as $t \to \infty$. Define $P : \mathcal{M} \to \mathcal{M}$ by $\phi \in \mathcal{M}$ implies that $(P\phi)(t) = \psi(t)$ on $[-L, 0]$ and for $t \geq 0$ then

$$\begin{aligned}(P\phi)(t) = h(\phi(t - L_1)) &+ \int_{t-L_2}^{t} g(\phi(s))\,ds \\ &+ \psi(0) - h(\psi(-L_1)) - \int_{-L_2}^{0} g(\psi(s))\,ds.\end{aligned}$$

We may note that $(P\phi)(0) = \psi(0)$ and that since $\phi(t) \to k$ we have $(P\phi)(t) \to k$ as $t \to \infty$. Moreover, P is a contraction since for $\phi, \eta \in \mathcal{M}$ it follows that

$$\begin{aligned}|(P\phi)(t) - (P\eta)(t)| &\leq K_1|\phi(t - L_1) - \eta(t - L_1)| \\ &\quad + \int_{t-L_2}^{t} K_2|\phi(s) - \eta(s)|\,ds \\ &\leq (K_1 + K_2L_2)\|\phi - \eta\|.\end{aligned}$$

This completes the proof.

Several authors have investigated this problem for systems, but the central condition of Cooke and Yorke that g be an arbitrary differentiable function, possibly bounded by $K|x|$, seems to have been pushed aside, except in Arino and Pituk (1998). Atkinson and Haddock (1983) and Atkinson, Haddock, and Staffans (1982) have looked at problems involving a system $x' = P(t)[x(t) - x(t - h)]$ and have found a variety of conditions, such as $P \in L^2[0, \infty)$ to ensure that solutions approach constants.

Minorsky designed an automatic steering device for the large ship the New Mexico which was governed by an equation

$$x''(t) + cx'(t) + g(x(t - h)) = 0. \tag{2.1.7}$$

Here, the rudder of the ship has angular position $x(t)$ and there is friction force proportional to the velocity. There is a direction indicating instrument which points in the actual direction of motion and there is an instrument pointing in the desired direction. These two are connected to a device which activates an electric motor producing a certain force to move the rudder so as to bring the ship onto the desired course. There is a time lag of amount h between the time the ship gets off course and the time the electric motor activates the restoring force. The object is to give conditions ensuring that $x(t)$ will stay near zero so that the ship closely follows its proper course. One may find a bit about the problem in Minorsky (1960) (1962; p. 517) and in Burton (1985; p. 149). Other work of this sort is found in Haddock, Krisztin, Terjéki, and Wu (1990), Jehu (1978), Kaplan, Sorg, and Yorke (1979), and Krisztin and Wu (1996), among so many others. These papers should lead us to a great many harmless perturbations in exactly the same way as the present presentation.

These kinds of control devices tend to over-correct the errors. The Cooke-Yorke type equation may do a much better job. If Minorsky had applied the opposite force L-time units later, say $g(x(t-h-L))$, then his ship would have eventually gone in a straight line, but on a course parallel to his desired course, provided that g satisfies a Lipschitz condition and that h and L are not too large.

Instead of Minorsky's equation we look at

$$x'' + cx'(t) + g(x(t-h)) - g(x(t-h-L)) = 0 \tag{2.1.8}$$

which we write as

$$x'' + cx' = -\frac{d}{dt}\int_{t-h-L}^{t-h} g(x(s))\,ds. \tag{2.1.8.1}$$

Given an initial function $\psi : [-h-L, 0] \to R$ and $x'(0)$ we can write (2.1.8.1) as

$$\begin{aligned} x'(t) + cx(t) &= x'(0) + c\psi(0) + \int_{-h-L}^{-h} g(\psi(s))\,ds - \int_{t-h-L}^{t-h} g(x(s))\,ds \\ &=: \Psi - \int_{t-h-L}^{t-h} g(x(s))\,ds. \end{aligned} \tag{2.1.8.2}$$

and then as

$$x(t) = \psi(0)e^{-ct} + \int_0^t e^{-c(t-s)}[\Psi - \int_{s-h-L}^{s-h} g(x(u))du]\,ds. \tag{2.1.8.3}$$

We ask that there is $K > 0$ such that for all real x, y we have

$$|g(x) - g(y)| \le K|x-y| \tag{2.1.8.4}$$

and there is an $\alpha < 1$ with

$$KL \leq \alpha c. \tag{2.1.8.5}$$

Theorem 2.1.8. *Let (2.1.8.4) and (2.1.8.5) hold, let ψ be a given continuous initial function on $[-h-L, 0]$, let $x'(0)$ be given, and let Ψ be defined in (2.1.8.2). Then there is a unique constant k satisfying*

$$ck = \Psi - g(k)L \tag{2.1.8.6}$$

and the unique solution $x(t, 0, \psi, x'(0))$ of (2.1.8.3) converges to k as $t \to \infty$.

Proof. Clearly, the mapping $Q : R \to R$ defined by $k \in R$ implies

$$Qk := (\Psi - g(k)L)/c \tag{2.1.8.7}$$

is a contraction with unique fixed point. Let $(\mathcal{M}, \|\cdot\|)$ be the complete metric space with supremum metric of continuous functions $\phi : [-h - L, \infty) \to R$ satisfying $\phi(t) \to k$ as $t \to \infty$, $\phi(t) = \psi(t)$ on $[-h-L, 0]$. Define $P : \mathcal{M} \to \mathcal{M}$ by $\phi \in \mathcal{M}$ implies that $(P\phi)(t) = \psi(t)$ on $[-h-L, 0]$ and for $t > 0$ then

$$(P\phi)(t) = \psi(0)e^{-ct} + \int_0^t e^{-c(t-s)}[\Psi - \int_{s-h-L}^{s-h} g(\phi(u))du]\, ds. \tag{2.1.8.8}$$

We have

$$\begin{aligned}(P\phi)(t) &= \psi(0)e^{-ct} + \int_0^t e^{-c(t-s)}\Psi\, ds \\ &\quad - \int_0^t e^{-c(t-s)} \int_{s-h-L}^{s-h} [g(\phi(u)) - g(k) + g(k)]\, du\, ds.\end{aligned}$$

Now, examine part of the last term. Let T be a large positive number and let

$$\|\phi - k\|_{[T,\infty)} = \sup_{t \geq T-h-L} |\phi(u) - k|.$$

Then

$$\begin{aligned}|H(t)| &:= |\int_0^t e^{-c(t-s)} \int_{s-h-L}^{s-h} [g(\phi(u) - g(k)]\, du\, ds| \\ &\leq \int_0^T e^{-c(t-s)} KL\|\phi - k\| ds + \int_T^t e^{-c(t-s)} KL\|\phi - k\|_{[T,\infty)} ds.\end{aligned}$$

The last term can be made as small as we please by taking T large since $\phi(t) \to k$. The other term tends to zero; hence, this whole term, $H(t)$, can be made as small as we please by taking t large.

Thus, we can say that

$$\begin{aligned}(P\phi)(t) = \psi(0)e^{-ct} + \frac{\Psi}{c}[1 - e^{-ct}] \\ - \frac{g(k)L}{c}[1 - e^{-ct}] + H(t)\end{aligned}$$

which tends to

$$\frac{\Psi}{c} - \frac{g(k)L}{c} = k$$

as $t \to \infty$.

Moreover, P is a contraction since $\phi, \eta \in M$ imply that

$$|(P\phi)(t) - (P\eta)(t)| \leq \int_0^t e^{-c(t-s)} LK\|\phi - \eta\|\, ds \leq \frac{LK}{c}\|\phi - \eta\|.$$

Remark 2.1.7. *We now remind the reader of Remark 2.1.3. Given a target k, we can solve (2.1.8.6) for Ψ. If we then select any initial function having the value in (2.1.8.2) of this number Ψ and manually steer the ship in the pattern of that initial function, then the solution will converge to k. This is a type of control problem. Recently, Zhang and Gao (2005) have solved a control problem with analogous techniques.*

We now construct some examples which are certainly contrived, but are devised so that the reader can see precisely how the right-hand-sides of equations designated as (2.1.n) are harmless perturbations. In all of the rest of the sections of this chapter we will show classical examples which are in no manner contrived for which these same expressions will greatly simplify the stability analysis yielding entirely new results.

Our first example is entirely pedestrian, yet it illustrates how the comparatively small term $-\frac{1}{t+1}x(t)$ can overpower the much larger term $.4x(t)$ in the presence of a term $-.4x(t-1)$ so that the equation

$$x' = -\frac{1}{t+1}x(t) + .4x(t) - .4x(t-1)$$

is asymptotically stable.

Example 2.1.1. Consider the scalar equation

$$x'(t) = -a(t)x(t) + g(x(t)) - g(x(t-L)) \tag{2.1.9}$$

where

$$|g(x) - g(y)| \leq K|x - y|$$

for a fixed constant K and all $x, y \in R$, while

$$\alpha := KL < 1/2$$

and

$$a(t) \geq 0, \quad \int_0^\infty a(s)ds = \infty.$$

Even though g can have arbitrary behavior and seem to be greatly destabilizing, so long as it is Lipschitz and $KL < 1/2$, stability analysis of the case $L = 0$ is correct for the problem with $KL < 1/2$. We will show that every solution tends to zero as $t \to \infty$.

Let an arbitrary continuous initial function $\psi : [0, L] \to R$ be given. There will then be a unique solution $x(t) = x(t, 0, \psi)$ which agrees with ψ on $[-L, 0]$. It is not difficult to prove that this solution exists on $[0, \infty)$.

Thus, for this initial function, our equation

$$x'(t) = -a(t)x(t) + \frac{d}{dt}\int_{t-L}^{t} g(x(s))\, ds$$

can be written as

$$\begin{aligned}
x(t) &= \psi(0)e^{-\int_0^t a(s)\,ds} + \int_0^t e^{-\int_s^t a(u)\,du} \frac{d}{ds}\int_{s-L}^{s} g(x(u))\,du\,ds \\
&= \psi(0)e^{-\int_0^t a(s)\,ds} + e^{-\int_s^t a(u)\,du}\int_{s-L}^{s} g(x(u))\,du\Big|_0^t \\
&\quad - \int_0^t e^{-\int_s^t a(u)\,du} a(s)\int_{s-L}^{s} g(x(u))\,du\,ds \\
&= \psi(0)e^{-\int_0^t a(s)\,ds} + \int_{t-L}^{t} g(x(u))\,du \\
&\quad - e^{-\int_0^t a(s)\,ds}\int_{-L}^{0} g(\psi(s))\,ds \\
&\quad - \int_0^t e^{-\int_s^t a(u)\,du} a(s)\int_{s-L}^{s} g(x(u))\,du\,ds.
\end{aligned}$$

Let

$$\mathcal{M} = \{\phi : [-L, \infty) \to R \,|\, \phi \in \mathcal{C}, \phi_0 = \psi, \phi(t) \to 0 \quad \text{as} \quad t \to \infty\}$$

and define $P : \mathcal{M} \to \mathcal{M}$ by $(P\phi)(t) = \psi(t)$ on $[-L, 0]$, and for $t > 0$ then

$$\begin{aligned}(P\phi)(t) &= \psi(0)e^{-\int_0^t a(s)\,ds} + \int_{t-L}^{t} g(\phi(s))\,ds \\ &\quad - e^{-\int_0^t a(s)\,ds}\int_{-L}^{0} g(\psi(s))\,ds \\ &\quad - \int_0^t e^{-\int_s^t a(u)\,du}a(s)\int_{s-L}^{s} g(\phi(u))\,du\,ds.\end{aligned}$$

It is easy to see that the second term on the right tends to $Lg(0)$ as $t \to \infty$. We will show that the last term on the right tends to $-Lg(0)$ so that it will follow that $(P\phi)(t) \to 0$ as $t \to \infty$. Let $\varepsilon > 0$ be given and for a fixed $\phi \in M$, find $T > 0$ such that $t \geq T - L$ implies that $|Lg(\phi(t)) - Lg(0)| < \varepsilon$. Then the last term can be written as

$$\begin{aligned}&\int_0^T e^{-\int_s^t a(u)\,du}a(s)\int_{s-L}^{s} g(\phi(u))\,du\,ds \\ &\qquad + \int_T^t e^{-\int_s^t a(u)\,du}a(s)\int_{s-L}^{s} g(\phi(u))\,du\,ds.\end{aligned}$$

The first term of this last expression tends to zero as $t \to \infty$. Then

$$\begin{aligned}&\left|\int_T^t e^{-\int_s^t a(u)\,du}a(s)\int_{s-L}^{s} [g(\phi(u)) - g(0)]\,du\,ds\right| \\ &\leq \varepsilon \int_T^t e^{-\int_s^t a(u)du}a(s)\,ds \\ &< \varepsilon.\end{aligned}$$

Note that

$$\begin{aligned}&\int_T^t e^{-\int_s^t a(u)\,du}a(s)\int_{s-L}^{s} g(0)\,du\,ds \\ &= Lg(0)\left[1 - e^{-\int_T^t a(u)\,du}\right] \\ &\to Lg(0).\end{aligned}$$

To see that P is a contraction, if $\phi, \eta \in \mathcal{M}$ then

$$\begin{aligned}|(P\phi)(t) - (P\eta)(t)| &\leq \int_{t-L}^{t} |g(\phi(u)) - g(\eta(u))|\, du \\ &\quad + \int_0^t e^{-\int_s^t a(u)\,du} a(s) \int_{s-L}^{s} |g(\phi(u)) - g(\eta(u))|\, du\, ds \\ &\leq 2KL\|\phi - \eta\| = \alpha\|\phi - \eta\|.\end{aligned}$$

so P is a contraction with fixed point $\phi \in \mathcal{M}$ so ϕ tends to zero as $t \to \infty$.

Our next example shows how two very different terms can work together seamlessly to yield asymptotic stability.

Example 2.1.2. The zero solution of

$$x'(t) = \frac{1}{2}x(t) - x^3(t) - \sin x(t - L) \tag{2.1.10}$$

is globally asymptotically stable when

$$L < 1/2.$$

This is an interesting equation. Near the equilibrium, $x' = -\sin x(t-L)$ is asymptotically stable and it dominates $\frac{1}{2}x(t) - x^3(t)$, producing local asymptotic stability of the whole equation. However, farther from equilibrium, the sine term becomes unstable; stability is then picked up by the cubic term. In the work here, we effectively simply remove the L and the terms work together to produce global stability in one step.

We can write this equation as

$$\begin{aligned}x' &= \frac{1}{2}x - x^3 - \sin x + \sin x - \sin x(t - L) \\ &= \frac{1}{2}x - x\left(x^2 + \frac{\sin x}{x}\right) + \frac{d}{dt}\int_{t-L}^{t} \sin x(s)\, ds.\end{aligned}$$

If we write out the series for $\sin x$ and divide by x, we have an alternating series and for $|x| \leq 1$ we can say that $\frac{\sin x}{x} \geq 1 - \frac{x^2}{6}$. For $1 \leq |x| \leq \pi$, then $x^2 + \frac{\sin x}{x} \geq 1$. Finally, for $|x| \geq \pi$, then by factoring out x^2 we see that the same conclusion holds.

Let $\psi : [-L, 0] \to R$ be a fixed but arbitrary continuous initial function. For that initial function our equation has a unique solution

$x_1(t) = x(t, 0, \psi)$ and it is defined on $[0, \infty)$. With a view to writing our equation as

$$x' = -a(t)x + \frac{d}{dt}\int_{t-L}^{t} \sin x(s)\, ds,$$

(which will be called exact linearization) we define

$$a(t) := -\frac{1}{2} + x_1^2(t) + \frac{\sin x_1(t)}{x_1(t)} \geq \frac{1}{2}$$

by the above calculations. Thus, for that fixed initial function and, hence, for the fixed solution $x_1(t)$, then $a(t)$ is fixed and our equation becomes

$$x' = -a(t)x + \frac{d}{dt}\int_{t-L}^{t} \sin x(u)\, du$$

with solution

$$x(t) = \psi(0)e^{-\int_0^t a(s)ds} + \int_0^t e^{-\int_s^t a(u)\, du}\frac{d}{ds}\int_{s-L}^{s} \sin x(u)\, du\, ds.$$

We construct the set $\mathcal{M}$ and the mapping P in exactly the same way as in Example 2.1.1. The set $\mathcal{M}$ is identical, while the mapping P has $\sin x(u)$ substituted for $g(x(u))$. The remainder of the details are identical.

The next example is contrived to produce the same result as that just given by using two terms, neither of which is asymptotically stable. But when we erase the L the terms fit together to produce asymptotic stability.

Example 2.1.3. Define two functions by

$$r(x) = \begin{cases} x, & \text{for } x \geq 0 \\ 0, & \text{for } x < 0, \end{cases}$$

and

$$g(x) = \begin{cases} 0, & \text{for } x \geq 0 \\ x, & \text{for } x < 0. \end{cases}$$

Both r and g are 1-Lipschitz. For $L < 1/2$ we consider the equation

$$\begin{aligned} x'(t) &= -r(x(t)) - g(x(t-L)) \\ &= -[r(x) + g(x)] + g(x) - g(x(t-L)) \\ &= -x + \frac{d}{dt}\int_{t-L}^{t} g(x(s))\, ds. \end{aligned} \tag{2.1.11}$$

The remainder of the problem is just like Example 2.1.1. We readily show that all solutions tend to zero.

The next example continues with the idea of exact linearization which is the basis of the entire next section.

Example 2.1.4. A form of Example 2.1.1 with the unperturbed part nonlinear offers a bit of difficulty which we circumvent here. Consider the equation

$$x'(t) = -x^3(t) + g(x(t)) - g(x(t-L)) \tag{2.1.12}$$

where

$$|g(x) - g(y)| \leq K|x - y|$$

for a fixed constant K and all $x, y \in R$, while

$$\alpha := KL < 1/2.$$

We will show that all solutions tend to zero. This is an autonomous equation and we lose nothing by considering an arbitrary continuous initial function $\psi : [-L, 0] \to R$. There will then be a unique solution $x_1(t) = x(t, 0, \psi)$ which agrees with ψ on $[-L, 0]$. It is not difficult to prove that this solution exists on $[0, \infty)$.

For that fixed ψ there is a unique function $a : [0, \infty) \to [0, \infty)$ such that the aforementioned solution $x_1(t)$ satisfies

$$x_1^3(t) =: a(t)x_1(t).$$

(This, again, is exact linearization.) Thus, for this initial function, our equation is

$$x'(t) = -a(t)x(t) + \frac{d}{dt}\int_{t-L}^{t} g(x(s))\, ds$$

so we can write

$$\begin{aligned}
x(t) &= \psi(0)e^{-\int_0^t a(s)\,ds} + \int_0^t e^{-\int_s^t a(u)du} \frac{d}{ds}\int_{s-L}^{s} g(x(u))\,du\,ds \\
&= \psi(0)e^{-\int_0^t a(s)\,ds} + e^{-\int_s^t a(u)du}\int_{s-L}^{s} g(x(u))du\Big|_0^t \\
&- \int_0^t e^{-\int_s^t a(u)\,du} a(s)\int_{s-L}^{s} g(x(u))\,du\,ds \\
&= \psi(s)e^{-\int_0^t a(s)\,ds} + \int_{t-L}^{t} g(x(u))\,du \\
&- \int_{-L}^{0} g(\psi(s))\,ds - \int_0^t e^{-\int_s^t a(u)du} a(s)\int_{s-L}^{s} g(x(u))\,du\,ds.
\end{aligned}$$

Let

$$\mathcal{M} = \{\phi : [-L, \infty) \to R \,|\, \phi \in \mathcal{C}, \phi_0 = \psi, \phi \quad \text{bounded}\}$$

and define $P : \mathcal{M} \to \mathcal{M}$ by $(P\phi)(t) = \psi(t)$ on $[-L, 0]$, and for $t > 0$ then

$$\begin{aligned}(P\phi)(t) = \psi(0)e^{-\int_0^t a(s)\,ds} &+ \int_{t-L}^{t} g(\phi(s))ds - \int_{-L}^{0} g(\psi(s))\,ds \\ &- \int_0^t e^{-\int_s^t a(u)\,du} a(s) \int_{s-L}^{s} g(\phi(u))\,du\,ds.\end{aligned}$$

Note that if we knew that $\int_0^t a(s)ds \to \infty$ as $t \to \infty$, we could show that $\phi(t) \to 0$ implies that $(P\phi)(t) \to 0$ and we could show that solutions tend to zero. We proceed in two steps, first showing that $x_1(t)$ is bounded.

To see that P is a contraction, if $\phi, \eta \in \mathcal{M}$ then

$$\begin{aligned}|(P\phi)(t) - (P\eta)(t)| &\leq \int_{t-L}^{t} |g(\phi(u)) - g(\eta(u))|\,du \\ &+ \int_0^t e^{-\int_s^t a(u)\,du} a(s) \int_{s-L}^{s} |g(\phi(u)) - g(\eta(u))|\,du \\ &\leq 2KL\|\phi - \eta\| = 2\alpha\|\phi - \eta\|.\end{aligned}$$

so P is a contraction with fixed point $\phi \in \mathcal{M}$ and $\phi = x_1$ is bounded.

Since $x_1(t)$ is bounded, it follows from the differential equation that $x_1'(t)$ is also bounded. Thus, if $x_1(t)$ does not tend to zero, then there is an $\varepsilon > 0$ and a sequence t_n with $|x_1(t_n)| > \varepsilon$; hence, there is a $\delta > 0$ with $|x_1(t)| > \varepsilon/2$ on $[t_n, t_n + \delta]$. Thus, $\int_0^t x_1^2(s)ds \to \infty$. This completes the proof.

Example 2.1.5. Consider the scalar equation

$$x' = -h(t, x) + \int_0^t C(t, s)g(x(s))\,ds \tag{2.1.13}$$

with a view to adding and subtracting a term $G(t)g(x)$ of such a nature that the integral, together with $-G(t)g(x)$, will constitute a harmless perturbation of the remaining ordinary differential equation

$$x' = -h(t, x) + G(t)g(x)$$

and that the solution is bounded. To that end, we ask that h, C, and $\int_{t-s}^{\infty} C(u+s,s)du$ be continuous for $0 \le s \le t < \infty$. Notice then that

$$\begin{aligned} &-\frac{d}{dt}\int_0^t \int_{t-s}^{\infty} C(u+s,s)\,du\; g(x(s))\,ds \\ &= -\int_0^{\infty} C(u+t,t)\,du\; g(x(t)) + \int_0^t C(t,s)g(x(s))\,ds \\ &=: -G(t)g(x(t)) + \int_0^t C(t,s)g(x(s))\,ds. \end{aligned}$$

We prepare the harmless term by asking that there exist $K > 0$ and $\alpha < 1/2$ such that

$$|g(x) - g(y)| \le K|x-y|$$

for all real x, y and that

$$K\int_0^t \int_{t-s}^{\infty} |C(u+s,s)|\,du\,ds \le \alpha.$$

Moreover, to make the remaining ode stable, we need to ask that

$$\frac{-h(t,x) + g(x)G(t)}{x} \le 0$$

for $x \neq 0$, and have a finite continuous limit, say $-a(t)$, as $x \to 0$.

We will start all our solutions at $t = 0$ so we need an initial point, x_0, to specify a fixed, but non-unique, solution $x_1(t) = x(t, 0, x_0)$. The conditions we are going to impose would allow us to prove that any such solution can be defined in the future; however, for this discussion we simply ask that this be the case. For that fixed $x_1(t)$ we define a function, $a(t)$, by

$$-a(t) := \frac{-h(t, x_1(t)) + G(t)g(x_1(t))}{x_1(t)}.$$

By an earlier assumption, $a(t) \ge 0$ and it is continuous.

We can now quickly show that under these conditions the solution $x_1(t)$ is bounded. Write the equation along that fixed solution $x_1(t)$ as

$$x' = -a(t)x - \frac{d}{dt}\int_0^t \int_{t-s}^{\infty} C(u+s,s)\,du\, g(x(s))\,ds$$

or

$$\begin{aligned}
x(t) &= x_0 e^{-\int_0^t a(s)\,ds} \\
&\quad - \int_0^t e^{-\int_s^t a(u)\,du} \frac{d}{ds} \int_0^s \int_{s-v}^\infty C(u+v,v)\,du g(x(v))\,dv\,ds \\
&= x_0 e^{-\int_0^t a(s)\,ds} \\
&\quad - e^{-\int_s^t a(u)\,du} \int_0^s \int_{s-v}^\infty C(u+v,v)\,du g(x(v))\,dv\,ds \Big|_0^t \\
&\quad + \int_0^t e^{-\int_s^t a(u)\,du} a(s) \int_0^s \int_{s-v}^\infty C(u+v,v)\,du g(x(v))\,dv\,ds \\
&= x_0 e^{-\int_0^t a(s)\,ds} - \int_0^t \int_{t-v}^\infty C(u+v,v) du g(x(v))\,dv \\
&\quad + \int_0^t e^{-\int_s^t a(s)\,ds} a(s) \int_0^s \int_{s-v}^\infty C(u+v,v)\,du g(x(v))\,dv\,ds.
\end{aligned}$$

Let

$$\mathcal{M} = \{\phi : [0,\infty) \to R \mid \phi(0) = x_0, \phi \in \mathcal{C}, \phi \quad \text{bounded}\}$$

and define P using the above equation for the solution $x(t)$. Clearly, $P : \mathcal{M} \to \mathcal{M}$. To see that P is a contraction, if $\phi, \eta \in \mathcal{M}$ then

$$|(P\phi)(t) - (P\eta)(t)| \le 2K \int_0^t \int_{t-v}^\infty |C(u+v,v)| du \|\phi - \eta\| \le 2\alpha \|\phi - \eta\|.$$

Example 2.1.6. Consider the scalar equation

$$x' = \frac{1}{4}x - x^3 - d\int_0^t \frac{\sin x(s)}{(1+(t-s))^3}\,ds \tag{2.1.14}$$

with $1/2 < d < 2$. This is an example of Example 2.1.5. We briefly show that the conditions of Example 2.1.5 hold and we also show that if $\phi \to 0$, so does $P\phi$; thus, our conclusion will be that all solutions tend to zero.

Here, $K = 1$ and

$$K \int_0^t \int_{t-s}^\infty |C(u+s,s)|\,du\,ds = d \int_0^t \int_{t-s}^\infty (1+u)^{-3}\,du\,ds \le d/2$$

while

$$G(t) = -d \int_0^\infty (1+u)^{-3} du = -d/2$$

and

$$\frac{-h(t,x)+g(x)G(t)}{x} = \frac{\frac{1}{4}x - x^3 - (d/2)\sin x}{x} \leq \frac{1}{4} - \frac{d}{2}$$

by calculations in Ex 2.1.2 so $a(t) \geq 1/4$.

Thus, if we let $\mathcal{M}$ be as stated in Example 2.1.5, but with $\phi(t) \to 0$ as $t \to \infty$, then $(P\phi)(t) \to 0$ as $t \to \infty$. The conditions of Example 2.1.5 are satisfied and all solutions tend to zero.

This will acknowledge that much of the material in this section was first published in *The Electronic Journal of Qualitative Theory of Differential Equations*, as described in Burton (2004d), which is published by the University of Szeged.

2.2 Equations of Volterra-Levin Type

All of the last section should be considered as the introduction to this section. We have seen a number of forms which were presented as candidates to be harmless perturbations. Those candidates will now be put to the test to yield new types of stability criteria for well-known classical problems. The work here is taken from Burton (2005b).

In Burton (2003a) (See Section 1.6.) we compared the techniques of fixed point theory and Liapunov theory in proving stability for functional differential equations. We found that sometimes one is better and sometimes the other is better. In the same vein, in this section we will treat important classical problems by means of fixed point theory, while different classical problems will be treated by Liapunov theory. All of these problems will be approached via the harmless perturbation ideas of the last section.

The problems we consider here using contractions are:

$$x' = -\int_{t-L}^{t} p(s-t)g(x(s))\,ds,$$

$$x' = -\int_{0}^{t} e^{-a(t-s)}\sin(t-s)g(x(s))\,ds,$$

$$x' = -\int_{-\infty}^{t} q(s-t)g(x(s))\,ds,$$

and

$$x' = -a(t)g(x(q(t))).$$

We always have $xg(x) > 0$ for $x \neq 0$, so that each of these equations can be written as

$$x' = -g(x) + \text{a harmless perturbation}$$

and then we use contraction mappings to show that the equation is stable. In this way we can show the fixed point technique working on a distributed bounded delay, a distributed unbounded delay, a distributed infinite delay, and a pointwise variable delay. Moreover, the fixed point arguments are simple and unified with promise of application to a very wide class of problems. In particular, we conjecture that all the papers which followed the Cooke and Yorke (1973) paper will generate additional harmless perturbations, stimulating a great resurgence of interest in stability theory.

In all of our examples in this section we will have a function $g : R \to R$ satisfying:

$$|g(x) - g(y)| \leq K|x - y| \tag{2.2.1}$$

for some $K > 0$ and all $x, y \in R$,

$$\frac{g(x)}{x} \geq 0 \quad \text{and} \quad \lim_{x \to 0} \frac{g(x)}{x} \quad \text{exits,} \tag{2.2.2}$$

and sometimes

$$\frac{g(x)}{x} \geq \beta \tag{2.2.3}$$

for some $\beta > 0$.

We have spoken in terms of stability. But each of our theorems claim that solutions are bounded and, with an additional assumption, that these solutions tend to zero. It is a simple, but lengthy, exercise to show that these statements could be extended to say that the zero solution is stable and globally asymptotically stable. One defines the mapping set in terms of the given $\varepsilon > 0$ in the stability argument. The reader is referred to type I, II exercises.

Existence theory is found in Chapter 3 of Burton (1985), for example. Briefly, for the type of initial function which we will give, owing to the continuity and the Lipschitz condition, there will be a unique solution. Because of the Lipschitz growth condition, that solution can be continued for all future time.

Consider the scalar equation

$$x' = -\int_{t-L}^{t} p(s-t)g(x(s))\,ds \tag{2.2.4}$$

with $L > 0$, p continuous,

$$\int_{-L}^{0} p(s)\,ds = 1, \tag{2.2.5}$$

and for the K of (2.2.1) let

$$2K\int_{-L}^{0} |p(v)v|\,dv =: \alpha < 1. \tag{2.2.6}$$

This is the Volterra-Levin equation motivating so much of this work. Note that (2.2.5) can be interpreted as only asking that $H := \int_{-L}^{0} p(s)ds > 0$. Then H is absorbed into K in (2.2.1) and the only functional condition on the size of p resides in (2.2.6). Note, then, that our condition is only an averaging one and in no manner is it pointwise. Moreover, the condition is in the form of a bound on magnitude. This is most crucial when applying the result to real-world problems in which there are always uncertainties.

Theorem 2.2.1. *If (2.2.1), (2.2.2), (2.2.5), and (2.2.6) hold, then every solution of (2.2.4) is bounded. Moreover, if (2.2.3) also holds, then every solution of (2.2.4) tends to zero.*

Proof. Let $\psi : [-L,0] \to R$ be a given continuous initial function and let $x_1(t) := x(t,0,\psi)$ be the unique resulting solution. By the growth condition on g, $x_1(t)$ exists on $[0,\infty)$. If we add and subtract $g(x)$ we can write the equation as

$$x' = -g(x) + \frac{d}{dt}\int_{-L}^{0} p(s)\int_{t+s}^{t} g(x(u))\,du\,ds.$$

Define a continuous non-negative function $a : [0,\infty) \to [0,\infty)$ by

$$a(t) := \frac{g(x_1(t))}{x_1(t)}.$$

Since a is the quotient of continuous functions it is continuous when assigned the limit at $x_1(t) = 0$, if such a point exists.

Thus, for the fixed solution, our equation is

$$x' = -a(t)x + \frac{d}{dt}\int_{-L}^{0} p(s)\int_{t+s}^{t} g(x(u))\,du\,ds$$

which, by the variation of parameters formula, followed by integration by parts, can then be written as

$$
\begin{aligned}
x(t) &= \psi(0)e^{-\int_0^t a(s)\,ds} \\
&+ \int_0^t e^{-\int_v^t a(s)ds} \frac{d}{dv} \int_{-L}^0 p(s) \int_{v+s}^v g(x(u))\,du\,ds\,dv \\
&= \psi(0)e^{-\int_0^t a(s)\,ds} + e^{-\int_v^t a(s)\,ds} \int_{-L}^0 p(s) \int_{v+s}^v g(x(u))\,du\,ds\Big|_0^t \\
&- \int_0^t a(v)e^{-\int_v^t a(s)\,ds} \int_{-L}^0 p(s) \int_{v+s}^v g(x(u))\,du\,ds\,dv \\
&= \psi(0)e^{-\int_0^t a(s)ds} + \int_{-L}^0 p(s) \int_{t+s}^t g(x(u))\,du\,ds \\
&- e^{-\int_0^t a(s)\,ds} \int_{-L}^0 p(s) \int_s^0 g(\psi(u))\,du\,ds \\
&- \int_0^t e^{-\int_v^t a(s)ds} a(v) \int_{-L}^0 p(s) \int_{v+s}^v g(x(u))\,du\,ds\,dv.
\end{aligned}
$$

Let

$$
\mathcal{M} = \{\phi : [-L,\infty) \to R \mid \phi_0 = \psi, \phi \in \mathcal{C}, \phi \text{ bounded}\}
$$

and define $P : \mathcal{M} \to \mathcal{M}$ using the above equation in $x(t)$. For $\phi \in \mathcal{M}$ define $(P\phi)(t) = \psi(t)$ if $-L \le t \le 0$. If $t \ge 0$, then define

$$
\begin{aligned}
(P\phi)(t) &= \psi(0)e^{-\int_0^t a(s)\,ds} + \int_{-L}^0 p(s) \int_{t+s}^t g(\phi(u))\,du\,ds \\
&- e^{-\int_0^t a(s)\,ds} \int_{-L}^0 p(s) \int_s^0 g(\psi(u))duds \\
&- \int_0^t e^{-\int_v^t a(s)ds} a(v) \int_{-L}^0 p(s) \int_{v+s}^v g(\phi(u))\,du\,ds\,dv.
\end{aligned}
$$

To see that P is a contraction, if $\phi, \eta \in \mathcal{M}$ then

$$
\begin{aligned}
&|(P\phi)(t) - (P\eta)(t)| \\
&\le \int_{-L}^0 |p(s)| \int_{t+s}^t |g(\phi(u)) - g(\eta(u))|\,du\,ds \\
&+ \int_0^t e^{-\int_v^t a(s)\,ds} a(v) \int_{-L}^0 |p(s)| \int_{v+s}^v |g(\phi(u)) - g(\eta(u))|\,du\,ds\,dv \\
&\le 2K\|\phi - \eta\| \int_{-L}^0 |p(s)s|\,ds \\
&\le \alpha\|\phi - \eta\|.
\end{aligned}
$$

Hence, there is a unique fixed point, a bounded solution.

If $\frac{g(x)}{x} \geq \beta > 0$, then add to $\mathcal{M}$ the condition that $\phi(t) \to 0$ as $t \to \infty$. We can show that $(P\phi)(t) \to 0$ whenever $\phi(t) \to 0$ and, hence, that the fixed point tends to zero.

Equation (2.2.4) has a long and interesting history under much different conditions than those listed in Theorem 2.2.1. Volterra (1928) used it as a model for a biological problem and suggested that a Liapunov functional might be derived in a certain way under very regular conditions on p. Ergen (1954) used (2.2.4) to model a circulating fuel nuclear reactor with x representing the neutron density; others continued that work. It has also been used as a model for a one-dimensional viscoelasticity where x is the strain and, with $a(t-s) := p(s-t)$, then a is the relaxation function. Burton (1993) used it to model a neural network. The equation has served as a model for many other physical systems. Taking

$$a(t-s) := p(s-t),$$

Levin (1963) used Volterra's suggestion to construct a Liapunov functional. We discussed his result in Chapter 0, but it is worth reviewing here.

Theorem 2.2.2. *Suppose that $a''(t)$ is not identically zero, that*

$$xg(x) > 0 \quad \text{if} \quad x \neq 0, \qquad \int_0^{\pm\infty} g(s)\,ds = \infty,$$

and that

$$a(t) \in C[0,\infty), \quad (-1)^k a^{(k)}(t) \geq 0 \quad (0 < t < \infty; k = 0,1,2,3).$$

Then the derivative of

$$V(t, x(\cdot)) = \int_0^x g(s)\,ds + (1/2)a(t)\Big[\int_0^t g(x(s))\,ds\Big]^2 - (1/2)\int_0^t a'(t-s)\Big[\int_s^t g(x(w))dw\Big]^2 ds \geq 0$$

along solutions of

$$x' = -\int_0^t a(t-s)g(x(s))\,ds$$

satisfies

$$V'(t, x(\cdot)) = (1/2)a'(t)\Big[\int_0^t g(x(s))\,ds\Big]^2 - (1/2)\int_0^t a''(t-s)\Big[\int_s^t g(x(w))dw\Big]^2 ds \leq 0.$$

If $x(t)$ is any solution which exists on $[0, \infty)$, then

$$\lim_{t\to\infty} x^{(j)}(t) = 0, \quad j = 0, 1, 2.$$

It is, of course, a beautiful result. The fact that he proved asymptotic stability without having the derivative of the Liapunov functional negative definite stimulated a great amount of interest. Krasovskii (1963) had developed some theory for just such cases and Levin's work was prominently featured there. Krasovskii (1963; p. 155) had shown that bounded solutions approach invariant subsets of the set where $V' = 0$ under regularity conditions. The next year, Levin and Nohel (1964) extended Levin's Liapunov functional to the case of (2.2.4), again with the regularity conditions on $a(t)$. Here is their result.

Theorem 2.2.3. *Suppose that $a''(t)$ is not identically zero, that*

$$xg(x) > 0 \quad \text{if} \quad x \neq 0, \qquad \int_0^{\pm\infty} g(s)\, ds = \infty,$$

and that

$$a(t) \in C[0, L], \quad a(L) = 0, \quad (-1)^k a^{(k)}(t) \geq 0 \quad (0 \leq t \leq L; k = 0, 1, 2, 3).$$

Then the derivative of

$$\begin{aligned} V(t, x(\cdot)) &= \int_0^x g(s)\, ds \\ &\quad - (1/2) \int_{t-L}^t a'(t-s) \Big[\int_s^t g(x(w))\, dw \Big]^2 ds \geq 0 \end{aligned}$$

along solutions of

$$x' = -\int_{t-L}^t a(t-s) g(x(s))\, ds$$

satisfies

$$\begin{aligned} V'(t, x(\cdot)) &= (1/2) a'(L) \Big[\int_{t-L}^t g(x(s))\, ds \Big]^2 \\ &\quad - (1/2) \int_{t-L}^t a''(t-s) \Big[\int_s^t g(x(w)) dw \Big]^2 ds \leq 0. \end{aligned}$$

If $x(t)$ is any solution which exists on $[0, \infty)$, then

$$\lim_{t\to\infty} x^{(j)}(t) = 0, \quad j = 0, 1, 2.$$

Theorems 2.2.1 and 2.2.3 serve as a fundamental non-contrived demonstration of relations between fixed point methods and Liapunov methods. The Levin-Nohel conditions on g are far less stringent than those in Theorem 2.2.1. They would allow the integral of a to be as large as we please, while Theorem 2.2.1 has a very definite bound on it. But Levin and Nohel require $a(L) = 0$ and very stringent conditions on the derivatives of $a(t)$ which one feels would virtually never be verifiable in a real-world problem. Indeed, it would seem to be fair to judge that those real-world problems mentioned above remain unsolved insofar as they depend on the Levin-Nohel result. Thus, the fixed point method and the Liapunov method unite to greatly extend the theory. Krasovskii (1963) also features Theorem 2.2.3, as does Hale (1965) in his theory of limit sets. Moreover, Hale (1965) continued working with the Levin Liapunov functional and applied it to an infinite delay problem we consider in this section.

In Theorem 2.2.1 we saw a kernel which was allowed to change sign in an arbitrary manner, so long as it was positive on average. There are other examples in the literature where the kernel is allowed to change sign, but usually it is tightly controlled in some way. The next example concerns a problem of Halanay (1965) and later of MacCamy and Wong (1972; p.16) in stability investigation using positive kernels. They mention that the positive kernel technique works on

$$x' = -\int_0^t e^{-a(t-s)} \cos(t-s) g(x(s))\, ds$$

but does not work on

$$x' = -\int_0^t e^{-a(t-s)} \sin(t-s) g(x(s))\, ds.$$

The idea of adding and subtracting the same thing which we use here works in exactly the same way for both of them. It would even work if the right-hand-side began with an unstable term, such as $+\gamma g(x)$ where $\gamma < \beta$, and is left as one of our standard exercises. The fixed point technique works in the aforementioned case where the positive kernel technique does not apply.

Consider the scalar equation

$$x' = -\int_0^t e^{-a(t-s)} \sin(t-s) g(x(s))\, ds \tag{2.2.7}$$

where $a > 0$, and for the K of (2.2.1) we have

$$\alpha := 2K \sup_{t\geq 0} \int_0^t \int_{t-u}^{\infty} e^{-av} |\sin v|\, dv\, du < 1. \tag{2.2.8}$$

Notice that

$$k := \int_0^\infty e^{-av} \sin v dv = \frac{1}{a^2+1}, \tag{2.2.9}$$

a condition parallel to (2.2.5); indeed, we could replace our kernel with a general kernel satisfying (2.2.9). Because of the Lipschitz condition on g we can show that for each $x(0)$ there is a unique solution $x(t, 0, x(0))$ defined for all future t.

Theorem 2.2.4. *If (2.2.1), (2.2.2), and (2.2.8) hold, then every solution of (2.2.7) is bounded. If, in addition, (2.2.3) holds then every solution tends to zero as $t \to \infty$.*

Proof. Let $x(0) = x_0$ be given, resulting in a unique solution $x_1(t)$. With the k defined above, write the equation as

$$x' = -kg(x) + \frac{d}{dt}\int_0^t \int_{t-s}^\infty e^{-av} \sin v \, dv \;\, g(x(s))\, ds.$$

Define a function $c(t)$ by

$$c(t) := \frac{g(x_1(t))}{x_1(t)}$$

so that the equation can be written as

$$x' = -kc(t)x + \frac{d}{dt}\int_0^t \int_{t-s}^\infty e^{-av} \sin v \, dv \;\, g(x(s))\, ds$$

which will still have the unique solution $x_1(t)$ for the given initial condition $x(0) = x_0$. We can then use the variation of parameters formula to write the solution as

$$\begin{aligned}
x(t) &= x_0 e^{-k\int_0^t c(s)\, ds} \\
&\quad + \int_0^t e^{-k\int_u^t c(s)\, ds} \frac{d}{du}\int_0^u \int_{u-s}^\infty e^{-av} \sin v \, dv \;\, g(x(s))\, ds\, du \\
&= x_0 e^{-k\int_0^t c(s) ds} + e^{-k\int_u^t c(s)\, ds} \int_0^u \int_{u-s}^\infty e^{-av} \sin v \, dv \, g(x(s))\, ds \Big|_0^t \\
&\quad - \int_0^t e^{-k\int_u^t c(s)\, ds} kc(u) \int_0^u \int_{u-s}^\infty e^{-av} \sin v \, dv \, g(x(s))\, ds\, du \\
&= x_0 e^{-k\int_0^t c(s) ds} + \int_0^t \int_{t-s}^\infty e^{-av} \sin v \, dv \, g(x(s))\, ds \\
&\quad - \int_0^t e^{-k\int_u^t c(s)\, ds} kc(u) \int_0^u \int_{u-s}^\infty e^{-av} \sin v \, dv \, g(x(s))\, ds\, du.
\end{aligned}$$

Let $\mathcal{M}$ be defined as the set of bounded continuous $\phi : [0, \infty) \to R$, $\phi(0) = x_0$, and define $P : \mathcal{M} \to \mathcal{M}$ using the above equation for $x(t)$ as we did in the proof of Theorem 2.2.1. To see that P is a contraction, if $\phi, \eta \in \mathcal{M}$ then

$$|(P\phi)(t) - (P\eta)(t)| \leq 2K \sup_{t \geq 0} \int_0^t \int_{t-u}^{\infty} e^{-av} |\sin v| \, dv \, du \; \|\phi - \eta\|.$$

Thus, P will have a unique fixed point, a bounded function satisfying the differential equation.

If $\frac{g(x)}{x} \geq \beta > 0$ then we can show that $(P\phi)(t) \to 0$ whenever $\phi(t) \to 0$, thereby concluding that all solutions tend to zero.

We next consider the equation

$$x'(t) = -\int_{-\infty}^{t} q(s-t) g(x(s)) \, ds \tag{2.2.10}$$

where

$$\int_{-\infty}^{0} q(s) \, ds = 1 \quad \text{and} \quad \int_{-\infty}^{0} \int_{-\infty}^{v} |q(u)| \, du \, dv \quad \text{exists,} \tag{2.2.11}$$

and there is a positive number $\alpha < 1$ with

$$2K \sup_{t \geq 0} \int_0^t \int_{-\infty}^{s-t} |q(u)| \, du \, ds \leq \alpha \tag{2.2.12}$$

where K is from (2.2.1).

Theorem 2.2.5. *Suppose that (2.2.1), (2.2.2), (2.2.11), and (2.2.12) hold. Then every solution of (2.2.10) with bounded continuous initial function $\psi : (-\infty, 0] \to R$ is bounded. If, in addition, (2.2.3) holds then those solutions tend to zero as $t \to \infty$.*

Proof. Write (2.2.10) as

$$x' = -g(x(t)) + \frac{d}{dt} \int_{-\infty}^{t} \int_{-\infty}^{s-t} q(u) \, du \, g(x(s)) \, ds.$$

For a given bounded continuous initial function ψ, let $x_1(t)$ be the resulting unique solution which will be defined on $[0, \infty)$. Define a unique continuous function by

$$a(t) := g(x_1(t))/x_1(t)$$

and write the equation as

$$x' = -a(t)x(t) + \frac{d}{dt}\int_{-\infty}^{t}\int_{-\infty}^{s-t} q(u)\,du\,g(x(s))\,ds$$

which, for the same initial function, still has the unique solution $x_1(t)$. Use the variation of parameters formula to write the solution as the integral equation

$$\begin{aligned}
x(t) &= \psi(0)e^{-\int_0^t a(s)\,ds} \\
&\quad + \int_0^t e^{-\int_v^t a(u)\,du}\frac{d}{dv}\int_{-\infty}^{v}\int_{-\infty}^{s-v} q(u)\,du\; g(x(s))\,ds\,dv \\
&= \psi(0)e^{-\int_0^t a(s)\,ds} + e^{-\int_v^t a(u)\,du}\int_{-\infty}^{v}\int_{-\infty}^{s-v} q(u)\,du\; g(x(s))\,ds\Big|_0^t \\
&\quad - \int_0^t a(v)e^{-\int_v^t a(s)\,ds}\int_{-\infty}^{v}\int_{-\infty}^{s-v} q(u)\,du\; g(x(s))\,ds\,dv \\
&= \psi(0)e^{-\int_0^t a(s)ds} + \int_{-\infty}^{t}\int_{-\infty}^{s-t} q(u)\,du\; g(x(s))\,ds \\
&\quad - e^{-\int_0^t a(u)\,du}\int_{-\infty}^{0}\int_{-\infty}^{s} q(u)\,du\; g(\psi(s))\,ds \\
&\quad - \int_0^t a(v)e^{-\int_v^t a(s)\,ds}\int_{-\infty}^{v}\int_{-\infty}^{s-v} q(u)\,du\; g(x(s))\,ds\,dv.
\end{aligned}$$

Let

$$\mathcal{M} = \{\phi : R \to R \,|\, \phi \in \mathcal{C}, \phi(t) = \psi(t) \quad \text{for} \quad t \le 0,\ \phi \quad \text{bounded}\}$$

and define $P : \mathcal{M} \to \mathcal{M}$ by $\phi \in \mathcal{M}$ implies $(P\phi)(t) = \psi(t)$ for $t \le 0$ and $P\phi$ be defined from the last equation above for x with x replaced by ϕ, as we have done before. To see that P is a contraction, if $\phi, \eta \in \mathcal{M}$, then

$$\begin{aligned}
&|(P\phi)(t) - (P\eta)(t)| \\
&\le \int_{-\infty}^{t}\int_{-\infty}^{s-t} |q(u)|\,du\,|g(\phi(s)) - g(\eta(s))|\,ds \\
&\quad + \int_0^t a(v)e^{-\int_v^t a(s)\,ds}\int_{-\infty}^{v}\int_{-\infty}^{s-v} |q(u)|\,du\,|g(\phi(s)) - g(\eta(s))|\,ds\,dv \\
&\le \int_0^t\int_{-\infty}^{s-t} |q(u)|\,du\,ds\; K\|\phi - \eta\| \\
&\quad + \int_0^t a(v)e^{-\int_v^t a(s)\,ds}\int_0^{v}\int_{-\infty}^{s-v} |q(u)|\,du\,ds\,dv\; K\|\phi - \eta\|
\end{aligned}$$

$$\leq 2K\|\phi - \eta\| \sup_{t \geq 0} \int_0^t \int_{-\infty}^{s-t} |q(u)| \, du \, ds$$

$$\leq \alpha \|\phi - \eta\|.$$

If $g(x)/x \geq \beta > 0$, then we modify $\mathcal{M}$ to include $\phi(t) \to 0$ and we show that this means that $P\phi$ also tends to zero. That will complete the proof.

Finally, we consider a scalar equation

$$x'(t) = -a(t)g(x(q(t))) \tag{2.2.13}$$

where $q : [0, \infty) \to R$ is continuous and strictly increasing to ∞, $q(t) < t$, q has the inverse function $h(t)$ so that $q(h(t)) = t$, and $a : [0, \infty) \to [0, \infty)$ is continuous. We suppose that there is an $\alpha < 1$ with

$$2K \sup_{t \geq 0} \int_t^{h(t)} a(u) \, du \leq \alpha < 1 \tag{2.2.14}$$

where K is from (2.2.1).

Theorem 2.2.6. *Let (2.2.1), (2.2.2), and (2.2.14) hold. Then every solution of (2.2.13) is bounded. If, in addition, (2.2.3) holds and*

$$\int_0^t a(s) \, ds \to \infty \quad \text{as} \quad t \to \infty \tag{2.2.15}$$

then every solution of (2.2.13) tends to 0 as $t \to \infty$.

Proof. Write (2.2.13) as

$$x'(t) = -a(h(t))h'(t)g(x(t)) - \frac{d}{dt} \int_{h(t)}^t a(s)g(x(q(s))) \, ds.$$

Given a continuous initial function $\psi : [q(0), 0] \to R$, let $x_1(t)$ denote the unique solution having that initial function and define a continuous function by

$$c(t) := a(h(t))h'(t)g(x_1(t))/x_1(t).$$

For that fixed solution $x_1(t)$ and that initial function ψ it follows that

$$x' = -c(t)x - \frac{d}{dt} \int_{h(t)}^t a(s)g(x(q(s))) \, ds$$

has the unique solution $x_1(t)$ and, by the Lipschitz condition, we can argue that it exits on $[0, \infty)$.

By the variation of parameters formula we have

$$\begin{aligned}
x(t) &= \psi(0)e^{-\int_0^t c(s)\,ds} - \int_0^t e^{-\int_s^t c(u)\,du} \frac{d}{ds}\int_{h(s)}^s a(u)g(x(q(u)))\,du\,ds \\
&= \psi(0)e^{-\int_0^t c(s)\,ds} - e^{\int_s^t c(u)\,du}\int_{h(s)}^s a(u)g(x(q(u)))\,du\Big|_0^t \\
&+ \int_0^t c(s)e^{-\int_s^t c(u)\,du}\int_{h(s)}^s a(u)g(x(q(u)))\,du\,ds \\
&= \psi(0)e^{-\int_0^t c(s)\,ds} - \int_{h(t)}^t a(u)g(x(q(u)))\,du \\
&+ e^{-\int_0^t c(u)\,du}\int_{h(0)}^0 a(u)g(x(q(u)))\,du \\
&+ \int_0^t c(s)e^{-\int_s^t c(u)\,du}\int_{h(s)}^s a(u)g(x(q(u)))\,du\,ds.
\end{aligned}$$

We would define the complete metric space of bounded continuous functions which agree with ψ and use the above equation for x to define a mapping. That mapping would be a contraction because of (3.14). We would complete the proof as before.

Remark 2.2.1. *This is a general method applied to four very different problems which have been studied closely by other methods for many years. Yet, new information is found in each case. In Theorem 2.2.1, far less is required on the kernel than in traditional approaches. Theorem 2.2.4 succeeds where the positive kernel method failed. Theorem 2.2.5 again requires less on the kernel than did the traditional Liapunov functional. Something intriguing occurs in Theorem 2.2.6. This problem has been studied intensively by many investigators for at least 54 years, using techniques devised specifically for it, going under the name of the 3/2−theorem, and discussed, for example, by Krisztin (1991) or Graef, Qian, and Zhang (2000). Thus, it is unreasonable to expect a general technique to compare favorably with the special techniques. Yet, something new does occur. Our measure is in the integral with limits from t to $h(t)$, while traditional techniques measure from t to $t + r(t)$. It is known that $h(t)$ is smaller than $t + r(t)$ when $r(t)$ is decreasing. In the linear case, we may let $a(t)$ change sign. But the real value of the technique is that it is simple, can be applied to many problems with little difficulty, and indicates once more that fixed point theory is a viable stability tool.*

Remark 2.2.2. *This section concerns stability by fixed point methods and condition (2.2.3) provides us with a simple way of showing that solu-*

tions tend to zero by endowing the mapping set with this property. But (2.2.3) can often be relaxed using an old technique from Liapunov theory called the annulus argument. See, for example, Burton (1985; p. 231). The idea works in all problems in which $x'(t)$ is bounded whenever $x(t)$ is a bounded function. Instead of (2.2.3), strengthen (2.2.2) to

$$\frac{g(x)}{x} > 0 \quad \text{for} \quad x \neq 0 \quad \text{and} \quad \lim_{x \to 0} \frac{g(x)}{x} \quad \text{exists.}$$

Consider the proof of Theorem 2.2.1 at the point where we have shown that $x_1(t)$ is bounded, say $|x_1(t)| \leq H$, and suppose this means that $|x_1'(t)| \leq L$ for some $L > 0$. We will now show that $x_1(t) \to 0$. By way of contradiction, if it does not, then there is an $\varepsilon > 0$ and sequence $\{t_n\} \uparrow \infty$ with $|x_1(t_n)| \geq \varepsilon$. Now, from $\frac{g(x)}{x} > 0$ for $x \neq 0$, there is a $\gamma > 0$ with $\frac{g(x)}{x} \geq \gamma$ for $\varepsilon/2 \leq |x| \leq H$. Moreover, since $|x'(t)| \leq L$, there is a $\mu > 0$ with $|x_1(t)| \geq \varepsilon/2$ for $|t_n - t| \leq \mu$. Hence, $\frac{g(x_1(t))}{x_1(t)} \geq \gamma$ for $|t_n - t| \leq \mu$. Thus, $\int_0^\infty a(t)dt = \infty$. Now, add to M the condition that $\phi(t) \to 0$ as $t \to \infty$. It follows that $(P\phi)(t) \to 0$ as $t \to \infty$. Hence, the fixed point tends to zero.

This is to acknowledge that much of this section was first published in *Fixed Point Theory and Applications*, as described in Burton (2005b), which is published by Hindawi Publishing Corporation.

2.3 Liénard's Equation and Liapunov Functions

In Section 2.1 we developed a theory of harmless perturbations using fixed point theory. In Section 2.2 we applied that theory to some classical problems using fixed point theory. Throughout this book we strive to point out that sometimes fixed point theory is more effective and sometimes Liapunov theory is more effective. Here, we come to a number of Liénard equations with delays. We attack them by first putting them in a form containing a harmless perturbation, as developed with the fixed point methods, and then prove stability using the well-developed Liapunov theory for Liénard equations. The two methods work together very well. This work may be found in Burton (2004b).

We present a technique which allows us to remove the delay in a functional differential equation and replace it with a stable term without delay plus a harmless perturbation. This is done with three delayed Liénard equations, as well as two related equations. Those equations are

$$x'' + f(x)x' + g(x(t-L)) = 0, \tag{2.3.1}$$

$$x'' + f(x)x' + \int_{t-L}^{t} p(s-t)g(x(s))\,ds = 0, \tag{2.3.2}$$

with

$$\int_{-L}^{0} p(s)\,ds = 1, \quad \int_{-L}^{0} |p(s)|\,ds =: K,$$

and

$$x'' + f(x)x' + \int_{-\infty}^{t} q(s-t)g(x(s))\,ds = 0, \tag{2.3.3}$$

with

$$\int_{-\infty}^{0} q(s)\,ds = 1, \quad \int_{-\infty}^{0}\int_{-\infty}^{v} |q(u)|\,du\,dv =: D$$

where L, K, D are all finite positive numbers.

It can be noted that the conditions

$$\int_{-L}^{0} p(s)\,ds = 1 \quad \text{and} \quad \int_{-\infty}^{0} q(s)\,ds = 1$$

are conveniences in notation. In fact, they ask only that

$$\int_{-L}^{0} p(s)\,ds > 0 \quad \text{and} \quad \int_{-\infty}^{0} q(s)\,ds > 0.$$

The exact values of the integrals can be incorporated into $g(x)$.

A change of variable will show that each of these equations is an autonomous functional differential equation; hence, stability implies uniform stability and asymptotic stability implies uniform asymptotic stability. We generally think of f as being the damping and g as being the restoring force. We ask that

$$f(x) \geq 0, \quad xg(x) > 0 \quad \text{if} \quad x \neq 0 \tag{2.3.4}$$

and denote

$$G(x) = \int_{0}^{x} g(s)\,ds, \quad F(x) = \int_{0}^{x} f(s)\,ds. \tag{2.3.5}$$

Existence theory is found in Chapter 3 of Burton (1985). The conditions given here are adequate for existence for a given continuous initial function.

The literature concerning (2.3.1) is massive when $L = 0$, while much work has also been done when $L > 0$, particularly by Zhang (1992, 1993, 1996, 1997). Some of his results are both necessary and sufficient for boundedness and stability.

It is assumed that f and g are continuous and in (2.3.3) it is convenient to ask that g be just a bit more than continuous so as to satisfy the following two conditions.

$$\text{If} \quad G(x) \quad \text{is bounded for} \quad x \geq 0, \qquad \text{then} \quad g(x) \quad \text{is bounded for} \quad x \geq 0. \tag{2.3.6}$$

$$\text{If} \quad G(x) \quad \text{is bounded for} \quad x \leq 0, \qquad \text{then} \quad g(x) \quad \text{is bounded for} \quad x \leq 0. \tag{2.3.7}$$

While these conditions are mild, they will prove to simplify analysis of (2.3.3) and to be of interest in themselves.

It is known that when (2.3.4) holds with $f(0) > 0$ and $L = 0$ then the zero solution of (2.3.1) is globally asymptotically stable if and only if

$$\int_0^{\pm\infty} [f(x) + |g(x)|]\, dx = \pm\infty. \tag{2.3.8}$$

That condition was derived in Burton (1965) and has played a central role in investigations of all these equations.

We now set up the problems as developed in Section 2.1. Beginning with (2.3.1), subtract and add $g(x)$ so that we can write it as

$$x'' + f(x)x' - \frac{d}{dt}\int_{t-L}^{t} g(x(s))\, ds + g(x) = 0.$$

Then write it as the Liénard system

$$\begin{aligned} x' &= y - F(x) + \int_{t-L}^{t} g(x(s))\, ds \\ y' &= -g(x). \end{aligned}$$

Next, write

$$\begin{aligned} x' &= y - F(x) + Lg(x) - Lg(x) + \int_{t-L}^{t} g(x(s))\, ds \\ y' &= -g(x). \end{aligned}$$

We are now in a position to identify the terms discussed in Section 2.1. Separate that system into two parts as follows. The first part consists of

$$\begin{aligned} x' &= y - F(x) + Lg(x) \\ y' &= -g(x). \end{aligned}$$

Note that under (2.3.4) the zero solution is asymptotically stable in case there is a $\delta > 0$ such that

$$Lg^2(x) - g(x)F(x) < 0 \quad \text{for} \quad 0 < |x| < \delta, \tag{2.3.9}$$

as may be argued without difficulty using the well-known Liapunov function

$$V(x, y) = y^2 + 2G(x).$$

The second part consists of $-Lg(x) + \int_{t-L}^{t} g(x(s))\, ds$ which has the form of a harmless perturbation, as discussed in Section 2.1 since each constant is a solution of

$$x' = -Lg(x) + \int_{t-L}^{t} g(x(s))\, ds = -\frac{d}{dt} \int_{-L}^{0} \int_{t+s}^{t} g(x(u))\, du\, ds.$$

Our discussion in Section 2.1 now tells us that (2.3.1) should be asymptotically stable when (2.3.4) and (2.3.9) hold. Moreover, the same type of preparation should work for (2.3.2) and (2.3.3). The perturbation $-Lg(x) + \int_{t-L}^{t} g(x(s))ds$ can be ignored.

We have mentioned before that the literature on the Liénard equation without a delay is large and will not be cited. The interested reader should consult the online Mathematical Reviews. The first result here can be compared with work of Zhang (1992, 1993) concerning (2.3.1). He obtained several results on boundedness and stability by asking for an $N > 1$ with $LNg^2(x) - g(x)F(x) < 0$ on certain intervals in order to get stability and boundedness. As mentioned before, some of his results are both necessary and sufficient. But the technique is sharper in that $N = 1$ suffices, while yielding the idea that (2.3.9) is the condition needed. Finally, this presentation readily covers the more sophisticated problems (2.3.2) and (2.3.3).

Actually, (2.3.9) can be weakened further. We can change (2.3.9) to $Lg^2(x) - g(x)F(x) \leq 0$ for $0 < |x| < \delta$ and then argue that the limit set of a solution does not intersect any value of x for which that expression is negative.

Theorem 2.3.1. *Let (2.3.4) hold for (2.3.1) and suppose there is a $\delta > 0$ with*

$$Lg^2(x) - g(x)F(x) < 0 \quad \text{for} \quad 0 < |x| < \delta. \tag{2.3.9}$$

Then the zero solution of (2.3.1) is asymptotically stable. If (2.3.4), (2.3.8), and (2.3.9) hold with $\delta = \infty$, then the zero solution of (2.3.1) is asymptotically stable in the large.

Proof. Subtract and add $g(x)$ to (2.3.1) so that it can be written as

$$x'' + f(x)x' - \frac{d}{dt}\int_{t-L}^{t} g(x(s))\,ds + g(x) = 0.$$

Then write it as the Liénard system

$$\begin{aligned} x' &= y - F(x) + \int_{t-L}^{t} g(x(s))\,ds \\ y' &= -g(x). \end{aligned}$$

Define the Liapunov functional

$$V(t, x(\cdot), y(t)) = (1/2)y^2 + G(x) + (1/2)\int_{-L}^{0}\int_{t+v}^{t} g^2(x(u))\,du\,dv$$

whose derivative along solutions of the system satisfies

$$\begin{aligned} V'(t, x(\cdot), y(t)) &= g(x)y - g(x)F(x) + g(x)\int_{t-L}^{t} g(x(s))\,ds \\ &\quad + (1/2)\int_{-L}^{0} [g^2(x(t)) - g^2(x(t+v))]\,dv - g(x)y \\ &\le -g(x)F(x) + (1/2)\int_{t-L}^{t} [g^2(x(t)) + g^2(x(s))]\,ds \\ &\quad + (L/2)g^2(x(t)) - (1/2)\int_{t-L}^{t} g^2(x(v))\,dv \\ &= -g(x)F(x) + Lg^2(x) \\ &< 0 \quad \text{for} \quad 0 < |x| < \delta \end{aligned}$$

by (2.3.9). From V and V' it follows that the zero solution is stable.

Let $(x(t), y(t))$ be any fixed solution of the system remaining in a region in which $|x(t)| < \delta$. Clearly, $y(t)$ is bounded. Let $V(t) := V(t, x(\cdot), y(t))$ and note that $V'(t) \le 0$ so $V(t) \to c$, where c is a non-negative constant. If

$c = 0$ then the solution tends to $(0,0)$. If $c > 0$, then it follows readily that $x(t) \to 0$ as $t \to \infty$; thus, $V(t) \to y^2(t)/2 \to c$. Assume that $y(t) \to \sqrt{2c}$. Then for large t we have $x'(t) \geq \sqrt{c}$, a contradiction to $x(t) \to 0$. As V is positive definite, we have shown that the zero solution is asymptotically stable.

Next, if $\delta = \infty$ and if $G(x) \to \infty$ as $|x| \to \infty$ then all solutions are bounded and, hence, converge to zero. Now, if $G(x)$ is bounded on the right and if $x(t)$ is any fixed solution, then it is clear from the Liapunov functional that $y(t)$ is bounded, say $|y(t)| \leq B_1$. By (2.3.8) we can conclude that $F(x) \to \infty$ with x. From $y' = -g(x)$ we obtain

$$|\int_{t-L}^{t} g(x(s))\,ds| \leq |y(t)| + |y(t-L)| \leq 2B_1.$$

Hence, we can find x_1 so large that $x'(t) < 0$ if $x(t) \geq x_1$; therefore, the solution is bounded on the right. A similar argument shows that $x(t)$ is bounded on the left. By the above argument, the solution tends to zero. This completes the proof.

Remark 2.3.1. *The next result contains a very interesting assumption. We require $xg(x) > 0$ but the weight $p(t)$ can change sign, so long as it is positive on average.*

Theorem 2.3.2. *Let (2.3.4) hold for (2.3.2) and suppose there is a $\delta > 0$ such that*

$$KLg^2(x) < g(x)F(x) \quad \text{if} \quad 0 < |x| < \delta. \tag{2.3.10}$$

Then the zero solution of (2.3.2) is asymptotically stable. If, in addition, (2.3.8) holds and $\delta = \infty$ then the zero solution of (2.3.2) is asymptotically stable in the large.

Proof. Add and subtract $g(x)$ to (2.3.2) so that it can be written as

$$x'' + f(x)x' - \frac{d}{dt}\int_{-L}^{0} p(s)\int_{t+s}^{t} g(x(u))\,du\,ds + g(x) = 0.$$

Then write it as a system

$$x' = y - F(x) + \int_{-L}^{0} p(s)\int_{t+s}^{t} g(x(u))\,du\,ds$$
$$y' = -g(x).$$

Define a Liapunov functional by

$$V(t, x(t), y(t)) = (1/2)y^2 + G(x) + (K/2)\int_{-L}^{0}\int_{t+v}^{t} g^2(x(u))\,du\,dv$$

so that if we denote the last term by Y, then the derivative along a solution of the system satisfies

$$\begin{aligned}
V' &= yg(x) - g(x)F(x) + g(x)\int_{-L}^{0} p(s)\int_{t+s}^{t} g(x(u))\,du\,ds \\
&\quad - yg(x) + Y' \\
&\leq -g(x)F(x) + |g(x)|\int_{-L}^{0} |p(s)|\,ds\int_{t-L}^{t} |g(x(u))|\,du + Y' \\
&\leq -g(x)F(x) + (K/2)\int_{-L}^{0} [g^2(x(t)) - g^2(x(t+v))]\,dv \\
&\quad + (K/2)\int_{t-L}^{t} (g^2(x(t)) + g^2(x(u)))\,du \\
&\leq -g(x)F(x) + KLg^2(x) < 0
\end{aligned}$$

if $0 < |x| < \delta$. The same arguments given in the proof of the last theorem now complete the proof here.

To prepare for (2.3.3) we notice that

$$\begin{aligned}
&\frac{d}{dt}\int_{-\infty}^{t}\int_{-\infty}^{s-t} q(u)\,du\; g(x(s))\,ds \\
&= \int_{-\infty}^{0} q(u)\,du g(x) - \int_{-\infty}^{t} q(s-t)g(x(s))\,ds \\
&= g(x(t)) - \int_{-\infty}^{t} q(s-t)g(x(s))\,ds.
\end{aligned}$$

While (2.3.6) and (2.3.7) are not essential, they are mild and greatly simplify the proof of the next result.

Theorem 2.3.3. *Let (2.3.4) hold for (2.3.3) and suppose there is a $\delta > 0$ such that*

$$-g(x)F(x) + Dg^2(x) < 0 \quad \text{for} \quad 0 < |x| < \delta.$$

Then the zero solution of (2.3.3) is asymptotically stable. If, in addition, (2.3.6)-(2.3.8) hold and $\delta = \infty$, then the zero solution of (2.3.3) is globally asymptotically stable.

Proof. Write (2.3.3) as

$$x'' + f(x)x' - \frac{d}{dt}\int_{-\infty}^{t}\int_{-\infty}^{s-t} q(u)\,du\; g(x(s))\,ds + g(x) = 0$$

and then as the system

$$\begin{aligned} x' &= y - F(x) + \int_{-\infty}^{t}\int_{-\infty}^{s-t} q(u)\,du\; g(x(s))\,ds \\ y' &= -g(x). \end{aligned}$$

Define

$$V(t,x,y) = \frac{1}{2}y^2 + G(x) + \frac{1}{2}\int_{-\infty}^{t}\int_{t-s}^{\infty}\int_{-\infty}^{-v} |q(u)|\,du\,dv g^2(x(s))\,ds$$

and call the last term Y. Then the derivative of V along the system satisfies

$$\begin{aligned} V' &= yg(x) - g(x)F(x) \\ &\quad + g(x)\int_{-\infty}^{t}\int_{-\infty}^{s-t} q(u)\,du\; g(x(s))\,ds - yg(x) + Y' \\ &\leq -g(x)F(x) \\ &\quad + (1/2)\int_{-\infty}^{t}\int_{-\infty}^{s-t} |q(u)|\,du\; (g^2(x(t)) + g^2(x(s)))\,ds + Y' \\ &= -g(x)F(x) + (1/2)g^2(x)\int_{-\infty}^{t}\int_{-\infty}^{s-t} |q(u)|\,du\,ds \\ &\quad + (1/2)\int_{-\infty}^{t}\int_{-\infty}^{s-t} |q(u)|\,du\; g^2(x(s))\,ds \\ &\quad + (1/2)\int_{0}^{\infty}\int_{-\infty}^{-v} |q(u)|\,du\,dv g^2(x(t)) \\ &\quad - (1/2)\int_{-\infty}^{t}\int_{-\infty}^{s-t} |q(u)|\,du\; g^2(x(s))\,ds \\ &= -g(x)F(x) + (1/2)\Bigg(\int_{0}^{\infty}\int_{-\infty}^{-v} |q(u)|\,du\,dv \\ &\quad + \int_{-\infty}^{t}\int_{-\infty}^{s-t} |q(u)|\,du\,ds\Bigg) g^2(x(t)) \\ &= -g(x)F(x) + Dg^2(x) < 0 \end{aligned}$$

for $0 < |x| < \delta$.

The remainder of the proof is just as that of Theorem 2.3.1 until we reach the last paragraph. We proceed as follows.

If $\delta = \infty$ and if $G(x) \to \infty$ as $|x| \to \infty$ then all solutions are bounded and, hence, converge to zero. If $G(x)$ is bounded on the right and if $x(t)$ is any fixed solution, then it is clear from the Liapunov functional that $y(t)$ is bounded and by (2.3.6) $g(x(t))$ is bounded. By (2.3.8) we can then find x_1 so large that $x'(t) < 0$ if $x(t) \geq x_1$; hence the solution is bounded on the right. A similar argument shows that $x(t)$ is bounded on the left. By the argument given in the proof of Theorem 2.3.1, the solution tends to zero. This completes the proof.

We look now at problems which are parallel to Liénard equations to see if our results can be extended to similar cases in which we can not locate a harmless perturbation quite so simply. Consider the scalar equation

$$x'' + f(x)x' + a(t)g(x(q(t))) = 0 \tag{2.3.11}$$

where $f : R \to [0, \infty)$ is continuous, $q : [0, \infty) \to R$ is continuous, strictly increasing, and $q(t) < t$. Let $h(t)$ be the inverse of q and let $a : [0, \infty) \to (0, \infty)$ be bounded and continuous. Assume that $g : R \to R$ is continuous,

$$xg(x) > 0 \tag{2.3.12}$$

for $x \neq 0$,

$$(a(h(t))h'(t))' > 0, \tag{2.3.13}$$

there is an $r > 0$ with

$$q(t) \geq t - r, \tag{2.3.14}$$

and for $F(x) = \int_0^x f(s)\,ds$ we suppose that there is a $\delta > 0$ for which

$$-g(x)F(x) + (1/2)\sup_{t\geq 0}\Big[a(h(t))h'(t)r + \int_t^{h(t)} a(s)\,ds\Big]g^2(x) < 0 \tag{2.3.15}$$

if $0 < |x| < \delta$. Note that (2.3.15) will require that $a(h(t))h'(t)$ be bounded above. A calculation will show that if $q(t) = t - L$ and if $a(t) = 1$ then (2.3.15) and (2.3.9) are the same.

Theorem 2.3.4. *Let (2.3.12)-(2.3.15) hold. Then the zero solution of (2.3.11) is asymptotically stable. If (2.3.8) holds and if (2.3.15) holds for $\delta = \infty$, then the zero solution is globally asymptotically stable.*

Proof. We can rewrite (2.3.11) as

$$x'' + f(x)x' + a(h(t))h'(t)g(x) + \frac{d}{dt}\int_{h(t)}^{t} a(s)g(x(q(s)))\,ds = 0. \quad (2.3.16)$$

Then write it as the system

$$\begin{aligned} x' &= y - F(x) - \int_{h(t)}^{t} a(s)g(x(q(s)))\,ds \\ y' &= -a(h(t))h'(t)g(x). \end{aligned}$$

We will define a Liapunov functional in two steps. First, let

$$V_1(t,x,y) = \frac{y^2}{2a(h(t))h'(t)} + G(x) \quad (2.3.17)$$

so that the derivative of V along a solution of the system satisfies

$$\begin{aligned} V_1' = g(x)y - g(x)F(x) - g(x)\int_{h(t)}^{t} a(s)g(x(q(s)))\,ds - g(x)y \\ - \frac{(a(h(t))h'(t))'y^2}{2(a(h(t))h'(t))^2}. \end{aligned} \quad (2.3.18)$$

Make a change of variable. Let $w = q(s)$ so that $s = h(w)$, $ds = h'(w)\,dw$, and $s = h(t)$ implies $w = q(h(t)) = t$, while $s = t$ implies that $w = q(t)$. Thus,

$$\int_{h(t)}^{t} a(s)g(x(q(s)))\,ds = \int_{t}^{q(t)} a(h(w))g(x(w))h'(w)\,dw.$$

Since $q(t) \geq t - r$ we have

$$\int_{t-r}^{t} a(h(w))h'(w)g^2(x(w))\,dw \geq \int_{t}^{h(t)} a(s)g^2(x(q(s)))\,ds.$$

Next, define the second part of the Liapunov functional by

$$2V_2(t,x(\cdot)) = \int_{-r}^{0}\int_{t+s}^{t} a(h(w))h'(w)g^2(x(w))\,dw\,ds \quad (2.3.19)$$

so that the derivative of V_2 along a solution of the system is

$$\begin{aligned} 2V_2' &= \int_{-r}^{0} [a(h(t))h'(t)g^2(x(t)) - a(h(t+s))h'(t+s)g^2(x(t+s))]\,ds \\ &= a(h(t))h'(t)rg^2(x) - \int_{t-r}^{t} a(h(s))h'(s)g^2(x(s))\,ds. \end{aligned} \quad (2.3.20)$$

Hence,

$$\begin{aligned}
&(V_1 + V_2)' \\
&\leq -g(x)F(x) + (1/2)\int_t^{h(t)} a(s)[g^2(x(t)) + g^2(x(q(s)))]\,ds \\
&+ (1/2)a(h(t))h'(t)rg^2(x) - (1/2)\int_{t-r}^{t} a(h(s))h'(s)g^2(x(s))\,ds \\
&\leq -g(x)F(x) + (1/2)[a(h(t))h'(t)r + \int_t^{h(t)} a(s)\,ds]g^2(x). \qquad (2.3.21)
\end{aligned}$$

By assumption (2.3.15), that is a negative definite function of x on $0 < |x| < \delta$. The remainder of the proof is just as before.

There is a very interesting classical problem concerning the equation

$$x'' + a(t)x(t) = 0$$

in which $a(t)$ is positive and increases monotonically. It is readily shown that each solution is bounded; there is then the problem of showing that solutions tend to zero. There is considerable history of the problem given in Cesari (1962; pp. 80 - 86). Much of the early work was by Armellini, Tonelli, and Sansone, as detailed by Cesari. Both linear and nonlinear versions are still of considerable interest. It is known that if $a(t) \to \infty$ in a regular way, then all solutions tend to zero. It was shown by example by Galbraith, McShane, and Parrish (1965) that both solutions need not tend to zero if $a(t)$ does most of its increasing on very short intervals. The solutions oscillate, with their derivatives becoming unbounded. It is known that delays can cause oscillations and destroy oscillations. Thus, with a delay there is considerable chance that even boundedness may be lost. We now show that boundedness can be preserved.

Our interest centers on the scalar equation

$$x'' + a(t)g(x(q(t))) = 0 \qquad (2.3.22)$$

in which $a : [0, \infty) \to (0, \infty)$ is differentiable, $q(t) = t - r(t)$ where $r : [0, \infty) \to (0, \infty)$ is continuous and q is strictly increasing to ∞ with inverse function $h(t)$. This is, of course, our previous problem (2.3.11) in which the damping has been eliminated. The problems are connected by transformations explained by Cesari. Indeed, the R. A. Smith problem of the next section may also be linked to this one by the transformations.

We suppose that

$$(a(h(t))h'(t))' \geq 0 \tag{2.3.23}$$

and that for

$$\alpha(t) := \int_t^{h(t)} a(s)\,ds \tag{2.3.24}$$

then

$$\int_0^\infty \alpha(s)\,ds =: \beta < \infty. \tag{2.3.25}$$

It is also supposed that $xg(x) > 0$ for $x \neq 0$, that g is odd and increasing, and sometimes that

$$\int_0^x g(s)\,ds/\beta g^2(x) \to \infty \quad \text{as} \quad |x| \to \infty. \tag{2.3.26}$$

Theorem 2.3.5. *Let (2.3.23) - (2.3.26) hold. Then every solution of (2.3.22) is bounded. If $g(x) = x$, then (2.3.26) can be replaced by $\beta < 1/2$.*

Proof. The proof is of Razumikhin type. Write (3.2.22) as

$$x'' + \frac{d}{dt}\int_{h(t)}^t a(s)g(x(q(s)))\,ds + a(h(t))h'(t)g(x(t)) = 0$$

and then as the system

$$x' = y - \int_{h(t)}^t a(s)g(x(q(s)))\,ds$$
$$y' = -a(h(t))h'(t)g(x).$$

Define a Razumikhin function

$$V(t, x, y) = \int_0^x g(s)\,ds + \frac{y^2}{2a(h(t))h'(t)} \tag{2.3.27}$$

with derivative along a solution of the system satisfying

$$V' = -g(x)\int_{h(t)}^t a(s)g(x(q(s)))\,ds - \frac{(a(h(t))h'(t))'y^2}{2(a(h(t))h'(t))^2}. \tag{2.3.28}$$

For a given solution $(x(t), y(t))$ suppose that t is any value with $|x(t)| \geq |x(s)|$ for $0 \leq s \leq t$. Integrate V', use the monotonicity and oddness of g, and obtain

$$\int_0^{x(t)} g(s)\,ds \leq V(t) \leq V(0) + g^2(x(t))\beta.$$

By (2.3.26) this shows that $x(t)$ is bounded.

In view of the literature, it would be very interesting to study what more is needed to show that solutions tend to zero.

This is to acknowledge that much of the material in this section was first published in *Fixed Point Theory*, as described in Burton (2004b), which is published by the House of the Book of Science.

2.4 A Problem of R. A. Smith Type

We will now look at natural extensions of the Liénard equations studied in the last section which have features which seem to have defied study by Liapunov's direct method. This work may be found in Burton (2005a).

Smith (1961) proved a beautiful result on asymptotic stability of a second order linear equation

$$x'' + h(t)x' + k^2 x = 0$$

where $h(t)$ is continuous for $t \geq 0$, k^2 is a positive constant, and h_0 is a positive constant with

$$h(t) \geq h_0.$$

He showed that the zero solution is asymptotically stable if and only if

$$\int_0^\infty e^{-H(t)} \int_0^t e^{H(s)}\, ds\, dt = \infty$$

where $H(t) := \int_0^t h(s)\, ds$. It is not hard to show that the condition holds for $h(t) = h_0 + t$, but fails for $h(t) = h_0 + t^2$.

There are many recent extensions of the result including Hatvani (1992) (1996), Hatvani and Krisztin (1997), Hatvani, Krisztin, and Totik (1995), Hatvani and Totik (1993), and Pucci and Serrin (1994). They studied the linear case and also a nonlinear version

$$x'' + f(t, x, x')x' + g(x) = 0$$

with variants of $f(t, x, x') \geq h(t) \geq 0$, obtaining results very close to those of Smith. The condition $h(t) \geq h_0$ was reduced (by Smith also), usually asking that $h(t)$ be integrally positive but, in any case, asking that the intervals on which $h(t)$ vanishes be bounded by a local Lipschitz constant for g. Those results were dictated in large measure by the technique of proof, often a Liapunov argument and/or differential inequalities. In particular, they did not extend well to the case of time dependent restoring force with a delay. One may find Liapunov treatment of

$$x'' + f(x)x' + g(x(t-L)) = 0$$

in Zhang (1992).

Here, we use fixed point theory to develop a close counterpart of the sufficient part of Smith's theorem for the delay equation

$$x'' + f(t,x,x')x' + b(t)g(x(t-L)) = 0$$

where $f(t,x,y) \geq a(t)$ for some continuous function a. Like Smith's result, our condition holds for $a(t) = t$ but fails for $a(t) = t^2$. And, like Smith's result again, actually verifying the conditions can be a challenging chore.

When we refer to (2.4.1) below, all properties in this paragraph are assumed. Consider the scalar equation

$$x'' + f(t,x,x')x' + b(t)g(x(t-L)) = 0 \tag{2.4.1}$$

in which $b : [0,\infty) \to [0,\infty)$ is continuous and bounded. We require that $g : R \to R$ and $f : R \times R \times R \to [0,\infty)$ be continuous. Also, we ask that there are continuous functions a and $c : [0,\infty) \to [0,\infty)$ and $F : R \times R \to [0,\infty)$ such that for all $t \geq 0, x \in R, y \in R$ we have

$$a(t) \leq f(t,x,y) \leq F(x,y)c(t), \qquad \int_0^\infty a(t)dt = \infty,$$
$$\frac{g(x)}{x} \geq \beta > 0, \quad \text{and} \quad \lim_{x\to 0} \frac{g(x)}{x} \quad \text{exists.} \tag{2.4.2}$$

We assume also that f satisfies a local Lipschitz condition and that there is a positive constant K with

$$|g(x) - g(y)| \leq K|x-y| \tag{2.4.3}$$

for all real x, y. Our objective is to give conditions under which each solution of (2.4.1) satisfies $(x(t), x'(t)) \to (0,0)$ as $t \to \infty$. Thus, we suppose that there is given an arbitrary continuous initial function $\psi : [-L,0] \to R$ and a slope $x'(0)$. This is sufficient to establish a unique solution $x_1(t)$. It will be crucial to have such solutions defined for all future time. Thus, we will suppose that either $L > 0$ and $b(t)$ is continuous, or that $L = 0$ and $b(t)$ is differentiable. We will return to the argument in a moment.

But first we need to define certain functions from $x_1(t)$. Let

$$A(t) := f(t, x_1(t), x_1'(t))) \tag{2.4.4}$$

(noting that $c(t)F(x_1(t), x_1'(t)) \geq A(t) \geq a(t) \geq 0$) and write (2.4.1) (retaining the initial condition) as

$$\begin{aligned} x' &= y \\ y' &= -A(t)y - b(t)g(x(t-L)). \end{aligned} \tag{2.4.5}$$

Notice that for the given initial condition, $(x_1(t), y_1(t))$ is the unique solution of (2.4.5). We have linearized (2.4.1), but with the initial condition it is exactly the same equation. This is exact linearization.

Concerning continuation of the solution $x_1(t)$, if $L > 0$ then in the second equation in (2.4.5) we take the last term as a forcing function, apply the variation of parameters formula, and conclude that $y(t)$ is bounded on any interval $[0, T)$ for which the solution is defined. We then use the first equation in (2.4.5) to see that $x(t)$ is also bounded on that interval. It then follows that the solution is defined for all future time, as may be seen in Chapter 3 of Burton (1985). Next, if $L = 0$ and $b(t)$ is differentiable, write $b(t) = \lambda(t)\mu(t)$ where $\lambda(t)$ is increasing and $\mu(t)$ is decreasing. Then define a function

$$W(t, x, y) = \frac{y^2}{2\lambda(t)} + \mu(t)\int_0^x g(s)\,ds$$

whose derivative along solutions of (2.4.5) is non-positive. We readily see that $y(t)$ is bounded on any finite interval. This, in the first equation of (2.4.5), yields $x(t)$ also bounded on any finite interval. This proves that solutions can be continued for all future time. In particular, the aforementioned $(x_1(t), y_1(t))$ is defined for all future time. Note carefully that this does not yield bounded solutions because $\lambda(t)$ can tend to ∞ and $\mu(t)$ can tend to zero.

Return now to (2.4.5) and use the variation of parameters formula on the second member to obtain

$$x'(t) = y(t) = x'(0)e^{-\int_0^t A(s)\,ds} - \int_0^t e^{-\int_u^t A(s)\,ds} b(u)g(x(u-L))\,du \tag{2.4.6}$$

which we now write as

$$x' = B(t) - \int_0^t C(t, u)g(x(u - L))\,du \tag{2.4.7}$$

where $B(t)$ is the first term on the right in (2.4.6). We have reduced (2.4.1) (retaining its initial condition) to a first order integro-differential equation.

If

$$\int_{t-s}^{\infty} C(u+s,s)\,du \tag{2.4.8}$$

exists, then we can write (2.4.7) as

$$x' = B(t) - g(x(t-L)) \int_0^{\infty} C(u+t,t)\,du + \frac{d}{dt}\int_0^t \int_{t-s}^{\infty} C(u+s,s)\,du\, g(x(s-L))\,ds. \tag{2.4.9}$$

Taking $D(t) = \int_0^{\infty} C(u+t,t)\,du$ and $E(t,s) = \int_{t-s}^{\infty} C(u+s,s)\,du$ we can write (2.4.9) as

$$x' = B(t) - D(t+L)g(x(t)) + \frac{d}{dt}\int_{t-L}^{t} D(s+L)g(x(s))\,ds + \frac{d}{dt}\int_0^t E(t,s)g(x(s-L))\,ds. \tag{2.4.10}$$

When $L=0$ this becomes

$$x' = B(t) - D(t)g(x(t)) + \frac{d}{dt}\int_0^t E(t,s)g(x(s))\,ds. \tag{2.4.11}$$

Notice that $x_1(t)$ is involved in every one of these new terms:

$$0 \le a(t) \le A(t) := f(t, x_1(t), x_1'(t)) \le F(x_1(t), x_1'(t))c(t), \tag{2.4.4}$$

$$C(t,u) := e^{-\int_u^t A(s)\,ds} b(u) \ge 0, \tag{2.4.12}$$

$$B(t) := x'(0)e^{-\int_0^t A(s)\,ds}, \tag{2.4.13}$$

$$D(t) := \int_0^{\infty} C(u+t,t)\,du = \int_0^{\infty} e^{-\int_t^{u+t} A(s)\,ds} b(t)\,du \ge 0, \tag{2.4.14}$$

and

$$E(t,s) := \int_{t-s}^{\infty} C(u+s,s)\,du = \int_{t-s}^{\infty} e^{-\int_s^{u+s} A(v)dv} b(s)\,du \ge 0. \tag{2.4.15}$$

Our task now is to obtain a stability relation for (2.4.9) when (2.4.2)-(2.4.4) hold, including the initial condition $(\psi, x'(0))$. By (2.4.2) and (2.4.14) we can define a continuous function

$$q(t) := \frac{D(t+L)g(x_1(t))}{x_1(t)} \ge 0 \tag{2.4.16}$$

so that for the given initial condition then from (2.4.16) and (2.4.10) we have

$$\begin{aligned} x' =& B(t) - q(t)x + \frac{d}{dt}\int_{t-L}^{t} D(s+L)g(x(s))\,ds \\ &+ \frac{d}{dt}\int_{0}^{t} E(t,s)g(x(s-L))\,ds \end{aligned} \tag{2.4.17}$$

with unique solution $x_1(t)$. Again, we have exact linearization.

It will help to understand the next result if we think of (2.4.18) as stipulating a minimum lower bound on $A(t)$, while (2.4.19) asks that the growth of $A(t)$ be controlled. Furthermore, the last part of the coming proof involving showing that $P\phi$ tends to zero becomes almost trivial if we replace (2.4.20) by $A(t) \geq a_0 > 0$.

In the theorem below g enters through K, f enters through $a(t)$ and $c(t)$, and all the functions in the equation are related through these integrals on $0 \leq t < \infty$. While $A(t)$ is approximated through $a(t)$ and $c(t)$, in the linear case

$$x'' + A(t)x' + b(t)x(t-L) = 0,$$

(2.4.18) and (2.4.19) with $a(t) = \gamma c(t) = A(t)$ would exactly reflect the equation. Moreover, this linear equation would serve as a guide to a local asymptotic stability result, which we do not formulate here.

Theorem 2.4.1. *Let the conditions in the paragraph containing (2.4.1) hold, including (2.4.2) and (2.4.3). Suppose also that there is an $\alpha < 1$ such that*

$$\begin{aligned} 2K\sup_{t\geq 0}\int_{t-L}^{t}\int_{0}^{\infty} e^{-\int_{s+L}^{u+s+L} a(v)\,dv} b(s+L)\,du\,ds \\ + 2K\sup_{t\geq 0}\int_{0}^{t}\int_{t-s}^{\infty} e^{-\int_{s}^{u+s} a(v)dv} b(s)\,du\,ds \leq \alpha, \end{aligned} \tag{2.4.18}$$

that for each $\gamma > 0$

$$\int_{0}^{\infty}\int_{0}^{\infty} e^{-\int_{t+L}^{u+t+L}\gamma c(s)\,ds} b(t+L)\,du\,dt = \infty, \tag{2.4.19}$$

and that there are numbers $a_0 > 0$ and $Q > 0$ such that for all $t \geq 0$ if $J \geq Q$ then

$$\int_{t}^{t+J} a(v)\,dv \geq a_0 J. \tag{2.4.20}$$

(Note that since $A(t) \geq a(t)$, it also follows that $\int_t^{t+J} A(v)\,dv \geq a_0 J$.) Then $(x_1(t), y_1(t)) \to (0,0)$ as $t \to \infty$.

Proof. By the variation of parameters formula applied to (2.4.17) we have

$$
\begin{aligned}
x(t) =&\psi(0)e^{-\int_0^t q(s)\,ds} + \int_0^t e^{-\int_u^t q(s)ds}B(u)\,du \\
&+ \int_0^t e^{-\int_u^t q(s)\,ds}\frac{d}{du}\int_{u-L}^u D(s+L)g(x(s))\,ds\,du \\
&+ \int_0^t e^{-\int_u^t q(s)ds}\frac{d}{du}\int_0^u E(u,s)g(x(s-L))\,ds\,du.
\end{aligned}
$$

If we integrate the last two terms by parts we obtain

$$
\begin{aligned}
&e^{-\int_u^t q(s)\,ds}\left[\int_{u-L}^u D(s+L)g(x(s))\,ds + \int_0^u E(u,s)g(x(s-L))\,ds\right]\Bigg|_0^t \\
&- \int_0^t q(u)e^{-\int_u^t q(s)ds}\left[\int_{u-L}^u D(s+L)g(x(s))\right. \\
&\left. + \int_0^u E(u,s)g(x(s-L))\,ds\right]du \\
&= \int_{t-L}^t D(s+L)g(x(s))\,ds + \int_0^t E(t,s)g(x(s-L))\,ds \\
&- e^{-\int_0^t q(s)\,ds}\left[\int_{-L}^0 D(s+L)g(\psi(s))\,ds\right] \\
&- \int_0^t q(u)e^{-\int_u^t q(s)\,ds}\left[\int_{u-L}^u D(s+L)g(x(s))\,ds\right. \\
&\left. + \int_0^u E(u,s)g(x(s-L))\,ds\right]du.
\end{aligned}
$$

This yields

$$
\begin{aligned}
x(t) =&\psi(0)e^{-\int_0^t q(s)\,ds} + \int_0^t e^{-\int_u^t q(s)\,ds}B(u)\,du \\
&+ \int_{t-L}^t D(s+L)g(x(s))\,ds + \int_0^t E(t,s)g(x(s-L))\,ds \\
&- e^{-\int_0^t q(s)\,ds}\left[\int_{-L}^0 D(s+L)g(\psi(s))\,ds\right] \\
&- \int_0^t q(u)e^{-\int_u^t q(s)\,ds}\left[\int_{u-L}^u D(s+L)g(x(s))\,ds\right. \\
&\left. + \int_0^u E(u,s)g(x(s-L))\,ds\right]du.
\end{aligned}
$$

Let $\mathcal{M}$ be the complete metric space with the supremum metric defined by

$$\mathcal{M} = \{\phi : [-L,\infty) \to R \,|\, \phi \in C, \phi_0 = \psi, \phi(t) \to 0 \quad \text{as} \quad t \to \infty\}.$$

Define $P : \mathcal{M} \to \mathcal{M}$ by $\phi \in \mathcal{M}$ implies $(P\phi)(t) = \psi(t)$ on $[-L,0]$, while $t \geq 0$ implies that

$$\begin{aligned}(P\phi)(t) =& \psi(0)e^{-\int_0^t q(s)\,ds} + \int_0^t e^{-\int_u^t q(s)\,ds} B(u)\,du \\ &+ \int_{t-L}^t D(s+L)g(\phi(s))\,ds + \int_0^t E(t,s)g(\phi(s-L))\,ds \\ &- e^{-\int_0^t q(s)\,ds}\left[\int_{-L}^0 D(s+L)g(\psi(s))\,ds\right] \\ &- \int_0^t q(u)e^{-\int_u^t q(s)\,ds}\left[\int_{u-L}^u D(s+L)g(\phi(s))\,ds\right. \\ &\left. + \int_0^u E(u,s)g(\phi(s-L))\,ds\right] du.\end{aligned}$$

We will show that $P : \mathcal{M} \to \mathcal{M}$ and that, under (2.4.18), P is a contraction with unique fixed point $\phi \in \mathcal{M}$ which is therefore the unique solution $x_1(t)$, inheriting the properties of $\mathcal{M}$. Then we will show from (2.4.6) that y_1 also has the desired properties. To ensure that $P : \mathcal{M} \to \mathcal{M}$ we must examine all terms of $P\phi$ and show that each term tends to zero as $t \to \infty$. But first we need to find a bound on $F(x_1(t), x_1'(t))$.

Lemma. *Let*

$$\mathcal{M}^* = \{\phi : [-L,\infty) \to R \,|\, \phi \in C, \phi_0 = \psi, \quad \phi \quad \text{bounded}\}.$$

Then $P : \mathcal{M}^* \to \mathcal{M}^*$ and is a contraction under the supremum metric.

Proof. We examine the terms of $P\phi$ for a fixed $\phi \in \mathcal{M}^*$.

The first term is bounded since $q(t) \geq D(t+L)\beta \geq 0$.

The absolute value of the second term with $t \geq Q$ is

$$\begin{aligned}\left|\int_0^t e^{-\int_u^t q(s)\,ds} B(u)\,du\right| &= |x'(0)| \int_0^t e^{-\int_u^t q(s)\,ds} e^{-\int_0^u A(s)\,ds}\,du \\ &\leq |x'(0)| \int_0^t e^{-\int_0^u a(s)\,ds}\,du \\ &\leq |x'(0)| \left[\int_0^Q du + \int_Q^t e^{-a_0 u}\,du\right]\end{aligned}$$

and that is bounded.

The third term is bounded by

$$\begin{aligned}
&\int_{t-L}^{t} D(s+L)|g(\phi(s))|\,ds \\
&\leq \int_{t-L}^{t}\int_{0}^{\infty} e^{-\int_{s+L}^{u+s+L} A(v)\,dv} b(s+L)\,du|g(\phi(s))|\,ds \\
&\leq \int_{t-L}^{t}\int_{0}^{\infty} e^{-\int_{s+L}^{u+s+L} a(v)\,dv} b(s+L)\,du|g(\phi(s))|\,ds
\end{aligned}$$

which is bounded by (2.4.18). The fourth term is bounded by

$$\int_{0}^{t}\int_{t-s}^{\infty} e^{-\int_{s}^{u+s} a(v)\,dv} b(s)|g(\phi(s-L))|\,du\,ds$$

which can be shown to be bounded using a lengthy argument which will be given in the proof of the theorem itself.

The fifth term is clearly bounded, while the sixth term is bounded for the same reasons given for the third term. The seventh term is like the fourth term.

Next, to see that P is a contraction on $\mathcal{M}^*$, if $\phi, \eta \in \mathcal{M}^*$, then

$$\begin{aligned}
&|(P\phi)(t)-(P\eta)(t)| \\
&\qquad \leq 2\sup_{t\geq 0}\int_{t-L}^{t} D(s+L)K|\phi(s)-\eta(s)|\,ds \\
&\qquad + 2\sup_{t\geq 0}\int_{0}^{t} E(t,s)K|\phi(s)-\eta(s)|\,ds \\
&\qquad \leq \Big[2\sup_{t\geq 0}\int_{t-L}^{t}\int_{0}^{\infty} e^{-\int_{s+L}^{u+s+L} a(v)\,dv} b(s+L)\,du\,ds \\
&\qquad + 2\sup_{t\geq 0}\int_{0}^{t}\int_{t-s}^{\infty} e^{-\int_{s}^{u+s} a(v)dv} b(s)\,du\,ds\Big]K\|\phi-\eta\| \\
&\qquad \leq \alpha\|\phi-\eta\|
\end{aligned}$$

by (2.4.18). Hence, P has a unique fixed point $x_1(t) \in \mathcal{M}^*$ and it is bounded.

Proof. We now return to the proof of the theorem and see that $x_1(t)$ bounded in (2.4.6) will yield $x_1'(t)$ bounded since $b(u)$ and $g(x_1(u-L))$ are bounded, while $A(s) \geq a(s)$. It suffices to note that for $t > Q$ we have

$$\begin{aligned}\int_0^t e^{-\int_u^t A(s)\,ds}\,du &= \int_0^{t-Q} e^{-\int_u^t A(s)\,ds}\,du + \int_{t-Q}^t e^{-\int_u^t A(s)\,ds}\,du\\ &\leq \int_0^{t-Q} e^{-a_0(t-u)}\,du + Q\end{aligned}$$

which is bounded. (See (2.4.20) for definition of Q.)

Hence there is a $\gamma > 0$ with

$$a(t) \leq A(t) = f(t, x_1(t), x_1'(t)) \leq F(x_1(t), x_1'(t))c(t) \leq \gamma c(t).$$

If we now consider (2.4.19) we have

$$\begin{aligned}\int_0^\infty D(t+L)\,dt &= \int_0^\infty \int_0^\infty e^{-\int_{t+L}^{u+t+L} A(s)\,ds} b(t+L)\,du\,dt\\ &\geq \int_0^\infty \int_0^\infty e^{-\int_{t+L}^{u+t+L} \gamma c(s)\,ds} b(t+L)\,du\,dt = \infty\end{aligned}$$

by (2.4.19).

Now we are ready to show that $\phi \in \mathcal{M}$ implies that $(P\phi)(t) \to 0$ as $t \to \infty$.

The first term of $P\phi$ is

$$\psi(0)e^{-\int_0^t q(s)\,ds}$$

and since $q(t) \geq D(t+L)\beta$ and $\int_0^\infty D(t+L)dt = \infty$ by (2.4.19) that term tends to 0.

The second term is

$$\int_0^t e^{-\int_u^t q(s)\,ds} B(u)\,du = x'(0) \int_0^t e^{-\int_u^t q(s)\,ds} e^{-\int_0^u A(s)\,ds}\,du.$$

Let $2\varepsilon > 0$ be given and find $T > Q$ so that $e^{-a_0 T} < \varepsilon a_0$. Since $T < t$ and $0 < Q < T \leq u$,

$$\begin{aligned}\int_T^t e^{-\int_0^u A(s)\,ds}\,du &\leq \int_T^t e^{-a_0 u}\,du\\ &\leq e^{-a_0 T}/a_0 < \varepsilon.\end{aligned}$$

Next, for $t > T$ we have

$$\int_0^T e^{-\int_u^t q(s)\,ds} e^{-\int_0^u A(s)\,ds} = e^{-\int_T^t q(s)\,ds} \int_0^T e^{-\int_u^T q(s)\,ds} e^{-\int_0^u A(s)\,ds}\,du.$$

The first factor tends to zero as $t \to \infty$, while the second factor is simply a fixed number.

The third term is

$$\int_{t-L}^{t} D(s+L)g(\phi(s))\,ds$$

which, by (2.4.18), tends to zero if $\phi \to 0$.

The fourth term is

$$\int_0^t E(t,s)g(\phi(s-L))\,ds = \int_0^t \int_{t-s}^{\infty} e^{-\int_s^{u+s} A(v)\,dv} b(s)g(\phi(s-L))\,du\,ds.$$

Let $2\varepsilon > 0$ be given and use boundedness of b and ϕ to find $T > 2Q$ with

$$\int_T^t b(t-w)g(\phi(t-w-L))e^{-wa_0}\,dw, < \varepsilon a_0.$$

for $t \geq T$. Then change variable below by $w = t - s$ and for $0 < T < t$ obtain

$$\begin{aligned}
&\int_0^t b(s)g(\phi(s-L)) \int_{t-s}^{\infty} e^{-\int_s^{u+s} A(v)\,dv}\,du\,ds \\
&= \int_0^t b(t-w)g(\phi(t-w-L)) \int_w^{\infty} e^{-\int_{t-w}^{u+t-w} A(v)\,dv}\,du\,dw \\
&= \int_0^T b(t-w)g(\phi(t-w-L)) \int_w^{\infty} e^{-\int_{t-w}^{u+t-w} A(v)\,dv}\,du\,dw \\
&+ \int_T^t b(t-w)g(\phi(t-w-L)) \int_w^{\infty} e^{-\int_{t-w}^{u+t-w} A(v)\,dv}\,du\,dw \\
&=: I_1 + I_2.
\end{aligned}$$

In I_2 take into account that $u \geq w \geq T \geq 2Q$ so that we have

$$\begin{aligned}
I_2 &\leq \int_T^t b(t-w)g(\phi(t-w-L)) \int_w^{\infty} e^{-a_0 u}\,du\,dw \\
&= \int_T^t b(t-w)g(\phi(t-w-L))e^{-a_0 w}/a_0\,dw \\
&< \varepsilon.
\end{aligned}$$

Now,

$$\begin{aligned}I_1 &= \int_0^T b(t-w)g(\phi(t-w-L)) \int_w^{w+Q} e^{-\int_{t-w}^{u+t-w} A(v)\,dv}\,du\,dw \\ &+ \int_0^T b(t-w)g(\phi(t-w-L)) \int_{w+Q}^{\infty} e^{-\int_{t-w}^{u+t-w} A(v)\,dv}\,du\,dw \\ &\leq \int_0^T b(t-w)g(\phi(t-w-L))Q\,dw \\ &+ \int_0^T b(t-w)g(\phi(t-w-L)) \int_{w+Q}^{\infty} e^{-a_0 u}\,du\,dw\end{aligned}$$

since $u \geq w+Q \geq Q$ in that last inner integral. Both tend to zero since $\phi \to 0$ as $t \to \infty$.

The fifth term is

$$e^{-\int_0^t q(s)\,ds}\Big[\int_{-L}^{0} D(s+L)g(\psi(s))\,ds\Big]$$

which tends to zero by (2.4.19).

The sixth term is

$$\int_0^t q(u)e^{-\int_u^t q(s)\,ds} \int_{u-L}^{u} D(s+L)g(\phi(s))\,ds$$

and it has the same bound as the third term.

The seventh term is

$$\int_0^t q(u)e^{-\int_u^t q(s)\,ds} \int_0^u E(u,s)g(\phi(s-L))\,ds\,du$$

and it has the same bound as the fourth term.

This verifies that $P : \mathcal{M} \to \mathcal{M}$.

To see that P is a contraction, if $\phi, \eta \in \mathcal{M}$ then

$$\begin{aligned}|(P\phi)(t) - (P\eta)(t)| &\leq 2\sup_{t\geq 0} \int_{t-L}^{t} D(s+L)K|\phi(s)-\eta(s)|\,ds \\ &+ 2\sup_{t\geq 0} \int_0^t E(t,s)K|\phi(s)-\eta(s)|\,ds\end{aligned}$$

and the contraction follows from (2.4.18). It now follows that P has a unique fixed point and it is the function $x_1(t)$. The fixed point tends to zero since it is in $\mathcal{M}$.

Now that we have shown that $x_1(t)$ tends to zero, refer back to (2.4.6) and apply the above arguments to show that $y_1(t)$ also tends to zero. Here are the details. Let the bound on $b(t)$ be 1, let $G(t) := |g(x_1(t-L))|$, let G^* be the maximum of $G(t)$, and let $2\varepsilon > 0$ be given. Fix $T > 2Q$ so that $e^{-a_0 T}G^* < \varepsilon a_0$. Take N so large that $TG(t) < \varepsilon$ if $t > N$. Then $t > N+T$ implies that

$$\begin{aligned}
\int_0^t & e^{-\int_u^t A(s)\,ds} b(u)|g(x_1(u-L))|\,du \\
&= \int_0^{t-T} e^{-\int_u^t A(s)\,ds} b(u)G(u)\,du \\
&\quad + \int_{t-T}^{t} e^{-\int_u^t A(s)\,ds} b(u)G(u)\,du \\
&\le \int_0^{t-T} e^{-a_0(t-u)} b(u)G(u)\,du + \int_{t-T}^{t} G(u)\,du \\
&\le G^* e^{-a_0 T}/a_0 + \varepsilon < 2\varepsilon.
\end{aligned}$$

This completes the proof.

Proposition 2.4.1. *Consider (2.4.19). Let $c(t) = h_0 + 2t$ for some $h_0 \ge 0$ and let $0 \le b(t) \le M$ for some $M > 0$. Then the integral condition in (2.4.19) is satisfied provided that $b(t)$ is large enough that*

$$\int_0^\infty \frac{b(t+L)}{(t+L)}\,dt = \infty.$$

Proof. The computations are virtually unchanged if we take $h_0 = 0$ and $\gamma = 1$. We have

$$\begin{aligned}
\int_0^\infty \int_0^\infty & e^{-\int_{t+L}^{u+t+L} 2s\,ds} b(t+L)\,du\,dt \\
&= \int_0^\infty b(t+L) \int_0^\infty e^{-(u+t+L)^2+(t+L)^2}\,du\,dt \\
&= \int_0^\infty b(t+L) \int_0^\infty e^{-u^2-2u(t+L)}\,du\,dt \\
&\ge \int_0^\infty b(t+L) \int_0^1 e^{-u^2-2u(t+L)}\,du\,dt \\
&\ge \int_0^\infty b(t+L)e^{-1} \int_0^1 e^{-2(t+L)u}\,du\,dt \\
&= \int_0^\infty \frac{b(t+L)}{-2e(t+L)} e^{-2(t+L)u}\Big|_0^1\,dt
\end{aligned}$$

$$
\begin{aligned}
&= \int_0^\infty \frac{b(t+L)}{2e(t+L)}\,dt - \int_0^\infty \frac{b(t+L)}{2e(t+L)} e^{-2(t+L)}\,dt \\
&= \infty,
\end{aligned}
$$

as required. In fact, we see that $b(t)$ can vanish on long intervals.

Proposition 2.4.2. *Consider (2.4.19). If $c(t) = 3t^2$ and if $0 \le b(t) \le \beta$ for some $\beta > 0$, then (2.4.19) is not satisfied.*

Proof. We have

$$
\begin{aligned}
&\int_0^\infty \int_0^\infty e^{-\int_{t+L}^{u+t+L} 3w^2\,dw} b(t+L)\,du\,dt \\
&\quad = \int_0^\infty \int_0^\infty e^{-(u+t+L)^3 + (t+L)^3} b(t+L)\,du\,dt \\
&\quad = \int_0^\infty \int_0^\infty e^{-u^3 - 3u^2(t+L) - 3u(t+L)^2} b(t+L)\,du\,dt \\
&\quad \le \int_0^\infty \int_0^\infty e^{-3(t+L)^2 u} b(t+L)\,du\,dt \\
&\quad = \int_0^\infty \frac{b(t+L)}{-3(t+L)^2} e^{-3(t+L)^2 u}\Big|_0^\infty\,dt \\
&\quad = \int_0^\infty \frac{b(t+L)}{3(t+L)^2}\,dt < \infty,
\end{aligned}
$$

as required.

We return to Proposition 2.4.1 now and see what must be done to satisfy (2.4.18). In fact, there are three simple ways to do so. We could take a large t_0 as a starting point for our solutions and go back through the calculations replacing 0 by the new t_0. We could take h_0 large; this is the easiest one for we will be able to approximate $a(t)$ by h_0 and the integrals will become of convolution type and very simple to evaluate. Or, we could let $b(t)$ be zero on an interval $[0, N]$ for large N. Since letting $b(t)$ vanish is a condition not seen before in the literature, we select that alternative.

Proposition 2.4.3. *Let $a(t) = 2t$, let $b(t) = 0$ on $[0, N]$ where L and N satisfy*

$$
K\left[\ln\left(1 + \frac{L}{N}\right) + \frac{2}{N^2}\right] < 1.
$$

and let $b(t) \le 1$. Then (2.4.18) is satisfied.

Proof. The second integral in (2.4.18) is

$$\begin{aligned}\int_0^t \int_{t-s}^{\infty} e^{-u^2-2us} b(s)\,du\,ds &\leq \int_N^t \int_{t-s}^{\infty} e^{-(u+s)u} b(s)\,du\,ds \\ &\leq \int_N^t \int_{t-s}^{\infty} e^{-(t-s+s)u} b(s)\,du\,ds \\ &= \int_N^t \frac{e^{-tu}}{-t}\bigg|_{t-s}^{\infty} ds \\ &= \frac{e^{-t^2}}{t} \int_N^t e^{ts}\,ds \\ &\leq 1/t^2.\end{aligned}$$

Taking the supremum in t we have the value $1/N^2$.

In the computation below, $b(s+L) = 0$ if $s \leq N - L$, so we take $t \geq N$. Moreover, $b(t) \leq 1$ so it exits the sequence of computations. The first integral in (2.4.18) is

$$\begin{aligned}&\int_{t-L}^t \int_0^{\infty} e^{-\int_{s+L}^{u+s+L} a(v)\,dv} b(s+L)\,du\,ds \\ &\qquad = \int_{t-L}^t \int_0^{\infty} e^{-v^2|_{s+L}^{u+s+L}} b(s+L)\,du\,ds \\ &\qquad \leq \int_{t-L}^t \int_0^{\infty} e^{-(u+s+L)^2+(s+L)^2}\,du\,ds \\ &\qquad = \int_{t-L}^t \int_0^{\infty} e^{-u^2-2u(s+L)}\,du\,ds \\ &\qquad \leq \int_{t-L}^t \int_0^{\infty} e^{-2(s+L)u}\,du\,ds \\ &\qquad = \int_{t-L}^t \frac{1}{2(s+L)}\,ds \\ &\qquad = (1/2)\ln(1+\frac{L}{t}).\end{aligned}$$

Because $b(t)$ is zero on $[0, N]$ the maximum is $(1/2)\ln(1+\frac{L}{N})$.

Putting both integrals together, (2.4.18) will be satisfied if

$$K\left[\ln\left(1+\frac{L}{N}\right)+\frac{2}{N^2}\right] < 1.$$

There is an intricate relation between $g(s)$, $b(t)$, L, and $a(t)$ on the whole interval $[0, \infty)$. We can construct examples such as $a(t) = |\sin 2\pi t| + \sin 2\pi t$ with crude approximations like

$$\int_t^{t+J} A(s)\, ds \geq (1/\pi)J$$

to fulfill (2.4.20). Condition (2.4.19) is readily satisfied. And the approximation

$$e^{-\int_s^{u+s} a(v)\, dv} \leq e^{-(1/\pi)u+(1/\pi)}$$

starts us on our way to verifying (2.4.18), but we are still left with finding bounds on

$$\pi e^{(1/\pi)} \int_{t-L}^{t} b(s+L)\, ds$$

and

$$\pi e^{(1/\pi)} \int_0^t b(s)e^{-(1/\pi)(t-s)}\, ds$$

in order to fulfill (2.4.18). For many tabulated functions $b(t)$, the value of the last integral can be read from Laplace transform tables with the aid of shift theorems.

This is to acknowledge that much of this section was first published in *Nonlinear Analysis*, as described in Burton (2005a), which is published by Elsevier Science.

2.5 A Nonconvolution Levin Equation

In Section 2.2 we studied several forms of Volterra-Levin equations, but always of convolution type. There is a vast gulf between convolution and nonconvolution equations. Given a linear delay equation,

$$x'(t) = ax(t) + bx(t-r)$$

with a, b, r being constants, $r > 0$, we can assume a solution of the form $x(t) = e^{kt}$ and obtain a characteristic quasi-polynomial having an infinite sequence of roots k_i with Re $k_i \to -\infty$. Determining stability is associated with showing that none of the roots have positive real parts.

In parallel with this, consider

$$x' = Ax + \int_0^t B(t-s)x(s)\,ds$$

with A being an $n \times n$ constant matrix and B being an $n \times n$ matrix of L^1–functions. Grossman and Miller (1973) show that all solutions are L^1 if and only if

$$\det[sI - A - L(B)]$$

does not vanish for Re $s \geq 0$ where $L(B)$ is the Laplace transform of B. Thus, constant coefficient problems and convolution problems can share important characteristics not seen in general problems not of convolution type.

In this section we will consider both a linear and nonlinear problem of the form studied in Section 2.2, but of nonconvolution type. This work may be found in Burton (2004a). This is a transition section. We have been considering problems which were globally Lipschitz and we obtained results concerning all solutions. Moreover, the spaces had the most straightforward metrics. Both of those will change here. For the nonlinear equation the mapping set will be bounded and we will have a mapping of that set into itself. Once that situation obtains there are several paths which we will develop throughout the remainder of the book. We may define a weighted norm so that a mapping which does not look like a contraction is a contraction. This requires great care concerning the question of completeness in the new metric. In a different vein, if the mapping fails to satisfy a Lipschitz condition, we may use a different fixed point theorem.

Here, we study a scalar linear equation closely related to one studied in Section 2.2 of the form

$$x'(t) = -\int_{t-r}^t a(t,s)x(s)\,ds, \tag{2.5.1}$$

as well as the nonlinear analogue

$$x'(t) = -\int_{t-r}^t a(t,s)g(x(s))\,ds, \tag{2.5.2}$$

where it is assumed that r is a positive constant, $a : [0,\infty) \times [0,\infty) \to R$ is continuous,

$$A(t,s) := \int_{t-s}^r a(u+s,s)\,du, \quad A(t,t) = \int_0^r a(u+t,t)\,du \geq 0, \tag{2.5.3}$$

and

$$\beta := \sup_{t\geq 0} \int_0^r \left| \int_w^r a(v+t-w, t-w)\, dv \right| dw < 1/2. \tag{2.5.4}$$

Conditions (2.5.3) and (2.5.4) are sufficient for stability of (2.5.1); asymptotic stability requires that the integral of $A(t,t)$ diverge. For (2.5.2) it is supposed that $xg(x) > 0$ for small $x \neq 0$, plus other conditions including (2.5.3) and (2.5.4), to ensure stability and asymptotic stability. In Section 2.2 we gave considerable history concerning equations of the form of (2.5.1), as well as nonlinear counterparts. Levin (1968) extended his Liapunov functional in the convolution case to a form of (2.5.4) with the lower limit zero, while Burton (1980) constructed a similar Liapunov functional for the case of $A(t,s)$ being an $n \times n$ matrix. With care, those results can apply to (2.5.3).

Theorem 2.5.1. *If (2.5.3) and (2.5.4) hold, then the zero solution of (2.5.1) is stable. If, in addition, $\int_0^t A(s,s)\, ds \to \infty$ as $t \to \infty$, then the zero solution of (2.5.1) is asymptotically stable.*

Proof. We offer proofs only for $t_0 = 0$ and we cover only the case of the asserted asymptotic stability. The reader is referred to type I and II exercises. Let $\psi : [-r, 0] \to R$ be a given continuous initial function, let $\mathcal{C}$ be the set of continuous functions, and let

$$\mathcal{M} = \{\phi : [-r, \infty) \to R : \phi_0 = \psi, \phi \in \mathcal{C}, \quad \phi(t) \to 0 \text{ as } t \to \infty\}$$

so that if $\|\cdot\|$ is the supremum metric then $(\mathcal{M}, \|\cdot\|)$ is a complete metric space. Recall that $\phi_0 = \psi$ means $\phi(t) = \psi(t)$ for $-r \leq t \leq 0$. It will cause no confusion to also let $\|\psi\|$ denote the supremum of ψ on $[-r, 0]$ even though we use $\|\phi\|$ as the supremum of ϕ on $[-r, \infty)$.

Write (2.5.1) as

$$x'(t) = -A(t,t)x(t) + \frac{d}{dt}\int_{t-r}^t A(t,s)x(s)\, ds. \tag{2.5.5}$$

By the variation of parameters formula followed by integration by parts, we have

$$\begin{aligned}
x(t) &= e^{-\int_0^t A(s,s)\,ds}\psi(0) + \int_0^t e^{-\int_s^t A(u,u)du}\frac{d}{ds}\int_{s-r}^{s} A(s,u)x(u)\,du\,ds \\
&= e^{-\int_0^t A(s,s)ds}\psi(0) + e^{-\int_s^t A(u,u)du}\int_{s-r}^{s} A(s,u)x(u)du\Big|_{s=0}^{s=t} \\
&- \int_0^t e^{-\int_s^t A(u,u)\,du}A(s,s)\int_{s-r}^{s} A(s,u)x(u)\,du\,ds \\
&= e^{-\int_0^t A(s,s)\,ds}\psi(0) - e^{-\int_0^t A(s,s)\,ds}\int_{-r}^{0} A(0,u)\psi(u)\,du \\
&+ \int_{t-r}^{t} A(t,u)x(u)\,du \\
&- \int_0^t e^{-\int_s^t A(u,u)du}A(s,s)\int_{s-r}^{s} A(s,u)x(u)\,du\,ds. \qquad (2.5.6)
\end{aligned}$$

Use (2.5.6) to define the operator $P : \mathcal{M} \to \mathcal{M}$ as follows: for $\phi \in \mathcal{M}$ let $(P\phi)(t) = \psi(t)$ if $-r \le t \le 0$, and if $t > 0$, let

$$\begin{aligned}
(P\phi)(t) &= e^{-\int_0^t A(s,s)ds}\psi(0) - e^{-\int_0^t A(s,s)\,ds}\int_{-r}^{0} A(0,u)\psi(u)\,du \\
&+ \int_{t-r}^{t} A(t,u)\phi(u)\,du \\
&- \int_0^t e^{-\int_s^t A(u,u)\,du}A(s,s)\int_{s-r}^{s} A(s,u)\phi(u)\,du\,ds. \qquad (2.5.7)
\end{aligned}$$

A fixed point is a solution of (2.5.1).

Clearly, $\phi \in \mathcal{M}$ implies that $P\phi$ is continuous. We can give the classical argument that the convolution of an L^1 function with a function tending to zero does itself tend to zero to show that $(P\phi)(t) \to 0$; use is made of the fact that $\int_{t-r}^{t} |A(t,u)|du < 1/2$ by (2.5.4) using a change of variable.

To see that P is a contraction, consider $\phi, \eta \in \mathcal{M}$. For $t > 0$, we have

$$
\begin{aligned}
&|(P\phi)(t) - (P\eta)(t)| \\
&\leq \int_{t-r}^{t} |A(t,u)(\phi(u) - \eta(u))|\, du \\
&+ \int_0^t e^{-\int_s^t A(u,u)\, du} A(s,s) \int_{s-r}^{s} |A(s,u)(\phi(u) - \eta(u))|\, du\, ds \\
&\leq \int_{t-r}^{t} |A(t,u)|\, du \|\phi - \eta\| + \sup_{t \geq 0} \int_{t-r}^{t} |A(t,u)|\, du \|\phi - \eta\|.
\end{aligned}
$$

It follows that $\|P\phi - P\eta\| \leq \alpha\|\phi - \eta\|$ for $t \in [-r, \infty)$, for some $\alpha > 0$. Thus, P is a contraction on $\mathcal{M}$ if there is an $\alpha < 1$ such that

$$
2 \sup_{t \geq 0} \int_{t-r}^{t} |A(t,u)|\, du \leq \alpha. \tag{2.5.4*}
$$

Observe that this does hold if (2.5.4) is satisfied since the left-hand side of (2.5.4*) is 2β. This is easily seen with the change of variable $u = t - w$ in (2.5.4). In other words, under assumption (2.5.4), associated with each continuous initial function ψ is a complete metric space $\mathcal{M}$ and a unique $\phi \in \mathcal{M}$ that is a fixed point of P. Equivalently, ϕ is the unique solution of (2.5.1) with $\phi(t) = \psi(t)$ for $-r \leq t \leq 0$.

We now want to find an $\varepsilon - \delta$ relation for stability. If $\phi \in \mathcal{M}$ then we have

$$
\begin{aligned}
|(P\phi)(t)| &\leq \|\psi\| \left(1 + \int_{-r}^{0} |A(0,u)| du\right) + 2 \sup_{t \geq 0} \int_{t-r}^{t} |A(t,u)|\, du \|\phi\| \\
&\leq \|\psi\|(1 + \int_{-r}^{0} |A(0,u)| du) + 2\beta\|\phi\|
\end{aligned}
$$

where β is defined in (2.5.4). As P is a contraction, for each ψ there is a unique ϕ with $P\phi = \phi$ so that

$$
\|\phi\|(1 - 2\beta) \leq \|\psi\|(1 + \int_{-r}^{0} |A(0,u)|\, du).
$$

Thus, we can find $\delta > 0$ such that $\|\psi\| < \delta$ implies that $\|\phi\| < \varepsilon$. Now ϕ is the unique solution and $\phi \in \mathcal{M}$ so $\phi(t) \to 0$; hence, $\phi_t \to 0$ as $t \to \infty$.

We return to Equation (2.5.2) which we rewrite for reference as

$$
x'(t) = -\int_{t-r}^{t} a(t,s) g(x(s))\, ds. \tag{2.5.2}
$$

Here, r is a positive constant and $a : [0, \infty) \times [0, \infty) \to R$ is continuous.

Remark 2.5.1. *We assume that there is an $L > 0$ such that g is continuous, odd, Lipschitz, and increasing on $[-L, L]$ and that $0 \leq x - g(x)$ is nondecreasing on $[0, L]$. These monotone conditions hold in case $\frac{d}{dx}g(x)$ is continuous and positive at $x = 0$ by redefining g and a. In this case, there is an $L > 0$ for which $\frac{d}{dx}g(x) > 0$ on $[-L, L]$, with a maximum of $D > 0$. Write $a(t,s)g(x) = a(t,s)D\frac{g(x)}{D}$ and note that $0 \leq x - \frac{g(x)}{D}$ is nondecreasing.*

Theorem 2.5.2. *Assume that there is an $L > 0$ such that g is continuous, odd, Lipschitz, and increasing on $[-L, L]$. If (2.5.3) and (2.5.4) hold and if $0 \leq x - g(x)$ is nondecreasing on $[0, L]$, then the zero solution of (2.5.2) is stable.*

Proof. Write (2.5.2) as

$$\begin{aligned} x'(t) &= -A(t,t)g(x(t)) + \frac{d}{dt}\int_{t-r}^{t} A(t,s)g(x(s))\,ds \\ &= -A(t,t)x(t) + A(t,t)[x(t) - g(x(t))] \\ &+ \frac{d}{dt}\int_{t-r}^{t} A(t,s)g(x(s))\,ds. \end{aligned}$$

By the variation of parameters formula, followed by integration by parts we have

$$\begin{aligned} x(t) &= e^{-\int_0^t A(s,s)\,ds}\psi(0) - e^{-\int_0^t A(s,s)\,ds}\int_{-r}^{0} A(0,u)g(\psi(u))\,du \\ &+ \int_0^t e^{-\int_s^t A(u,u)\,du}A(s,s)[x(s) - g(x(s))]\,ds \\ &+ \int_{t-r}^{t} A(t,u)g(x(u))\,du \\ &- \int_0^t e^{-\int_s^t A(u,u)\,du}A(s,s)\int_{s-r}^{s} A(s,u)g(x(u))\,du\,ds. \end{aligned} \tag{2.5.8}$$

Define

$$\mathcal{M} = \{\phi : [-r,\infty) \to R : \phi_0 = \psi, \phi \in \mathcal{C}, |\phi(t)| \leq L\}$$

where ψ will be restricted in magnitude later. Use (2.5.8) to define $P : \mathcal{M} \to \mathcal{M}$ by $\phi \in \mathcal{M}$ implies that $(P\phi)(t) = \psi(t)$ if $-r \leq t \leq 0$ and for $t \geq 0$ then

$$\begin{aligned}(P\phi)(t) &= e^{-\int_0^t A(s,s)\,ds}\psi(0) - e^{-\int_0^t A(s,s)\,ds}\int_{-r}^{0} A(0,u)g(\psi(u))\,du \\ &\quad + \int_0^t e^{-\int_s^t A(u,u)\,du}A(s,s)[\phi(s) - g(\phi(s))]\,ds \\ &\quad + \int_{t-r}^{t} A(t,u)g(\phi(u))\,du \\ &\quad - \int_0^t e^{-\int_s^t A(u,u)\,du}A(s,s)\int_{s-r}^{s} A(s,u)g(\phi(u))\,du\,ds. \qquad (2.5.9)\end{aligned}$$

Since g is increasing and $x - g(x)$ is nondecreasing on $[0, L]$ and odd, we have

$$\begin{aligned}|(P\phi)(t)| &\leq \|\psi\| + \|g(\psi)\|\int_{-r}^{0} |A(0,u)|\,du \\ &\quad + L - g(L) + 2g(L)\sup_{t\geq 0}\int_{t-r}^{t} |A(t,u)|\,du \\ &= \|\psi\| + \|g(\psi)\|\int_{-r}^{0} |A(0,u)|\,du + L - g(L) + 2\beta g(L)\end{aligned}$$

where β is defined in (2.5.4). By (2.5.4) again, we can find $\alpha < 1$ with $L - g(L) + 2\beta g(L) = \alpha L$. As g is Lipschitz and $g(0) = 0$ there is a K with $|g(\psi(t))| \leq K|\psi(t)|$. Thus, we can find $\delta > 0$ such that $\|\psi\| < \delta$ implies that

$$\|\psi\| + \|g(\psi)\|\int_{-r}^{0} |A(0,u)|\,du < (1-\alpha)L.$$

This shows that $P : \mathcal{M} \to \mathcal{M}$.

In the proposition below we show that there is an exponentially weighted metric under which P is a contraction with unique fixed point ϕ for each ψ.

To see that we have proved stability, for a given $\varepsilon > 0$, $\varepsilon < L$, substitute ε for L in the above argument and conclude that $\|\psi\| < \delta$ yields $|(P\phi)(t)| \leq \varepsilon$; if $P\phi = \phi$, then $|\phi(t)| \leq \varepsilon$.

Remark 2.5.2. *We only conclude stability, not asymptotic stability here even if we ask that the integral of $A(t,t)$ diverges. And the reason for that* *lies in our change of metric. If we were to add the condition to $\mathcal{M}$ that $\phi(t) \to 0$ then under our new metric $\mathcal{M}$ would no longer be a complete metric space.*

Proposition 2.5.1. *Under the conditions of Theorem 2.5.2, there is a metric ρ such that $(\mathcal{M}, \rho)$ is a complete metric space and $P : \mathcal{M} \to \mathcal{M}$ is a contraction in that space.*

Proof. Let $B : [0, \infty) \to [0, \infty)$ be defined by $B(s) := \int_0^r |a(u+s, s)|\,du$ and let $d > 3$. Denote by K the common Lipschitz constant for $g(x)$ and $x - g(x)$ on $[-L, L]$. Define

$$h(t) = dK \int_0^t [A(u,u) + B(u)]\,du$$

and define a metric ρ on $\mathcal{M}$ by $\phi, \eta \in \mathcal{M}$ implies

$$\rho(\phi, \eta) = |\phi - \eta|_K := \sup_{0 \le t < \infty} |\phi(t) - \eta(t)|e^{-h(t)}.$$

Then $(\mathcal{M}, \rho)$ is a complete metric space. Let $\phi, \eta \in \mathcal{M}$ and use (2.5.8) to obtain

$$\begin{aligned}
&|(P\phi)(t) - (P\eta)(t)|e^{-h(t)} \\
&\le \int_0^t e^{-\int_s^t A(u,u)\,du} A(s,s)K|\phi(s) - \eta(s)|e^{-h(t)+h(s)-h(s)}\,ds \\
&+ \int_{t-r}^t |A(t,u)|K|\phi(u) - \eta(u)|e^{-h(t)+h(u)-h(u)}\,du \\
&+ \int_0^t e^{-\int_s^t A(u,u)\,du} A(s,s) \\
&\times \int_{s-r}^s |A(s,u)|K|\phi(u) - \eta(u)|e^{-h(t)+h(u)-h(u)}\,du\,ds \\
&\le \int_0^t e^{-(dK+1)\int_s^t A(u,u)\,du} A(s,s)K|\phi(s) - \eta(s)|e^{-h(s)}\,ds \\
&+ \int_{t-r}^t e^{-dK\int_u^t B(s)\,ds}|A(t,u)|K|\phi(u) - \eta(u)|e^{-h(u)}\,du \\
&+ \int_0^t e^{-\int_s^t A(u,u)\,du} A(s,s) \\
&\times \int_{s-r}^s e^{-dK\int_u^s B(v)\,dv}|A(s,u)|K|\phi(u) - \eta(u)|e^{-h(u)}\,du\,ds
\end{aligned}$$

$$\leq |\phi - \eta|_K K \Big[\int_0^t e^{-(dK+1)\int_s^t A(u,u)\,du} A(s,s)\,ds$$
$$+ \int_{t-r}^t e^{-dK \int_u^t B(s)\,ds} |A(t,u)|\,du$$
$$+ \int_0^t e^{-\int_s^t A(u,u)\,du} A(s,s) \int_{s-r}^s e^{-dK \int_u^s B(v)dv} |A(s,u)|\,du\,ds\Big].$$

For each fixed t notice that in

$$\int_{t-r}^t e^{-dK \int_u^t B(s)\,ds} |A(t,u)|\,du$$

we have $|A(t,u)| \leq \int_0^r |a(v+u,u)|dv = B(u)$ so that

$$\int_{t-r}^t e^{-dK \int_u^t B(s)ds} |A(t,u)|\,du \leq \int_{t-r}^t e^{-dK \int_u^t B(s)\,ds} B(u)\,du \leq \frac{1}{dK}.$$

This also applies to the last integral in our array and we arrive at

$$|(P\phi)(t) - (P\eta)(t)|e^{-h(t)}$$
$$\leq K|\phi - \eta|_K \Big[\frac{1}{dK+1} + \frac{1}{dK} + \frac{1}{dK}\Big]$$
$$\leq \frac{3}{d}|\phi - \eta|_K$$

which is a contraction since $d > 3$. This completes the proof.

In our next result we abuse notation by letting $g'(x) = \frac{d}{dx}g(x)$.

Corollary. *Let the conditions of Theorem 2.5.2 hold and suppose that $\int_0^t A(s,s)\,ds \to \infty$ as $t \to \infty$. If $g'(x)$ is continuous on $(-L, L)$ and if $g'(0) \neq 0$, then the zero solution of (2.5.2) is asymptotically stable.*

Proof. It can be seen that if Theorem 2.5.2 holds for one $L > 0$ then it holds for all smaller positive L. For a fixed $L > 0$ of Theorem 2.6.2, let $q = \min g'(x)$, $Q = \max g'(x)$ for $-L \leq x \leq L$. As $2\beta < 1$ by (2.5.4), if L is sufficiently small then q, Q will satisfy $2\beta Q < q$. That is, as $L \to 0$ we may let $q, Q \to g'(0)$. This means that

$$\mu := (1 - q) + 2\beta Q < 1. \tag{2.5.10}$$

For L, q, Q chosen so that (2.5.10) holds, define

$$\mathcal{M} = \{\phi : [-r, \infty) \to R : \phi_0 = \psi, \phi \in \mathcal{C}, |\phi(t)| \leq L, \phi(t) \to 0\}.$$

We have P defined in (2.5.9) and it readily follows that $\phi \in \mathcal{M}$ implies $(P\phi)(t) \to 0$ so $P : \mathcal{M} \to \mathcal{M}$.

To see that P is a contraction using the supremum metric, for $f(x) = x - g(x)$ on $(-L, L)$ we have $0 \le f'(x) = 1 - g'(x) \le 1 - q$ since, by hypothesis, f is increasing. We also have $0 \le g'(x) \le Q$. Thus, for $\phi, \eta \in \mathcal{M}$ we obtain

$$\begin{aligned}
&|(P\phi)(t) - (P\eta)(t)| \\
&\le \int_0^t e^{-\int_s^t A(u,u)\,du} A(s,s)|f(\phi(s)) - f(\eta(s))|\,ds \\
&+ \int_{t-r}^t |A(t-u)(g(\phi(u)) - g(\eta(u))|du \\
&+ \int_0^t e^{-\int_s^t A(u,u)\,du} A(s,s) \int_{s-r}^s |A(s,u)(g(\phi(u)) - g(\eta(u))|\,du\,ds \\
&\le \Big[(1-q) + 2\sup_{t\ge 0} \int_{t-r}^t |A(t,u)|\,duQ\Big]\|\phi - \eta\| \\
&\le [(1-q) + 2\beta Q]\|\phi - \eta\| \\
&= \mu\|\phi - \eta\|.
\end{aligned}$$

This completes the proof.

This is to acknowledge that much of the material in this section was first published in the *Proceeding of the American Mathematical Society*, as described in Burton (2004a), which is published by the American Mathematical Society.

2.6 The Bernoulli Equation

It is through the lowly Bernoulli equation that we get one of our rare glimpses into the consequences of nonlinearities. The Bernoulli equation is a first order scalar ordinary differential equation written as

$$x' + p(t)x = q(t)x^n.$$

It can be reduced to a linear equation through the substitution

$$v = x^{1-n},$$

but the substitution and the mechanics become much more clear if we first divide by x^n so that we can write it as

$$x^{-n}x' + p(t)x^{1-n} = q(t).$$

Nowadays, it is usually relegated to the end of a problem set in elementary texts on differential equations where the equation is given, the substitution is given, and several forms are stipulated, somewhat as follows:

$$x' = rx - kx^2$$

for $r > 0$ and $k > 0$ is a logistic equation in population theory;

$$x' = rx - kx^3$$

for $r > 0$ and $k > 0$ occurs in the study of stability of fluid flow;

$$x' = (r \cos t + k)x - x^3$$

for constants r and k is also used in the study of fluid flow. The average student may not even see the equation.

But the Bernoulli equation is also important in that it shows us that theorems like the ones we studied in Section 1.3 of the Liapunov-Perron-Krasovskii type can be greatly extended. When we study the behavior of solutions of the Bernoulli equation for $0 < n < 1$, for $1 < n < \infty$, and for $q(t) = e^{\pm Jt}$ with J a positive constant, we find some very interesting behavior which motivates the hypotheses of this section, as well as those of the next two. These are hypotheses which are in marked contrast to those of the Liapunov theorem discussed in Section 1.3. Recall that the Liapunov theorem states that if

$$x' = Dx + G(t, x) \tag{2.6.1}$$

with D a matrix, all of whose characteristic roots have negative real parts, and $\lim_{x\to 0} |G(t, x)|/|x| = 0$ uniformly for $0 \leq t < \infty$, then $x = 0$ is uniformly asymptotically stable. In contrast to that result, the solutions of the Bernoulli equation tell us that

$$x' + 2x = e^{-t}x^{3/5} \tag{2.6.2}$$

and

$$x' + 2x = e^{t}x^3 \tag{2.6.3}$$

may both have solutions tending to zero. Moreover, the case of $p(t)$ large and $n < 1$ leads us to understand that a fading memory, which was believed to be necessary for asymptotic stability, is not always required. Those are the ideas motivating this and the next two sections, as well as Section 5.1. The following work is found in Burton and Furumochi (2002b).

Here, we study the idea that there is a general theorem concerning asymptotic stability of the zero solution of

$$x' = f(t,x) + G(t,x) \tag{2.6.4}$$

when f satisfies a Lipschitz condition with $y' = f(t,y)$ uniformly asymptotically stable and, for example, when $|G(t,x)| \leq q(t)|x|^\alpha$ where $0 < \alpha < 1$ and q is small in some sense. Moreover, it seems that the following modification of Krasnoselskii's fixed point theorem may be a proper vehicle for the proof. The theorem was given in Section 1.3 and is repeated here for reference.

Theorem 1.2.9. *Let $\mathcal{M}$ be a closed, convex, and nonempty subset of a Banach space $(\mathcal{S}, \|\cdot\|)$. Suppose that $A : \mathcal{M} \to \mathcal{S}$ and $B : \mathcal{S} \to \mathcal{S}$ such that:*

(i) *B is a contraction with constant $\alpha < 1$,*
(ii) *A is continuous, $A\mathcal{M}$ resides in a compact subset of $\mathcal{S}$,*
(iii) *$[x = Bx + Ay, y \in \mathcal{M}] \Rightarrow x \in \mathcal{M}$.*

Then there is a $y \in \mathcal{M}$ with $Ay + By = y$.

As pointed out in Chapter 1, this result differs from one by Krasnoselskii in that the former requires that $Bx + Ay$ always resides in M. We will see that this is a crucial change in the present application.

We begin the construction with a simple equation to guide us in the construction of our theorem and then return to a similar problem as an example. Consider the scalar equation

$$x' = -2x + G(t,x) \tag{2.6.5}$$

where G is continuous,

$$|G(t,x)| \leq Ke^{-t}|x^{3/5}|, \tag{2.6.6}$$

and K is a positive constant. Let

$$M = \{\psi : [0,\infty) \to R | \psi \in \mathcal{C}, |\psi(t)| \leq e^{-t}\}, \tag{2.6.7}$$

where $\mathcal{C}$ denotes the set of continuous functions, and let $(\mathcal{S}, \|\cdot\|)$ be the Banach space of bounded continuous function on $[0,\infty) \to R$ with the supremum norm.

Lemma 1. *If $|x_0| + (5/2)K < 1$ and if $x(t) = x(t,0,x_0)$ is the solution of*

$$x' = -2x + G(t,\psi(t)), \psi \in M, \tag{2.6.8}$$

then $x \in M$.

Proof. We have

$$\begin{aligned}|x(t)| &\le |x_0|e^{-2t} + \int_0^t e^{-2(t-s)} K e^{-s} e^{-(3/5)s}\, ds \\ &\le |x_0|e^{-2t} + Ke^{-2t}\int_0^t e^{(2/5)s}\, ds \\ &\le |x_0|e^{-2t} + (5/2)Ke^{-t} < e^{-t}.\end{aligned}$$

Hence, $x \in M$.

Lemma 2. *If for $\psi \in M$ we define*

$$(A\psi)(t) = \int_0^t G(s, \psi(s))\, ds, \quad t \ge 0, \tag{2.6.9}$$

then AM resides in a compact subset of S.

Proof. It is clear that the integrals exist and we readily verify that AM is an equicontinuous set. Moreover, AM is bounded. If we have a sequence $\{A\psi_n\}$, then by Ascoli's theorem and a diagonalization process there is a subsequence, say $\{A\psi_n\}$ again, converging uniformly on compact subsets of $[0, \infty)$. We will now show that $\{A\psi_n\}$ is a Cauchy sequence on $[0, \infty)$.

Given $\varepsilon > 0$, fix $T > 0$ so that $\int_T^\infty 2Ke^{-s}\, ds < \epsilon/2$. Then find N such that $n, m > N$ implies that

$$\sup_{0 \le p \le T} \left| \int_0^p [G(s, \psi_n(s) - G(s, \psi_m(s))]\, ds \right| < \varepsilon/2.$$

Thus, if $n, m > N$ then

$$\begin{aligned}&\sup_{0\le t<\infty} \left| \int_0^t [G(s, \psi_n(s)) - G(s, \psi_m(s))]\, ds \right| \\ &\le \sup_{0\le p\le T} \left| \int_0^p [G(s, \psi_n(s)) - G(s, \psi_m(s))]\, ds \right| \\ &+ \int_T^\infty [|G(s, \psi_n(s))| + |G(s, \psi_m(s))|]\, ds \\ &< \varepsilon.\end{aligned}$$

As AM is contained in $\mathcal{S}$ and $\mathcal{S}$ is complete, AM is contained in a compact subset of $\mathcal{S}$.

The following result is known, but we supply the details for reference. We saw a similar result in the last section and we will see more along the same lines in the next chapter.

Lemma 3. *Let $b : R^{d+1} \to R^d$ be continuous and suppose there is an $L > 0$ so that $|b(t,x) - b(t,y)| \leq L|x - y|$. With the norm*

$$|\phi|_L = \sup_{0 \leq s < \infty} \{|e^{-2Ls}\phi(s)|\}$$

on the Banach space $\mathcal{U}$ of bounded continuous functions $\phi : [0,\infty) \to R^d$, then the operator H defined by

$$(Hx)(t) = x_0 + \int_0^t b(s, x(s))\, ds, \quad t \geq 0,$$

is a contraction with contraction constant $1/2$.

Proof. We have

$$\begin{aligned}
|Hx_1 - Hx_2|_L &= \sup_{0 \leq s < \infty} |e^{-2Ls} \int_0^s (b(u, x_1(u)) - b(u, x_2(u))\, du| \\
&\leq \sup_{0 \leq s < \infty} \int_0^s e^{-2Ls} L|x_1(u) - x_2(u)|\, du \\
&= \sup_{0 \leq s < \infty} \int_0^s e^{-2Ls} L|x_1(u) - x_2(u)| e^{-2Lu} e^{2Lu}\, du \\
&\leq |x_1 - x_2|_L \sup_{0 \leq s < \infty} \int_0^s e^{-2Ls} L e^{2Lu}\, du \\
&\leq (1/2)|x_1 - x_2|_L
\end{aligned}$$

a contraction.

In the proof of Lemma 2, the norm $|\cdot|_L$ works as well as the supremum norm.

With this example in mind we now consider a general theorem. Let $a, b : [0,\infty) \times R^d \to R^d$ be continuous and consider

$$x' = b(t,x) + a(t,x), \quad x(0) = x_0 \tag{2.6.10}$$

where

$$|b(t,x) - b(t,y)| \leq L|x - y| \text{ on } [0,\infty) \times R^d. \tag{2.6.11}$$

Thus, (2.6.10) has a solution.

Let $(\mathcal{U}, \|\cdot\|)$ denote a Banach space of bounded continuous functions $\phi : [0, \infty) \to R^d$ and M denote a closed convex nonempty subset of $\mathcal{U}$. Let the operator $A : M \to \mathcal{U}$ defined by $\psi \in M$ implies that

$$(A\psi)(t) = \int_0^t a(s, \psi(s))\, ds, \quad t \geq 0, \tag{2.6.12a}$$

be continuous and define the operator B by

$$(B\phi)(t) = x_0 + \int_0^t b(s, \phi(s))\, ds, t \geq 0, \tag{2.6.12b}$$

for each $\phi \in \mathcal{U}$.

Theorem 2.6.1. *Let B be a contraction with constant $\alpha < 1$ on the space $(\mathcal{U}, \|\cdot\|)$ and suppose that AM resides in a compact subset of that space. Suppose also that for each $\psi \in M$ the unique solution ϕ of*

$$\phi'(t) = b(t, \phi(t)) + a(t, \psi(t)), \quad \phi(0) = x_0 \tag{2.6.13}$$

is in M. Then a solution of (2.6.10) is in M.

Proof. Notice first that if $\phi \in M$ is a fixed point of P, where P is defined by

$$(P\phi)(t) = x_0 + \int_0^t b(s, \phi(s))\, ds + \int_0^t a(s, \phi(s))\, ds, \quad t \geq 0, \tag{2.6.14}$$

then ϕ is a solution of (2.6.10).

Now, for fixed $\psi \in M$ and all $\phi \in \mathcal{U}$, define Q by

$$(Q\phi)(t) = x_0 + \int_0^t b(s, \phi(s))\, ds + \int_0^t a(s, \psi(s))\, ds, \quad t \geq 0. \tag{2.6.15}$$

If $Q\phi = \phi$ for some $\phi \in \mathcal{U}$, then ϕ is the unique solution of

$$\phi' = b(t, \phi) + a(t, \psi(t)), \quad \phi(0) = x_0. \tag{2.6.16}$$

By assumption, that unique solution of (2.6.16) is in M. By Theorem 1.2.9, P itself has a fixed point ϕ in M.

Corollary. *If, in addition to the assumptions of Theorem 2.6.1, all functions in M tend to 0 as $t \to \infty$, then a solution of (2.6.10) tends to zero as $t \to \infty$.*

Remark 2.6.1. *Solutions of (2.6.10) may not be uniquely determined by their initial conditions. Theorem 2.6.1 only claims that one solution is in M.*

The following example is parallel in content, but different in technique, to the results in Coddington and Levinson (1955; pp. 314, 327) and Lakshmikantham and Leela (1969; p. 115). It will then be followed by a nonlinear example.

Example 2.6.1. Let D be a $d \times d$ constant matrix, all of whose characteristic roots have negative real parts; thus, there exist $\alpha > 0$ and $k > 0$ with

$$|e^{Dt}| \leq ke^{-\alpha t}, \quad t \geq 0. \tag{2.6.17}$$

Next, let $G : [0, \infty) \times R^d \to R^d$ be continuous and suppose there is a constant $\gamma > 0$, a continuous function $q : [0, \infty) \to [0, \infty)$ with $q(t) \to 0$ as $t \to \infty$ and $q \in L^1[0, \infty)$ so that

$$|G(t, x)| \leq Kq(t)|x|^{\gamma}. \tag{2.6.18}$$

We will show that the conditions of Theorem 2.6.1 are satisfied for

$$x' = Dx + G(t, x)$$

when K is sufficiently small.

To this end, if we let

$$r(t) := \int_0^t e^{-\alpha(t-s)} q(s)\, ds \tag{2.6.19}$$

then $r(t) \to 0$ as $t \to \infty$ and $r \in L^1[0, \infty)$ since r is the convolution of appropriate functions.

Define

$$h(t) = \max[r(t), e^{-\alpha t}] \tag{2.6.20}$$

and note that $h(t) \leq |r(t)| + e^{-\alpha t} \in L^1[0, \infty)$; moreover $h(t) \to 0$ as $t \to \infty$. By redefining q and K we may assume without loss of generality that

$$h(t) \leq 1, \quad t \geq 0. \tag{2.6.21}$$

Define

$$M = \{\psi : [0, \infty) \to R^n \mid \psi \in C, |\psi(t)| \leq h(t)\}. \tag{2.6.22}$$

Thus, M is closed and convex.

For arbitrary $\psi \in M$, consider

$$x' = Dx + G(t, \psi(t)), \quad x(0) = x_0. \tag{2.6.23}$$

Then

$$\begin{aligned} |x(t)| &\leq |x_0|ke^{-\alpha t} + \int_0^t kKe^{-\alpha(t-s)}q(s)|\psi(s)|^\gamma \, ds \\ &\leq |x_0|kh(t) + kKr(t) \\ &\leq [|x_0|k + kK]h(t) \\ &\leq h(t) \end{aligned} \tag{2.6.24}$$

provided that

$$[|x_0| + K]k \leq 1.$$

Hence, $x(t) \in M$.

Exactly as in the proof of Lemma 2, if A is defined by (2.6.9) then any sequence $\{A\psi_n\}$ with $\psi_n \in M$ is equicontinuous and so we obtain a subsequence converging uniformly on compact sets. The norm $|\cdot|_L$ works just like the supremum norm in the convergence proof.

Next, we consider a perturbed Liénard equation

$$x'' + f(x)x' + g(x) = Kh(t, x, x') \tag{2.6.25}$$

which we write as the system

$$\begin{aligned} x' &= y \\ y' &= -f(x)y - g(x) + Kh(t, x, y) \end{aligned} \tag{2.6.26}$$

or in vector form as

$$X' = b(X) + a(t, X) \tag{2.6.27}$$

where

$$a(t, X) = (0, Kh(t, x, y))^T.$$

We assume that for any $\alpha > 0$ and for any $J > 0$, if $\psi : [0, \infty) \to R^2$ and $|\psi(t)| \leq Je^{-\alpha t}$ then

$$a(t, \psi(t)) \in L^1[0, \infty), \tag{2.6.28}$$

that for each $J > 0$ and for each $\alpha > 0$ there is a $D > 0$ such that $|\psi(t)| \le Je^{-\alpha t}$ implies that

$$\begin{aligned} |\int_{t_1}^{\infty} a(s,\psi(s))\,ds - \int_{t_2}^{\infty} a(s,\psi(s))\,ds| \\ = |\int_{t_1}^{t_2} a(s,\psi(s))\,ds| \le D|t_1 - t_2|, \end{aligned} \tag{2.6.29}$$

and that there are positive L_1, L_2, L_3, L_4 so that if $X_i \in R^2$ then

$$\begin{aligned} &|b(X_1) - b(X_2)| \le L_1|X_1 - X_2|, \\ &L_4 \ge f(x) \ge L_2, \\ \text{and} \quad &g(x)\int_0^x f(s)\,ds \ge L_3 x^2. \end{aligned} \tag{2.6.30}$$

Now, for J, α to be determined, let

$$M = \{\psi : [0,\infty) \to R^2 \mid \psi \in \mathcal{C}, |\psi(t)| \le Je^{-\alpha t}\}$$

and for each $\psi \in M$ consider the system

$$\begin{aligned} x' &= y \\ y' &= -f(x)y - g(x) + e(t) \end{aligned} \tag{2.6.31}$$

where $e(t) = Kh(t, \psi(t))$.

Lemma 4. *If (2.6.27)-(2.6.30) hold and if we define*

$$V(x,y) = (1/2)y^2 + 2\int_0^x g(s)ds + (1/2)\left(y + \int_0^x f(s)\,ds\right)^2 \tag{2.6.32}$$

then there is an $\eta > 0$ so that the derivative of V along a solution of (2.6.31) satisfies

$$V'(x(t), y(t)) \le -\eta V(x,y) + 2\sqrt{V(x,y)}|e(t)| \tag{2.6.33}$$

and there is a $k_1 > 0$ with

$$k_1(x^2 + y^2) \le V(x,y). \tag{2.6.34}$$

Proof. We have

$$\begin{aligned}
V'(x,y) &= 2g(x)y - f(x)y^2 - yg(x) + ye(t) \\
&+ \Big(y + \int_0^x f(s)ds\Big)\big(f(x)y - f(x)y - g(x) + e(t)\big) \\
&= -f(x)y^2 + ye(t) \\
&- g(x)\int_0^x f(s)\,ds + \Big(y + \int_0^x f(s)\,ds\Big)e(t) \\
&\leq -f(x)y^2 - g(x)\int_0^x f(s)\,ds \\
&+ |y||e(t)| + \Big|y + \int_0^x f(s)\,ds\Big||e(t)| \\
&\leq -L_2y^2 - L_3x^2 + \Big[\sqrt{2}(|y|/\sqrt{2}) \\
&+ \sqrt{2}\Big(\Big|y + \int_0^x f(s)\,ds\Big|\Big)/\sqrt{2}\Big]|e(t)| \\
&\leq -L_2y^2 - L_3x^2 + 2|e(t)|\sqrt{V(x,y)}.
\end{aligned}$$

But if we use (2.6.30), in particular g is Lipschitz, then we have

$$\begin{aligned}
V(x,y) &\leq (1/2)y^2 + (L_1)x^2 + y^2 + \Big(\int_0^x f(s)\,ds\Big)^2 \\
&\leq (3/2)y^2 + (L_1)x^2 + L_4^2x^2
\end{aligned}$$

and so there is an $\eta > 0$ with

$$V'(x,y) \leq -\eta V(x,y) + 2|e(t)|\sqrt{V(x,y)}.$$

To find k_1, we have

$$L_3x^2 \leq g(x)\int_0^x f(s)\,ds \leq |g(x)|L_4|x|$$

or

$$|g(x)| \geq L_3|x|/L_4$$

and so

$$\int_0^x g(s)\,ds \geq L_3x^2/(2L_4).$$

From these we can find k_1.

Theorem 2.6.2. *Suppose there are α, β, J, and S with $0 < \alpha \leq \beta < \eta/2$ so that*

$$|\psi(t)| \leq Je^{-\alpha t} \implies |h(t, \psi(t))| \leq Se^{-\beta t}, \; t \geq 0 \tag{2.6.35}$$

and

$$J((\eta/2) - \beta)\sqrt{k_1} > SK.$$

If

$$\mathcal{M} = \{\psi : [0, \infty) \to R^2 \mid \psi \in C, |\psi(t)| \leq Je^{-\alpha t}\}$$

and if $|(x_0, y_0)|$ is small, then the solution of (2.6.31) through (x_0, y_0) for any $t_0 \geq 0$ is in $\mathcal{M}$.

Proof. Select $\psi \in M$ and (x_0, y_0, t_0) so that $(x(t), y(t))$ is fixed, and hence, $V(t) := V(x(t), y(t))$ is determined in (2.6.32). In

$$V'(t) \leq -\eta V(t) + 2|e(t)|\sqrt{V(t)},$$

we first obtain

$$V(t) \leq V(0)e^{-\eta t} + 2\int_0^t e^{-\eta(t-s)}|e(s)|\sqrt{V(s)}ds$$

or

$$e^{\eta t}V(t) \leq V(0) + 2\int_0^t e^{(1/2)\eta s}|e(s)|\sqrt{e^{\eta s}V(s)}\, ds$$

which we write as

$$u(t) \leq u(0) + 2\int_0^t e^{(1/2)\eta s}|e(s)|\sqrt{u(s)}\, ds.$$

By Bihari's inequality (1956) and Hartman (1964; p. 29) we have $u(t) \leq w(t)$ where $w(t)$ is the maximal solution of

$$w(t) = u(0) + 2\int_0^t e^{(1/2)\eta s}|e(s)|\sqrt{w(s)}\, ds.$$

Thus, letting $v(t) = \sqrt{w(t)e^{-\eta t}}$ we obtain $2v'(t) + \eta v(t) = 2|e(t)|$ or $v' + (\eta/2)v = |e(t)|$. We then have

$$\begin{aligned}
v(t) &= v_0 e^{-(\eta/2)t} + \int_0^t e^{-(\eta/2)(t-s)}|e(s)|\, ds \\
&\leq v_0 e^{-(\eta/2)t} + \int_0^t SKe^{-(\eta/2)(t-s)-\beta s}\, ds \\
&= v_0 e^{-(\eta/2)t} + SKe^{-(\eta/2)t}[(\eta/2) - \beta]^{-1} e^{[(\eta/2)-\beta]s}\Big|_0^t \\
&\leq (v_0 + [(\eta/2) - \beta]^{-1}SK]e^{-\beta t}.
\end{aligned}$$

Hence,

$$\begin{aligned}\sqrt{k_1(x^2(t)+y^2(t))} &\le \sqrt{V(t)}\\ &\le \left[\sqrt{V(x_0,y_0)}+[(\eta/2)-\beta]^{-1}SK\right]e^{-\beta t}.\end{aligned} \tag{2.6.36}$$

Thus, $(x(t),y(t))$ is in $\mathcal{M}$ provided that

$$J_0 := \sqrt{V(x_0,y_0)/k_1}+[((\eta/2)-\beta)\sqrt{k_1}]^{-1}SK < J, \tag{2.6.37}$$

as required.

Remark. Notice that (2.6.35) is an interesting relation. For example, let $h(t,x,y)=Kp(t)x^n$. Thus, if $|\psi(t)|<Je^{-\alpha t}$, then

$$|h(t,\psi(t))|\le KJp(t)e^{-\alpha nt}<Se^{-\alpha t}$$

provided that

$$p(t)<(S/KJ)e^{-\alpha(1-n)t}:$$

(i) If $n=1$, $p(t)$ must be bounded.
(ii) If $n>1$, then $p(t)$ can be exponentially unbounded.
(iii) If $n<1$, then $p(t)$ must tend to 0 exponentially.

Now for a local result we look at (2.6.36) and (2.6.37). Let D be the set of (x_0,y_0) for which (2.6.37) holds. For any such (x_0,y_0) and any $t_0\ge 0$, the solution $(x(t),y(t))$ remains in a set

$$\Omega(J_0)=\{(x,y)|x^2+y^2\le J_0^2\}.$$

Theorem 2.6.3. *If (2.6.30) holds in $\Omega(J_0)$ and if (x_0,y_0) satisfies (2.7.37) then the solution of (2.6.31) through (x_0,y_0) for $t_0\ge 0$ is in M and the corresponding solution of (2.7.26) is in M.*

Proof. Notice that $\Omega(J_0)$ is convex. Write (2.6.31) as

$$x'=F(X)+E(t) \tag{2.6.31}$$

with $E(t)=(0,Kh(t,\psi(t)))^T$ and define a new system

$$X'=G(X)+E(t) \tag{2.6.31*}$$

by $G(X)=F(X)$ for $X\in\Omega(J_0)$ and if X is in the complement of $\Omega(J_0)$ then the line from (0,0) to X intersects the boundary of $\Omega(J_0)$ at a unique point X^*. In the latter case, define $G(X)=F(X^*)$. Then G is continuous and globally Lipschitz. Any solution of (2.6.31*) with initial values in $\Omega(J_0)$ lies in M. Krasnoselskii's theorem will now say that (2.6.27) has a solution in M.

This is to acknowledge that much of the material in this section was first published in *Nonlinear Analysis*, as described in Burton and Furumochi (2002b), which is published by Elsevier Science.

2.7 Liapunov-Perron-Krasovskii Equations

This section is named like Section 1.3 in which we discussed perturbed linear systems using Liapunov's direct method. We will now look at some generalizations of those problems which we will treat with fixed point theory.

Let S be an $n \times n$ constant matrix, all of whose characteristic roots have negative real parts and let $F : [0, \infty) \times R^n \to R^n$ be continuous. As we discussed in some detail in Section 1.3, Liapunov, and later Perron (1930), showed that uniform asymptotic stability of the zero solution of a vector system

$$x'(t) = Sx(t) + F(t, x(t))$$

is inherited from the uniform asymptotic stability of

$$y' = Sy$$

whenever

$$\lim_{|x| \to 0} \frac{|F(t, x)|}{|x|} = 0$$

uniformly for $0 \le t < \infty$. Bellman (1953) gives many associated criteria, as do Lakshmikantham and Leela (1969; p. 115). Krasovskii (1963) extended these to functional differential equations. The material of this section may be found in Burton (2003b).

We are interested in similar results for a system

$$x'(t) = Sx(t) + Px(t - r) + \frac{d}{dt}Q(t, x_t) + G(t, x_t)$$

where r is a positive constant and x_t is an element of the Banach space $(\mathcal{C}, \|\cdot\|)$ of continuous functions $\phi : [-\gamma, 0] \to R^n$, $\gamma > 0$, with $x_t(s) = x(t + s)$, $-\gamma \le s \le 0$. For the vector case we find that we need to take $P = 0$, but for the scalar case both S and P can be functions of t. Solutions of the Bernoulli equation will guide us in the hypotheses on these functions of t.

The above condition on F might be expressed as

$$|F(t,x)| \le h(|x|)|x|$$

where h is continuous with $h(0) = 0$. We ask a variant of this on both Q and G , allowing them to grow rapidly in t. We also ask strong continuity properties.

The nice thing about Liapunov's result is that in so many problems we simply glance at a very formidable equation and see that Liapunov's result applies. For example, in

$$x'' + (\cos x)x' + \sin x = 0 \tag{E_1}$$

we recognize that the linear part is the asymptotically stable equation

$$x'' + x' + x = 0$$

and that the rest of the terms are of higher order. Hence, (E_1) is uniformly asymptotically stable.

It would be very interesting to have a fairly simple criterion which would tell us that the zero solution of

$$x'' + (\cos x)x' + \sin x = \frac{d}{dt}t^3x^2(t - r(t)) + t^5x^3(t - r(t)) \tag{E_2}$$

is also asymptotically stable even when r is a bounded continuous function which is not differentiable. Our first result (Theorem 2.7.1) gives us just such a criterion and (E_2) is asymptotically stable. The technique will also work with a term of the form $\frac{d}{dt}tg(x(t - r(t)))$ where g satisfies a Lipschitz condition with sufficiently small Lipschitz constant. This is done in Theorem 2.7.2, but it requires careful calculations, while Theorem 2.7.1 relies on general order of growth.

In the scalar first order case we consider equations having a linear part of the form

$$x'(t) = -a(t)x(t) - b(t)x(t - r) \tag{E_3}$$

in which

$$a(t) + b(t + r) \ge 0, \qquad \int_0^\infty [a(t) + b(t + r)]\, dt = \infty,$$

and

$$\sup_{t\ge 0} \int_{t-1}^t [a(s) + b(s + r)]\, ds$$

is bounded. A simple example of that is

$$x'(t) = -(\sin t)x(t) - x(t-r).$$

We show that the equation is asymptotically stable under perturbations of the type in (E_2).

Neutral equations arise in natural ways such as in describing the vibration of masses on an elastic bar. Often it is necessary to convert a functional differential equation of the form $x'(t) = f(t, x_t)$ into a neutral equation in order to secure a tractable linear term from which to launch a stability investigation. That type of work is seen in the following pages.

The methods used here include the Krasnoselskii fixed point theorem which we stated earlier and will be repeated here for ready reference. Krasnoselskii's theorem is a natural tool for neutral equations being motivated by the fact that, frequently, the inversion of the perturbed differential operator yields the sum of a contraction and compact map. Neutral equations offer a perfect example of that property.

Compactness requirements in Krasnoselskii's theorem depend on good estimates of the rate of decay of solutions. Study of Bernoulli equations leads us to those estimates.

We have mentioned before advantages of Liapunov theory over fixed point theory and the reverse. In neutral theory with Liapunov's direct method it is frequently necessary that the neutral term, $\frac{d}{dt}Q(t, x_t)$, satisfy $|Q(t, x_t)| \leq \alpha \|x_t\|$, $\alpha < 1$, since we often show that $x(t) + Q(t, x_t) \to 0$ and wish to infer that $x(t) \to 0$. That problem does not occur with fixed point theory.

Our first result concerns a general neutral equation. Let $\gamma > 0$ and $(\mathcal{C}, \|\cdot\|)$ be the Banach space of continuous functions $\phi : [-\gamma, 0] \to R^n$ with the supremum norm, $\mathcal{C}_H$ the H-ball in $\mathcal{C}$. We suppose also that Q and G map $[0, \infty) \times \mathcal{C}_H \to R^n$ and are at least continuous. Finally, let S be an $n \times n$ real constant matrix, all of whose characteristic roots have negative real parts. We then choose positive constants K and α with

$$|e^{St}| \leq Ke^{-2\alpha t}, \ t \geq 0.$$

With this K and α in mind, consider the neutral functional differential equation

$$x'(t) = Sx(t) + \frac{d}{dt}Q(t, x_t) + G(t, x_t) \tag{2.7.1}$$

with continuous initial function $\psi : [-\gamma, 0] \to R^n$, $|e^{St}| \leq Ke^{-2\alpha t}, t \geq 0$.

In order to prove asymptotic stability by fixed point theory, write (2.7.1) as

$$\frac{d}{dt}[x(t) - Q(t, x_t)] = S[x(t) - Q(t, x_t)] + SQ(t, x_t) + G(t, x_t)$$

so that

$$\begin{aligned} x(t) = Q(t, x_t) &+ e^{St}[\psi(0) - Q(0, \psi)] \\ &+ \int_0^t e^{S(t-s)}[SQ(s, x_s) + G(s, x_s)]ds. \end{aligned} \tag{2.7.2}$$

For reference here we again state Krasnoselskii's theorem, discussed earlier in Chapter 1.

Theorem 1.2.7 (Krasnoselskii). *Let $\mathcal{M}$ be a closed convex non-empty subset of a Banach space $(\mathcal{S}, \|\cdot\|)$. Suppose that A and B map $\mathcal{M}$ into $\mathcal{S}$ such that*

(i) $Ax + By \in \mathcal{M}$ for all $x, y \in \mathcal{M}$,
(ii) *A is continuous and $A\mathcal{M}$ is contained in a compact set,*
(iii) *B is a contraction with constant $\alpha < 1$.*

Then there is a $y \in \mathcal{M}$ with $Ay + By = y$.

Looking at the theorem and at (2.7.2) we see that we must define a mapping from (2.7.2) and, hence, must separate it into two operators, one a contraction and one a compact mapping. The choice here is fairly clear since Q is exposed and does not smooth. Our only question concerns whether or not that Q should be joined with the Q under the integral (which does get smoothed); it turns out to make little difference.

Study of simple Bernoulli equations leads us to conjecture that our solution should reside in

$$\mathcal{M} = \{\phi : [-\gamma, \infty) \to R^n \mid \phi_0 = \psi, |\phi(t)| \leq Le^{-\alpha t}\} \tag{2.7.3}$$

where $L > 0$ is to be determined, ψ is the initial function with $\|\psi\|$ small. Now, from (2.7.2) we define two operators $A, B : \mathcal{M} \to \mathcal{M}$ by $\phi \in \mathcal{M}$ implies that

$$(B\phi)(t) = Q(t, \phi_t) + e^{St}[\psi(0) - Q(0, \psi)] \tag{2.7.4}$$

and

$$(A\phi)(t) = \int_0^t e^{S(t-s)}[SQ(s, \phi_s) + G(s, \phi_s)]\, ds. \tag{2.7.5}$$

There is great latitude in the conditions which can be placed on Q and G. From the following discussions it is hoped that the reader can see how to modify our conditions to fit a wide variety of problems.

Discussion of Q

If B is to define a contraction, Q will need to be Lipschitz with a very small Lipschitz constant. But we have in mind functions such as $Q(t, \phi) = r(t)\phi^n$ so, for example, if $Q(t, \phi) = r(t)\phi^2$ then we have

$$|Q(t, \phi_1) - Q(t, \phi_2)| = |r(t)||\phi_1 + \phi_2||\phi_1 - \phi_2|.$$

The fact that we are working entirely in $\mathcal{M}$ means that this will define a contraction with small contraction constant, provided only that $|r(t)|Le^{-\alpha(t-\gamma)}$ is small; we have the freedom to take L as small as we please, $L > 0$. The following assumption illustrates this.

Assume that there is a continuous increasing function $\beta : [0, \infty) \to [0, \infty)$, $\beta(0) = 0$, a continuous function $q : [0, \infty) \to [0, \infty)$, and a positive constant q^* such that $\phi, \eta \in C_H$ implies that

$$\begin{aligned} &|Q(t, \phi) - Q(t, \eta)| \leq \beta(q(t)[\|\phi\| + \|\eta\|])\|\phi - \eta\| \\ &\qquad \text{and} \qquad q(t)e^{-\alpha(t-\gamma)} < q^*. \end{aligned} \tag{2.7.6}$$

Assume also that $Q(t, 0) = 0$ so that

$$|Q(t, \phi)| \leq \beta(q(t)[\|\phi\|])\|\phi\|. \tag{2.7.7}$$

Discussion of the function G

We need three things. The growth of G must be controlled so that $A : \mathcal{M} \to \mathcal{M}$. A must map $\mathcal{M}$ into an equicontinuous set. A must be continuous. To those ends, for the function G we suppose there is a continuous function $f : [0, \infty) \to [0, \infty)$ and a continuous function $h : [0, H] \to [0, \infty)$ which is increasing.

Making $A : \mathcal{M} \to \mathcal{M}$

It will take the next two conditions to ensure that $A : \mathcal{M} \to \mathcal{M}$. Assume that for $\phi \in C_H, t \geq 0$ then

$$|G(t, \phi)| \leq f(t)h(\|\phi\|)\|\phi\|, \quad h(0) = 0, \tag{2.7.8}$$

and

$$\int_0^t e^{-\alpha(t-s)} f(s)h(Le^{-\alpha(s-\gamma)})\, ds \to 0 \tag{2.7.9}$$

as $L \to 0$ uniformly for $0 \leq t < \infty$.

Continuity of A

As an example, we will give a sample condition to ensure continuity of A based on a prototype of $G(t,\phi) = f(t)\phi^n$. To see how continuity proceeds, take $G(t,\phi) = f(t)\phi^2$ so that for a $\delta > 0$ and $\phi, \eta \in \mathcal{M}$ with $\|\phi - \eta\| < \delta$ we have

$$\begin{aligned} |G(t,\phi) - G(t,\eta)| &= |f(t)[\phi^2 - \eta^2]| \\ &\le |f(t)|[\|\phi\| + \|\eta\|]\|\phi - \eta\| \\ &\le 2\delta|f(t)|Le^{-\alpha(t-\gamma)}. \end{aligned}$$

Thus, the part of $|(A\phi)(t) - (A\eta)(t)|$ from Q offers no difficulty, while the part from G satisfies

$$\int_0^t e^{-2\alpha(t-s)} 2\delta|f(s)|Le^{-\alpha(s-\gamma)}\,ds \le 2\delta e^{\alpha\gamma}L\int_0^t e^{-2\alpha(t-s)}|f(s)|e^{-\alpha s}\,ds,$$

yielding the simple requirement that the last integral be bounded, which is an averaging condition on f.

Thus, to ensure that A is continuous we ask that for each $\varepsilon > 0$ there is a $\delta > 0$ such that $\phi_1, \phi_2 \in \mathcal{M}$ with $\|\phi_1 - \phi_2\| < \delta$ imply that

$$|G(t,\phi_1) - G(t,\phi_2)| \le \varepsilon f(t)(\|\phi_1\| + \|\phi_2\|) \tag{2.7.10}$$

and

$$\int_0^t e^{-2\alpha(t-s)} f(s)e^{-\alpha(s-\gamma)}\,ds \tag{2.7.11}$$

is bounded, another averaging condition on f.

A maps $\mathcal{M}$ into an equicontinuous set

Finally, to ensure that the operator A smooths, we may note that $(A\phi)(t)$ has a bounded derivative provided that

$$f(t)h(Le^{-\alpha(t-\gamma)})Le^{-\alpha(t-\gamma)} \text{ is bounded .} \tag{2.7.12}$$

This is not an averaging condition, but now we have the factor $h(Le^{-\alpha(t-\gamma)})$ which can go to zero very quickly.

Theorem 2.7.1. *If (2.7.6)-(2.7.12) hold then for L and ψ sufficiently small there is a solution $x(t,0,\psi) \in \mathcal{M}$ which satisfies (2.7.2) and $x(t) - Q(t,x_t)$ is continuously differentiable. In particular, $x(t) \to 0$.*

Proof. First, we will show that for L and ψ small enough, if $\phi, \eta \in \mathcal{M}$ then $A\phi + B\eta \in \mathcal{M}$. We have

$$\begin{aligned}
|(B\phi)(t)| &\leq \beta(q(t)\|\phi_t\|)\|\phi_t\| \\
&\quad + Ke^{-2\alpha t}[|\psi(0)| + \beta(q(0)\|\psi\|)\|\psi\|] \\
&\leq \beta(Lq(t)e^{-\alpha(t-\gamma)})Le^{-\alpha(t-\gamma)} \\
&\quad + Ke^{-2\alpha t}\|\psi\|[1 + \beta(q(0)\|\psi\|)] \\
&\leq \beta(Lq^*)Le^{-\alpha(t-\gamma)} \\
&\quad + Ke^{-2\alpha t}\|\psi\|[1 + \beta(q(0)\|\psi\|)] \\
&\leq \frac{1}{3}Le^{-\alpha t}
\end{aligned}$$

if L and $\|\psi\|$ are sufficiently small since $\beta(0) = 0$, β is continuous, and $q(t)e^{-\alpha(t-\gamma)} < q^*$.

Find K^* with $|S| \leq K^*$. Then $\phi \in \mathcal{M}$ implies that

$$\begin{aligned}
|(A\phi)(t)| &\leq K\int_0^t e^{-2\alpha(t-s)}[K^*\beta(q(s)\|\phi_s\|)\|\phi_s\| \\
&\quad + f(s)h(\|\phi_s\|)\|\phi_s\|]\,ds \\
&\leq K\int_0^t e^{-2\alpha(t-s)}[K^*\beta(Lq^*)Le^{\gamma\alpha}e^{-\alpha s} \\
&\quad + f(s)h(Le^{\alpha\gamma-\alpha s})Le^{\alpha\gamma}e^{-\alpha s}]\,ds \\
&= KK^*\beta(Lq^*)Le^{\alpha\gamma}e^{-\alpha t}\int_0^t e^{-\alpha(t-s)}\,ds \\
&\quad + KLe^{\alpha\gamma}e^{-\alpha t}\int_0^t e^{-\alpha(t-s)}f(s)h(Le^{-\alpha(s-\gamma)})\,ds \\
&\leq \frac{1}{3}Le^{-\alpha t}
\end{aligned}$$

if L is sufficiently small since $\beta(0) = h(0) = 0$, both are continuous, and (2.7.9) holds.

Next, we can show that $A\mathcal{M}$ is equicontinuous by noting that $\frac{d}{dt}(A\phi)(t)$ is bounded.

Also, we can use (2.7.6), (2.7.8), and (2.7.10) to prove that A is continuous.

Now $\mathcal{M}$ resides in the Banach space of bounded continuous functions with the supremum norm on the interval $[-\gamma, \infty)$. To see that B is a contraction, if $\phi, \eta \in \mathcal{M}$ then

$$\begin{aligned}|(B\phi)(t) - (B\eta)(t)| &\le |Q(t, \phi_t) - Q(t, \eta_t)| \\ &\le \beta(q(t)[\|\phi_t\| + \|\eta_t\|])\|\phi_t - \eta_t\| \\ &\le \beta(2q(t)Le^{-\alpha(t-\gamma)})\|\phi_t - \eta_t\| \\ &\le \beta(2q^*L)\|\phi_t - \eta_t\| \\ &\le d\|\phi - \eta\|_{[-\gamma,\infty)}\end{aligned}$$

for some $d < 1$ if L is small.

The conditions of Krasnoselskii's theorem are satisfied and there is a $\phi \in \mathcal{M}$ with $A\phi + B\phi = \phi$.

Example 1.7.1. Consider once more the equation

$$x'' + (\cos x)x' + \sin x = \frac{d}{dt}t^3x^2(t - r(t)) + t^5x^3(t - r(t)) \qquad (E_2.)$$

Theorem 2.7.1 was constructed so that the conditions were all based on the order of magnitude of terms, thereby avoiding tedious computations. This equation does satisfy the conditions of Theorem 2.7.1. We will write it as a system and separate the terms. First, take $\gamma = 1$ so that we have $0 \le r(t) \le 1$. Next, write

$$\begin{aligned}x' &= y \\ y' &= -\sin x - (\cos x)y + \frac{d}{dt}t^3x^2(t - r(t)) + t^5x^3(t - r(t))\end{aligned}$$

or

$$\begin{aligned}\begin{pmatrix} x' \\ y' \end{pmatrix} = \begin{pmatrix} 0 & 1 \\ -1 & -1 \end{pmatrix}\begin{pmatrix} x \\ y \end{pmatrix} + \begin{pmatrix} 0 \\ \frac{d}{dt}t^3x^2(t - r(t)) \end{pmatrix} \\ + \begin{pmatrix} 0 \\ x - \sin x + y(1 - \cos x) + t^5x^3(t - r(t)) \end{pmatrix}.\end{aligned}$$

It is an elementary exercise to show that these functions satisfy the conditions of Theorem 2.7.1. We showed how to obtain the continuity requirements in the development of (2.7.6) and (2.7.10).

We now turn to the scalar case which goes beyond the vector case in two ways. First, we add a delay term so that the basic equation being perturbed is a delay equation. Next, stability of the equation being perturbed is shown by an averaging process which averages in two ways.

Let r and γ be positive constants, $(\mathcal{C}, \|\cdot\|)$ be the Banach space of continuous functions $\phi : [-\gamma, 0] \to R$, $\mathcal{C}_H$ the H-ball in $\mathcal{C}$ for some $H > 0$, and let $Q, G : [0, \infty) \times \mathcal{C}_H \to R$ be continuous. Suppose also that $a, b : [0, \infty) \to R$ are continuous and define a function c by

$$2c(t) := a(t) + b(t+r), \quad t \geq 0,$$

$$c(t) = 0, \quad t < 0.$$

Thus, c is the average of $a(t)$ and $b(t+r)$.

Consider the equation

$$x'(t) = -a(t)x(t) - b(t)x(t-r) + \frac{d}{dt}Q(t, x_t) + G(t, x_t) \tag{2.7.13}$$

which we can write as

$$\begin{aligned}
&\frac{d}{dt}[x - \int_{t-r}^{t} b(u+r)x(u)\,du - Q(t, x_t)] \\
&- [a(t) + b(t+r)][x(t) - \int_{t-r}^{t} b(u+r)x(u)\,du - Q(t, x_t)] \\
&- [a(t) + b(t+r)][\int_{t-r}^{t} b(u+r)x(u)\,du + Q(t, x_t)] + G(t, x_t).
\end{aligned}$$

For an initial function ψ, by the variation of parameters formula we can write this as

$$\begin{aligned}
x(t) = &\int_{t-r}^{t} b(u+r)x(u)\,du + Q(t, x_t) \\
&+ e^{-2\int_0^t c(s)\,ds}[\psi(0) - \int_{-r}^{0} b(u+r)\psi(u)\,du - Q(0, \psi)] \\
&- \int_0^t e^{-\int_s^t 2c(u)\,du} 2c(s)[\int_{s-r}^{s} b(u+r)x(u)\,du + Q(s, x_s)]\,ds \\
&+ \int_0^t e^{-\int_s^t 2c(u)\,du} G(s, x_s)\,ds.
\end{aligned}$$

Let

$$\int_0^\infty c(u)\,du = \infty \text{ and } c(t) \geq 0 \tag{2.7.14}$$

and for some $L > 0$ let

$$\mathcal{M} = \{\phi : [-\gamma, \infty) \to R : \phi_0 = \psi, |\phi(t)| \leq Le^{-\int_0^t c(s)\,ds} \text{ for } t \geq 0\}. \tag{2.7.15}$$

For $\phi \in \mathcal{M}$ define operators A and B by

$$
\begin{aligned}
(B\phi)(t) &:= \int_{t-r}^{t} b(u+r)\phi(u)\,du + Q(t,\phi_t) \\
&+ e^{-2\int_0^t c(s)\,ds}\big[\psi(0) - \int_{-r}^{0} b(u+r)\psi(u)\,du - Q(0,\psi)\big] \\
&- \int_0^t e^{-\int_s^t 2c(u)\,du} 2c(s)\big[\int_{s-r}^{s} b(u+r)\phi(u)\,du + Q(s,\phi_s)\big]\,ds \qquad (2.7.16)
\end{aligned}
$$

and

$$
(A\phi)(t) = \int_0^t e^{-\int_s^t 2c(u)\,du} G(s,\phi_s)\,ds. \qquad (2.7.17)
$$

To get a bound on the first term in B we suppose there are positive constants α_1 and J with

$$
\begin{aligned}
&\sup_{t\geq 0} \int_{t-r}^{t} |b(s+r)| e^{\int_s^t c(u)du}\,ds \leq \alpha_1 \\
&\text{and} \quad \sup_{t\geq 0} \int_{t-\gamma}^{t} c(u)\,du \leq J. \qquad (2.7.18)
\end{aligned}
$$

To prepare the second term of B for a contraction argument we suppose there is a continuous increasing function $\beta : [0,\infty) \to [0,\infty)$, a continuous function $q : [0,\infty) \to [0,\infty)$, and a continuous increasing function $\alpha_2(L)$ such that for $\phi, \eta \in \mathcal{M}$ we have

$$
\begin{aligned}
&\beta\big(2q(t)Le^{-\int_0^{t-\gamma} c(s)\,ds}\big) \leq \alpha_2(L) \\
&|Q(t,\phi_t) - Q(t,\eta_t)| \leq \beta(q(t)[\|\phi_t\| + \|\eta_t\|])\|\phi_t - \eta_t\|, \\
&Q(t,0) = 0. \qquad (2.7.19)
\end{aligned}
$$

To ensure that $A : \mathcal{M} \to \mathcal{M}$ we suppose there is a continuous function $f : [0,\infty) \to [0,\infty)$ and a continuous function $h : [0,H] \to [0,\infty)$ which is increasing and for which $\phi \in C_H, t \geq 0$ imply that

$$
|G(t,\phi)| \leq f(t)h(\|\phi\|)\|\phi\|, \;\; h(0) = 0, \qquad (2.7.20)
$$

and that

$$
\int_0^t e^{-\int_s^t c(u)\,du} f(s)h(Le^{-\int_0^s c(u)du})\,ds \to 0 \qquad (2.7.21)
$$

as $L \to 0$ uniformly for $0 \leq t < \infty$.

To prove continuity of A we ask that for each $\varepsilon > 0$ there is a $\delta > 0$ such that $\phi_1, \phi_2 \in \mathcal{C}_H$ with $\|\phi_1 - \phi_2\| < \delta$ imply that

$$|G(t,\phi_1) - G(t,\phi_2)| \le \varepsilon f(t)(\|\phi_1\| + \|\phi_2\|). \tag{2.7.22}$$

In addition, let

$$\int_0^t e^{-\int_s^t 2c(u)du} f(s) e^{\int_0^{s-\gamma} c(u)\,du}\, ds \tag{2.7.23}$$

be bounded.

Theorem 2.7.2. *Suppose that (2.7.14) and (2.7.18) to (2.7.23) hold. Let $L^* > 0$ be fixed with*

$$\mu := 2\alpha_1 + 3\alpha_2(L) \sup_{t\ge 0} e^{\int_{t-\gamma}^t c(s)\,ds} < 1$$

and

$$f(t)h(Le^{-\int_0^{t-\gamma} c(u)\,du})Le^{-\int_0^{t-\gamma} c(u)\,du} \qquad \text{be bounded}$$

if $L \le L^*$. Then for ψ sufficiently small, (2.7.13) has a solution in $\mathcal{M}$.

The proof will be given in a series of lemmas. For $\phi, \eta \in \mathcal{M}$ and for $-\gamma \le \lambda_1 < \lambda_2 \le \infty$ we let $\|\phi - \eta\|_{[\lambda_1,\lambda_2]} = \sup_{\lambda_1 \le t \le \lambda_2} |\phi(t) - \eta(t)|$. Note that $(\mathcal{M}, \|\cdot\|_{[-\gamma,\infty)})$ is a complete metric space.

Lemma 1. *If (2.7.19) holds and if for $L \le L^*$ we have*

$$\mu = 2\sup_{t\ge 0} \int_{t-r}^t |b(u+r)|\,du + 3\alpha_2(L) < 1$$

then B is a contraction on $\mathcal{M}$.

Proof. For $\phi, \eta \in \mathcal{M}$ we have

$$\begin{aligned} |(B\phi)(t) - (B\eta)(t)| \le & \int_{t-r}^t |b(u+r)|\,du \|\phi_t - \eta_t\| \\ & + |Q(t,\phi_t) - Q(t,\eta_t)| \\ & + \sup_{s\ge 0} \int_{s-r}^s |b(u+r)||\phi(u) - \eta(u)|\,du \\ & + \sup_{0\le s\le t} |Q(s,\phi_s) - Q(s,\eta_s)|. \end{aligned}$$

Note that

$$\begin{aligned}|Q(t,\phi_t)-Q(t,\eta_t)| &\le \beta(q(t)[\|\phi_t\|+\|\eta_t\|])\|\phi_t-\eta_t\|\\ &\le \beta(2q(t)Le^{-\int_0^{t-\gamma}c(s)\,ds})\|\phi_t-\eta_t\|\\ &\le \alpha_2(L)\|\phi-\eta\|_{[-\gamma,\infty)}.\end{aligned}$$

Thus,

$$\begin{aligned}&|(B\phi)(t)-(B\eta)(t)| \le 2\sup_{t\ge 0}\int_{t-r}^{t}|b(u+r)|\,du\|\phi-\eta\|_{[-\gamma,\infty)}\\ &+2\alpha_2(L)\|\phi_t-\eta_t\|_{[-\gamma,\infty)}\\ &\le [2\sup_{t\ge 0}\int_{t-r}^{t}|b(u+r)|\,du+2\alpha_2(L)]\|\phi-\eta\|_{[-\gamma,\infty)}\\ &\le \mu\|\phi-\eta\|_{[-\gamma,\infty)}.\end{aligned}$$

Lemma 2. *If G satisfies (2.7.20) to (2.7.22) and if*

$$f(t)h(Le^{-\int_0^{t-\gamma}c(u)\,du})Le^{-\int_0^{t-\gamma}c(u)du}+c(t)e^{-\int_0^t c(s)ds}$$

is bounded for $L\le L^*$, then A maps $\mathcal{M}$ into an equicontinuous set.

We note that A has a bounded derivative on $\mathcal{M}$ when we see at the end of the proof of Lemma 4 that $|(A\phi)(t)|\le Le^{-\int_0^t c(s)ds}$.

Lemma 3. *If (2.7.18) and (2.7.19) hold then for L and ψ sufficiently small, $\phi\in\mathcal{M}$ implies that*

$$|(B\phi)(t)|\le \mu^* Le^{-\int_0^t c(s)\,ds}\quad \text{for}\quad \mu^*<1.$$

Proof. Taking $\Gamma(\psi)=e^{-2\int_0^t c(s)\,ds}[|\psi(0)|+\int_{-r}^0|b(u+r)\psi(u)|+|Q(0,\psi)|]$, we have

$$\begin{aligned}&|(B\phi)(t)|\\ &\le 2\sup_{t\ge 0}\int_{t-r}^{t}|b(s+r)|Le^{-\int_0^s c(u)\,du}\,ds+\alpha_2(L)Le^{-\int_0^{t-\gamma}c(s)\,ds}\\ &+\Gamma(\psi)+\int_0^t e^{-2\int_s^t c(u)\,du}2\alpha_2(L)c(s)\|\phi_s\|\,ds.\end{aligned}$$

The last term is bounded by

$$\int_0^t e^{-2\int_s^t c(u)\,du} 2Lc(s)\alpha_2(L)e^{-\int_0^{s-\gamma} c(u)\,du}\,ds$$
$$\le \int_0^t e^{-\int_s^t c(u)\,du} 2Lc(s)\alpha_2(L)e^{\int_{s-\gamma}^s c(u)\,du}\,ds\,e^{-\int_0^t c(u)\,du}.$$

From this we have

$$\begin{aligned}|(B\phi)(t)| &\le 2L\sup_{t\ge0}\int_{t-r}^t |b(u+r)|e^{\int_s^t c(u)\,du}\,ds\,e^{-\int_0^t c(u)\,du}\\ &+\alpha_2(L)Le^{-\int_0^{t-\gamma} c(s)\,ds}\\ &+\alpha_2(L)Le^{-\int_0^t c(u)\,du}\int_0^t 2c(s)e^{-\int_s^t c(u)\,du}e^{\int_{s-\gamma}^s c(u)\,du}\,ds\\ &+\Gamma(\psi)\\ &\le (2\alpha_1+3\alpha_2(L)\sup_{t\ge0}e^{\int_{t-\gamma}^t c(u)\,du})Le^{-\int_0^t c(s)\,ds}+\Gamma(\psi)\\ &\le \mu Le^{-\int_0^t c(s)\,ds}+\Gamma(\psi).\end{aligned}$$

Thus, if $L\le L^*$ and ψ is sufficiently small, then

$$|(B\phi)(t)|\le \mu^* Le^{-\int_0^t c(s)\,ds}$$

where $\mu^*<1$.

Lemma 4. *If (2.7.20) - (2.7.22) hold and if $0<\theta<1$, then there is an $L_1>0$ such that $\phi\in\mathcal{M}$ and $L\le L_1$ implies that $|(A\phi)(t)|\le\theta Le^{-\int_0^t c(s)\,ds}$. Thus, for L small enough and under the conditions of Lemma 3, if $\phi,\eta\in\mathcal{M}$ then $A\phi+B\eta\in\mathcal{M}$.*

Proof. We have

$$\begin{aligned}|(A\phi)(t)| &\le \int_0^t e^{-\int_s^t 2c(u)\,du} f(s)h(\|\phi_s\|)\|\phi_s\|\,ds\\ &\le \int_0^t e^{-\int_s^t 2c(u)\,du} f(s)h(Le^{-\int_0^{s-\gamma} c(u)\,du})Le^{-\int_0^{s-\gamma} c(u)\,du}\,ds\\ &\le \int_0^t e^{-\int_s^t 2c(u)\,du} f(s)h(Le^{-\int_0^s c(u)\,du}e^{\int_{s-\gamma}^s c(u)\,du})\\ &\times Le^{-\int_0^s c(u)\,du}e^{\int_{s-\gamma}^s c(u)\,du}\,ds\\ &\le \int_0^t e^{-\int_s^t c(u)\,du} f(s)h(Le^J e^{-\int_0^s c(u)\,du})e^J\,ds\,Le^{-\int_0^t c(s)\,ds}\\ &\le \theta Le^{-\int_0^t c(s)\,ds}\end{aligned}$$

if $L \leq L_1 \leq L^*$ is chosen so that

$$\int_0^t e^{-\int_s^t c(u)\,du} f(s) h(Le^J e^{-\int_0^s c(u)\,du}) e^J \, ds < \theta$$

by (2.7.21).

When we take $\mu^* + \theta < 1$ the conditions of Krasnoselskii's theorem are satisfied and there is a $\phi \in \mathcal{M}$ with $A\phi + B\phi = \phi$, a solution of (2.7.13).

This will acknowledge that much of the material of this section was first published in *Nonlinear Analysis*, as detailed in Burton (2003b), which is published by Elsevier Science.

2.8 Seifert's Example of Memory

Seifert (1973) presented a very simple example of a functional differential equation having a memory which did not fade. That, therefore, prevented the zero solution from being asymptotically stable. It was a stern warning and it influenced the research in asymptotic stability. In this section we take a step in the direction of proving asymptotic stability independent of the memory.

Many processes of interest possess a fading memory. Some studies show that a child of taller than average parents tends to be taller than average, but not as much taller than average than the parents. The process continues with grandchildren being taller than average, but approaching average height with each successive generation. Such ideas are prime examples of fading memory. Many elastic materials have memory. The Levin equation discussed earlier is used to describe one-dimensional viscoelasticity. Torsion in a wire is an example of memory discussed by Picard almost one hundred years ago. Volterra spent much of his life studying memory in biological problems.

A delay in an equation represents a memory. In

$$x'(t) = -x(t) + x(t-2)$$

the behavior of the solution is forcefully determined by all values of $x(s)$ for $t - 2 \leq s \leq t$ and the earlier history is totally forgotten, aside from any implications of continuity of solutions in initial conditions. Seldom does real world memory behave in that way. The effect of past behavior is closely related to the remoteness of that past. The memory in the last equation does not fade, it seemingly vanishes. A more realistic idea is shown in

$$x'(t) = \int_{t-r}^{t} e^{-(t-s)} x(s)\, ds.$$

At $s = t$ there is perfect memory. At $s = t - r$ then the integrand has value $e^{-r}x(t-r)$. If r is large, then little of the behavior of $x(t-r)$ is remembered at time t. While this would seem to be more realistic, it still is unreasonable that all behavior of $x(s)$ for $s < t - r$ should be totally forgotten. The memory fades in a realistic way, but then it vanishes. Even during the roaring stock market of the 1990's, few investors totally forgot the crash of 1929. Finally, if C is an L^1-function, then

$$x'(t) = \int_{-\infty}^{t} C(t-s)x(s)\, ds$$

offers a very reasonable fading memory. For any fixed $r > 0$, write

$$\int_{-\infty}^{t} C(t-s)x(s)\, ds \doteq \int_{-\infty}^{r} C(t-s)x(s)\, ds + \int_{r}^{t} C(t-s)x(s)\, ds$$

and notice that

$$\int_{-\infty}^{r} C(t-s)x(s)\, ds = \int_{t-r}^{\infty} C(u)x(t-u)\, du.$$

Thus, if x is a bounded solution then that integral fades to zero as $t \to \infty$. The solution slowly forgets everything which has happened long ago, but the memory is never totally gone.

In this section we will study an equation

$$x' = -a(t)x^3(t) + b(t)x^3(t - r(t)) \tag{2.8.1}$$

which will feature many of the difficulties encountered in Liapunov's direct method. It will also feature difficulties encountered in questions of contractions. Finally, it will illustrate that our ideas concerning the necessity of a fading memory for asymptotic stability are not always correct. The results of this section are found in Burton (2002).

This equation has been discussed under a variety of assumptions and we will present two such discussions here as a contrast to our work. Suppose that r is a positive constant and that

$$b(t) \text{ is bounded and } a(t) \geq |b(t+r)| + k$$

for some $k > 0$. If we define the Liapunov functional

$$V(t,x_t) = |x(t)| + \int_{t-r}^{t} |b(s+r)||x^3(s)|\,ds$$

then a calculation yields

$$V'(t,x_t) \leq -k|x^3|.$$

An argument of Burton (1978) will yield asymptotic stability.

An alternative is to ask that

$$-a(t) + (1/4)|b(t)|^2 + 1 \leq -k$$

for some $k > 0$. This time we define a Liapunov functional as

$$V(t,x) = (1/4)x^4(t) + \int_{t-r}^{t} x^6(s)ds$$

and obtain

$$V'(t,x_t) \leq -kx^6$$

from which we can again conclude uniform asymptotic stability. The two Liapunov functions and the two sets of conditions are astonishingly different and one begins to see why this equation attracts such wide interest. Formidable challenges are presented when r becomes a function of t.

These are both global results, while our upcoming fixed point results are local under a different set of conditions. Thus, once again we see a balance between the two methods. The union is better than either separately.

We now proceed to a fixed point solution. For that we need a mapping and will depend on the variation of parameters formula. Since (2.8.1) has no linear term we resort to adding and subtracting a linear term. In principle one believes that this should not work because the extraneous linear term would strongly dominate these nonlinear terms near zero, effectively destroying the original stability properties. In fact, that does not happen and the problem seems made to order for the following fixed point theorem which we discussed in Chapter 1. We review a definition, an example, and two theorems from Chapter 1 for reference here.

Definition 1.2.6. *Let $(\mathcal{M},\rho)$ be a metric space and $B : \mathcal{M} \to \mathcal{M}$. B is said to be a large contraction if ϕ, $\psi \in \mathcal{M}$, with $\phi \neq \psi$ then $\rho(B\phi, B\psi) < \rho(\phi,\psi)$ and if for each $\varepsilon > 0$ there exists $\delta < 1$ such that $[\phi,\psi \in \mathcal{M}, \rho(\phi,\psi) \geq \varepsilon] \Rightarrow \rho(B\phi, B\psi) \leq \delta\rho(\phi,\psi)$.*

Theorem 1.2.4. *Let $(\mathcal{M}, \rho)$ be a complete metric space and B be a large contraction. Suppose there is an $x \in \mathcal{M}$ and an $L > 0$, such that $\rho(x, B^n x) \leq L$ for all $n \geq 1$. Then B has a unique fixed point in $\mathcal{M}$.*

Example 1.2.7. If $\|\cdot\|$ is the supremum metric, if

$$\mathcal{M} = \{\phi : [0, \infty) \to R | \phi \in \mathcal{C}, \|\phi\| \leq \sqrt{3}/3\},$$

and if $(H\phi)(t) = \phi(t) - \phi^3(t)$, then H is a large contraction of the set $\mathcal{M}$.

Theorem 1.2.8. *Let $(\mathcal{S}, \|\cdot\|)$ be a Banach space, $\mathcal{M}$ a bounded convex nonempty subset of $\mathcal{S}$. Suppose that A, $B : \mathcal{M} \to \mathcal{M}$ and that*

(i) $x, y \in \mathcal{M} \Rightarrow Ax + By \in \mathcal{M}$,
(ii) *A is continuous and $A\mathcal{M}$ is contained in a compact subset of $\mathcal{M}$,*
(iii) *B is a large contraction.*

Then there exists $y \in \mathcal{M}$ with $Ay + By = y$.

Consider again

$$x' = -a(t)x^3 + b(t)x^3(t - r(t)) \tag{2.8.1}$$

in which $r(t) \geq 0$, the functions a, b, r are continuous, and there is a $J > 1$ with

$$J|b(t)| \leq a(t) \text{ and } \int_0^\infty a(s)\, ds = \infty. \tag{2.8.2}$$

In order to show that solutions of (2.8.1) with small initial functions tend to zero using Theorem 1.2.8, we define a mapping equation by first writing

$$x' + a(t)x = a(t)x - a(t)x^3(t) + b(t)x^3(t - r(t)) \tag{2.8.3}$$

and then using the variation of parameters formula to write

$$\begin{aligned} x(t) = x_0 e^{-\int_0^t a(s)\,ds} &+ \int_0^t e^{-\int_s^t a(u)\,du} a(s)[x(s) - x^3(s)]\, ds \\ &+ \int_0^t e^{-\int_s^t a(u)\,du} b(s)x^3(s - r(s))\, ds, \end{aligned} \tag{2.8.4}$$

where we take the first two terms on the right as Bx and the last as Ax, so that (2.8.4) is expressed as the operator equation

$$x = Bx + Ax. \tag{2.8.5}$$

We will define B and A more precisely in a moment. There is an assumed continuous initial function ψ on some initial interval $[-D, 0]$ with

$\psi(0) = x_0$. Here, $[-D, 0] = \{u \leq 0 | u = t - r(t), t \geq 0\}$. A solution is then denoted by $x(t, 0, \psi)$ which is continuous, agrees with ψ on the initial interval, satisfies (2.8.1) for $t \geq 0$, and which may have a discontinuity in its derivative whenever $t - r(t) = 0$, as discussed later.

Definitions of $\mathcal{S}$, $\mathcal{M}$, A, and B

We will do the work in two steps. First, we show that solutions starting in a certain set tend to zero. That work will use the supremum norm and will assume that $b(t)/a(t) \to 0$ as $t \to \infty$ which allows us to show that a certain set is compact. Then we will work in a space with weighted norm and prove a Liapunov stability result without that condition on $b(t)/a(t)$ since compactness will rest on the equi-continuity.

With Theorem 1.2.8 in mind we let $\mathcal{S}$ be the Banach space of bounded continuous functions $\phi : [0, \infty) \to R$ with the supremum norm $\|\cdot\|$ and define the set

$$\mathcal{M} = \{\phi \in \mathcal{S} \,|\, |\phi(t)| \leq L, |\phi(t)| \to 0 \text{ as } t \to \infty\} \tag{2.8.6}$$

where $L = \sqrt{3}/3$. Denote the initial function by ψ and its maximum on $[-D, 0]$ by $\|\psi\|$, which should not cause confusion with the same symbol denoting the norm in S. Then we require that ψ be chosen so that

$$\|\psi\| + (2\sqrt{3}/9) + (\sqrt{3}/[9J]) \leq \sqrt{3}/3. \tag{2.8.7}$$

We will use this relation in Lemma 2 and define a $\delta > 0$ so that if $\varepsilon = \sqrt{3}/3$ and if $\|\psi\| < \delta$, then the solution satisfies $|x(t, 0, \psi)| < \varepsilon$. We treat the more general $\varepsilon - \delta$ stability relation later.

Now, with $\mathcal{M}$ defined, if $\phi \in \mathcal{M}$ and if $x_0 = \psi(0)$ then

$$(B\phi)(t) = x_0 e^{-\int_0^t a(s)\,ds} + \int_0^t e^{-\int_s^t a(u)\,du} a(s)(\phi(s) - \phi^3(s))\,ds. \tag{2.8.8}$$

Note that B would have been a contraction for small ϕ had $a(t)\phi(t)$ not been present. But we will show that B is a large contraction for small ϕ. The reader should take careful note of the changes in contraction properties; this is exactly the typical occurrence when a linear term is added to both sides of an equation which had a *stable* local contraction before the addition.

Next, for $\phi \in M$, extend ϕ to $[-D, \infty)$ by defining $\phi(t) = \psi(t)$ on $[-D, 0)$ so that

$$(A\phi)(t) = \int_0^t e^{-\int_s^t a(u)\,du} b(s)\phi^3(s - r(s))\,ds \tag{2.8.9}$$

is defined for $t \geq 0$.

If there is a fixed point ϕ for the mapping $P\phi = A\phi + B\phi$, then $x(t,0,\psi) = \phi(t)$ for $t \geq 0$, $x(t,0,\psi) = \psi(t)$ on $[-D,0]$, and $x(t,0,\psi)$ satisfies (2.8.1) for $t > 0$, whenever its derivative exists.

NOTE. For functional differential equations we always expect a discontinuity in the derivative of the continuous solution at $t = t_0$; but here we would expect a possible discontinuity in the derivative of the solution at each point where

$$t - r(t) = 0. \tag{2.8.10}$$

We say *possible discontinuity* because $r(t) = t$ for all t, for example, would not yield discontinuities for $t > 0$.

Fulfillment of (i), (ii), (iii) in Theorem 1.2.8

Here is our main result. It will require the additional assumption that

$$b(t)/a(t) \to 0 \text{ as } t \to \infty. \tag{2.2.11}$$

Earlier we had discussed this central requirement and we now add some substance to it. Seifert (1973) studies asymptotic stability of functional differential equations by means of Razumikhin techniques which can be very fruitful in proving Liapunov stability for problems in the general class of (2.8.1). But he warns of the difficulties in proving asymptotic stability when $t - r(t)$ fails to tend to infinity; and his main result on asymptotic stability asks that along each solution the main action of the delay be on $[t-r,t]$ for r constant. He has an example of great difficulties encountered when $r(t) = t$. If we look at a linear problem parallel to (2.8.1) with $r(t) = t$ we would have

$$x' = -a(t)x + b(t)x(0)$$

with solution

$$\begin{aligned} x(t) &= x(0)e^{-\int_0^t a(s)\,ds} + \int_0^t e^{-\int_s^t a(u)\,du} b(s)x(0)\,ds \\ &= x(0)e^{-\int_0^t a(s)\,ds} + x(0)\int_0^t e^{-\int_s^t a(u)\,du} a(s)(b(s)/a(s))\,ds. \end{aligned}$$

It is readily shown that $x(t) \to 0$ as $t \to \infty$ in case $b(t)/a(t) \to 0$ as $t \to \infty$. On the other hand, if $a(t)/b(t)$ is bounded away from zero, then $x(t)$ does not tend to zero. Hence, (2.8.11) is critical to our proof and it is critical to the validity of the theorem. There are cases in between: we might get by with $\liminf_{t\to\infty} b(t)/a(t) = 0$.

This is more than simply a justification of the condition. All of this work was motivated by the difficulties encountered in studying this problem via Liapunov's direct method. In that method the real difficulties occur because of unboundedness of the delay and the properties of the derivative of $r(t)$; those difficulties with the derivative are not seen at all in our work here.

Theorem 2.8.1. *Let (2.8.2) and (2.8.11) hold. If $L = \sqrt{3}/3$ and if ψ is the initial function satisfying (2.8.7), $\psi \in C$, then there is a solution $x(t,0,\psi)$ of (2.8.1) with $|x(t,0,\psi)| < L$ for $t \geq 0$ and $x(t,0,\psi) \to 0$ as $t \to \infty$.*

The proof is based on four lemmas.

Lemma 1. *For A defined in (2.8.5), if $\phi \in \mathcal{M}$ then $|(A\phi)(t)| \leq L^3/J \leq L$. Moreover, $(A\phi)(t) \to 0$ as $t \to \infty$.*

Proof. We have

$$\int_0^t e^{-\int_s^t a(u)\,du}|b(s)||\phi^3(s-r(s))|\,ds$$
$$\leq (L^3/J)\int_0^t e^{-\int_s^t a(u)\,du}a(s)\,ds \leq L^3/J,$$

as required.

Let $\phi \in M$ be fixed. We will show that $(A\phi)(t) \to 0$. For a given $\varepsilon > 0$ we can find T such that $|\phi^3(t-r(t))| < \varepsilon$ for $t \geq T$. We then have

$$\begin{aligned}|(A\phi)(t)| &\leq \int_0^T e^{-\int_s^t a(u)\,du}a(s)\,ds(L^3/J)\\ &\quad + \int_T^t (\varepsilon/J)e^{-\int_s^t a(u)\,du}a(s)\,ds\\ &\leq (L^3/J)e^{-\int_T^t a(u)du} + (\varepsilon/J).\end{aligned}$$

The result follows from this.

Lemma 2. *For A, B defined in (2.8.5) and ψ satisfying (2.8.7), if $y \in \mathcal{M}$ is fixed, but arbitrary, then the mapping $Bx + Ay : \mathcal{M} \to \mathcal{M}$ and B is a large contraction on $\mathcal{M}$ with a unique fixed point in $\mathcal{M}$.*

Proof. Using the definition of B, the result of Lemma 1, and the fact that $|x| \leq \sqrt{3}/3$ implies $|x - x^3| \leq (2\sqrt{3})/9$ we obtain

$$\begin{aligned}|Bx + Ay| &\leq |x_0| + (2\sqrt{3}/9)\int_0^t e^{-\int_s^t a(u)\,du}a(s)\,ds + L^3/J \\ &\leq \|\psi\| + (2\sqrt{3}/9) + (\sqrt{3}/[9J]) \leq \sqrt{3}/3\end{aligned}$$

by (2.8.8). Note that $0 \in \mathcal{M}$ so we have proved that $B : \mathcal{M} \to \mathcal{M}$.

To see that $B : \mathcal{M} \to \mathcal{M}$ is a large contraction on $\mathcal{M}$, we note that our example after Theorem 1.2.8 showed that $\phi - \phi^3$ is a large contraction within the integrand; when that term is taken outside the integral as a supremum, then the resulting integral is bounded by 1. Thus for the ε of the proof of that example, we have

$$\begin{aligned}\|B\phi_1 - B\phi_2\| &\leq \int_0^t e^{-\int_s^t a(u)\,du}a(s)\,ds\|\phi_1 - \phi_2\|\delta \\ &\leq \|\phi_1 - \phi_2\|\delta,\end{aligned}$$

as required.

We already know that $Ay \to 0$ as $t \to \infty$ and the proof that $Bx \to 0$ is just the same.

Lemma 3. *The mapping $A : \mathcal{M} \to \mathcal{M}$ is continuous in the supremum norm.*

Proof. If $\phi_1, \phi_2 \in M$ then there are positive constants J and K with

$$\begin{aligned}&|(A\phi_1)(t) - (A\phi_2)(t)| \\ &\quad = \left|\int_0^t e^{-\int_s^t a(u)\,du}b(s)[\phi_1^3(s - r(s)) - \phi_2^3(s - r(s))]\,ds\right| \\ &\quad \leq \|\phi_1 - \phi_2\|(1/J)\int_0^t e^{-\int_s^t a(u)du}a(s)\,ds \\ &\quad \leq K\|\phi_1 - \phi_2\|.\end{aligned}$$

Lemma 4. *The operator A maps $\mathcal{M}$ into a compact subset of $\mathcal{M}$.*

Proof. Let $\phi \in \mathcal{M}$ and let $0 \le t_1 < t_2$ so that

$$\begin{aligned}
&|(A\phi)(t_2) - (A\phi)(t_1)| \\
&\quad = \left| \int_0^{t_2} e^{-\int_s^{t_2} a(u)\,du} b(s)\phi^3(s-r(s))\,ds \right. \\
&\qquad \left. - \int_0^{t_1} e^{-\int_s^{t_1} a(u)\,du} b(s)\phi^3(s-r(s))\,ds \right| \\
&\quad = \left| \int_0^{t_1} e^{-\int_s^{t_1} a(u)\,du} b(s)\phi^3(s-r(s))[e^{-\int_{t_1}^{t_2} a(u)\,du} - 1]\,ds \right. \\
&\qquad \left. + \int_{t_1}^{t_2} e^{-\int_s^{t_2} a(u)\,du} b(s)\phi^3(s-r(s))\,ds \right| \\
&\quad \le (L^3/J) \int_0^{t_1} e^{-\int_s^{t_1} a(u)du} a(s)\,ds \left| e^{-\int_{t_1}^{t_2} a(u)\,du} - 1 \right| \\
&\qquad + (L^3/J) \left| e^{-\int_{t_1}^{t_2} a(u)\,du} - 1 \right| \\
&\quad \le 2(L^3/J) \left| e^{-\int_{t_1}^{t_2} a(u)\,du} - 1 \right|.
\end{aligned}$$

Hence, $A\mathcal{M}$ is equi-continuous. Next, we notice that for arbitrary $\phi \in \mathcal{M}$ we have

$$|(A\phi)(t)| \le \int_0^t e^{-\int_s^t a(u)\,du} a(s)|b(s)/a(s)|L^3\,ds =: c(t)$$

where $c(t) \to 0$ as $t \to \infty$ by a proof like that of Lemma 1, because of the assumption that $|b(t)|/a(t) \to 0$ as $t \to \infty$. This, and the equi-continuity will imply the conclusion (see Theorem 1.2.2).

The conditions of Theorem 1.2.8 are satisfied and there is a $\phi \in \mathcal{M}$ with $\phi = A\phi + B\phi$, a solution of (2.8.1).

NOTE. The set $A\mathcal{M}$ contains the solution and so from $c(t)$ we see that we actually prove equi-asymptotic stability. If $b(t)/a(t) \to 0$ in a certain uniform way, we could prove uniform asymptotic stability.

Stability and compactness

We have mentioned before that Razumikhin techniques can be very effective in proving Liapunov stability for problems like (2.8.1). While one may show that the zero solution of (2.8.1) is Liapunov stable using a Razumikhin technique, we know of no way to prove that solutions tend to zero except by the method presented. As we are trying to give a general discussion of how fixed point theory will yield stability, it is still worth the space to give a short discussion here.

If all we ask is stability (not asymptotic stability) then we can avoid (2.8.11) and still use Krasnoselskii's theorem. In that case we would achieve the compactness by using a different norm. Any C_g−space will work.

Let $g(t) = t+1$ for $t \geq 0$ and define $(S, |\cdot|_g)$ to be the Banach space of continuous functions $\phi : [0, \infty) \to R$ for which

$$|\phi|_g := \sup_{t \geq 0} |\phi(t)|/g(t)$$

exists. Continue to use $\|\cdot\|$ as the supremum norm of any $\phi \in S$, provided ϕ is bounded. Also, continue to use $\|\psi\|$ as the bound on an initial function, as before.

Theorem 2.8.2. *If the conditions of Theorem 2.8.1 hold, except for (2.8.11), then the zero solution of (2.8.1) is stable.*

Proof. We give the proof for solutions starting at $t_0 = 0$. Refer to (2.8.7) and the proof of Lemma 2. If $\varepsilon > 0$ is given with $0 < \varepsilon < \sqrt{3}/3$, then for $|x| \leq \varepsilon$, find a δ^* with $|x - x^3| \leq \delta^*$; then we will ask that

$$\|\psi\| + \delta^* + [\varepsilon^3/J] \leq \varepsilon.$$

To verify that the last inequality allows $\|\psi\| > 0$, note that for $0 \leq x \leq \epsilon < \sqrt{3}/3$, the function $x - x^3$ is increasing so $0 \leq x - x^3 \leq \varepsilon - \varepsilon^3 =: \delta^*$. This will yield a $\delta > 0$ so that $\|\psi\| < \delta$ is the requirement on the initial function.

We now construct a set

$$\mathcal{M} = \{\phi \in S \mid \|\phi\| \leq \varepsilon\}.$$

Define A and B as before. We easily verify that B is a large contraction on $\mathcal{M}$ and that $Ax + By : \mathcal{M} \to \mathcal{M}$, just as before. $A\mathcal{M}$ is an equi-continuous set and in the $g-norm$ it is contained in a compact subset of $\mathcal{M}$. Moreover, both A and B are continuous in the g−norm. The mapping has a fixed point satisfying (2.8.1) and lying in $\mathcal{M}$.

This will acknowledge that much of the material from this section was first published in *Nonlinear Studies*, as detailed in Burton (2002), which is published by I and S Publishers.

2.9 The Problem of Lurie

The classical problem of Lurie (1951) is given as an $n+1$-dimensional system of ordinary differential equations written as

$$\begin{aligned} x' &= Ax + bf(\sigma) \\ \sigma' &= c^T x - rf(\sigma) \end{aligned}$$

in which A is an $n \times n$ constant real matrix, b and c are real constant vectors, f and σ are scalars with $\sigma f(\sigma) > 0$ if $\sigma \neq 0$, and r is a positive constant. In that form it is discussed by Lefschetz and LaSalle (1961) where it is stated that one would like to guarantee asymptotic stability for more or less arbitrary functions $f(\sigma)$ (with $\sigma f(\sigma) > 0$) and any initial conditions. The entire book by Lefschetz (1965) is devoted to the problem. The literature now is enormous, as may be seen by consulting the online Mathematical Reviews. Much work is available only in the Chinese language.

Investigators are far more interested in the problem when there is a delay in the control of either pointwise or distributed type. Moreover, the connection b can be a function of both t and σ in realistic problems. In the three problems listed below the reader will see in the proofs that the same analysis can be given if the term $c^T x$ is replaced with $c^T x(t-H))$ for a nonnegative constant H. There is a *plant*, the x-component, and *control*, the σ component. It takes some time to tell the control the location of x and it takes some time to apply the control back to the plant. Early work on the system with a delay is found in Somolinos (1977) who uses a converse Liapunov theorem very effectively in that work.

Here, we will look at three problems:

$$\begin{aligned} x' &= Ax + bf(\sigma(t-L)) \\ \sigma' &= c^T x - rf(\sigma), \end{aligned} \tag{2.9.1}$$

$$\begin{aligned} x' &= Ax + \int_{t-L}^{t} b(s-t) f(\sigma(s))\, ds \\ \sigma' &= c^T x - rf(\sigma), \end{aligned} \tag{2.9.2}$$

and

$$\begin{aligned} x' &= Ax + \int_{-\infty}^{t} b(s-t) f(\sigma(s))\, ds \\ \sigma' &= c^T x - rf(\sigma). \end{aligned} \tag{2.9.3}$$

One may notice in the subsequent work that if we let the elements of $b(t)$ change sign, then sometimes the control is promoting stability and sometimes it hinders the stability, as so often does occur in real problems. The critical requirement is that "on average" stability is promoted. In the original problem investigators used a Liapunov functional whose derivative is negative in case $c^T A^{-1} b + r > 0$. Thus, one readily sees the effect on stability as b changes sign.

We begin with

$$\begin{aligned} x' &= Ax + bf(\sigma(t - L)) \\ \sigma' &= c^T x - rf(\sigma) \end{aligned} \tag{2.9.4}$$

in which A is an $n \times n$ real constant matrix all of whose characteristic roots have negative real parts, x is an n-vector, σ is a scalar, r and L are real positive constants, b and c are constant n-vectors.

In the present discussion it is assumed that

$$\sigma f(\sigma) > 0 \text{ if } \sigma \neq 0, \quad \lim_{\sigma \to 0} \frac{f(\sigma)}{\sigma} \quad \text{exists,} \tag{2.9.5}$$

there are positive constants k and β with

$$|f(x) - f(y)| \leq K|x - y| \quad \text{and} \quad (\beta/r)|\sigma| \leq |f(\sigma)|. \tag{2.9.6}$$

Thus, (2.9.6) states that our control function lies in a certain sector

$$(\beta/r)|\sigma| \leq |f(\sigma)| \leq K|\sigma|.$$

Finally, we suppose that there is an $\alpha < 1$ with

$$\sup_{t \geq 0} \int_0^t e^{-\beta(t-u)} K \int_0^u |c^T c^{A(u-s)} b| \, ds \, du \leq \alpha. \tag{2.9.7}$$

Theorem 2.9.1. *If (2.9.5)-(2.9.7) hold then the zero solution of (2.9.4) is asymptotically stable.*

Proof. Let $\psi : [-L, 0] \to R$ be a given continuous initial function and x_0 a given vector. This will specify a unique solution

$(x(t,0,x_0,\psi),\sigma(t,0,x_0,\psi)) =: (x_1(t),\sigma_1(t))$. For any solution starting at $t_0 = 0$, we have from (2.9.4) and the variation of parameters formula that

$$x(t) = e^{At}x_0 + \int_0^t e^{A(t-s)}bf(\sigma(s-L))\,ds. \tag{2.9.8}$$

Substituting this into the second equation in (2.9.4) yields

$$\sigma' = -rf(\sigma) + c^T\left[e^{At}x_0 + \int_0^t e^{A(t-s)}bf(\sigma(s-L))\,ds\right]. \tag{2.9.9}$$

Now, for the fixed solution $(x_1(t),\sigma_1(t))$, define

$$d(t) := r\frac{f(\sigma_1(t))}{\sigma_1(t)} \tag{2.9.10}$$

which is continuous and non-negative by (2.9.2). Thus, along that solution (2.9.9) may be written as

$$\sigma'(t) = -d(t)\sigma + c^T\left[e^{At}x_0 + \int_0^t e^{A(t-s)}bf(\sigma(s-L))\,ds\right]$$

so that by the variation of parameters formula we have

$$\begin{aligned}\sigma(t) &= \sigma(0)e^{-\int_0^t d(s)\,ds}\\ &+ \int_0^t e^{-\int_u^t d(s)\,ds}c^T\left[e^{Au}x_0 + \int_0^u e^{A(u-s)}bf(\sigma(s-L))\,ds\right]du.\end{aligned} \tag{2.9.11}$$

Let

$$H = \{\phi : [-L,\infty) \to R \mid \phi \in C, \phi_0 = \psi, \phi(t) \to 0 \text{ as } t \to \infty\}$$

and define $P : H \to H$ by $(P\phi)(t) = \psi(t)$ on $[-L,0]$ and

$$\begin{aligned}(P\phi)(t) &= \sigma(0)e^{-\int_0^t d(s)\,ds}\\ &+ \int_0^t e^{-\int_u^t d(s)\,ds}c^T\left[e^{Au}x_0 + \int_0^u e^{A(u-s)}bf(\phi(s-L))\,ds\right]du.\end{aligned}$$

Now $\int_0^u e^{A(u-s)}bf(\phi(s-L))\,ds$ is the convolution of an L^1-function with a function tending to zero so it tends to zero. Likewise, as $|f(\sigma)/\sigma| \geq \beta/r$, then $d(t) \geq \beta$. Hence,

$$\int_0^t e^{-\int_u^t d(s)\,ds}\int_0^u e^{A(u-s)}bf(\phi(s-L))\,ds\,du$$

also tends to zero as $t \to \infty$. In fact, $P : H \to H$.

Moreover, P is a contraction if (2.9.7) holds. Thus, P has a unique fixed point σ tending to zero. Going back to (2.9.8) we see that $x(t)$ also tends to zero. We could readily give a stability proof as well.

Remark 2.9.1. *Many generalizations are easily obtained. Clearly, c and b can be functions of t and even functions of σ. Also, f need not be the same in both equations; (2.9.5) is needed in the second equation, while (2.9.6) is needed in the first equation of the system (2.9.4). Such changes can be important. Changes of the type mentioned here would be fatal in the standard Liapunov theory arguments, but fixed point theory is much more flexible in this respect.*

Remark 2.9.2. *The sector condition*

$$(\beta/r)|\sigma| \leq |f(\sigma)| \leq K|\sigma|$$

is a direct result of the delay. The function $f(\sigma(t))$ in the second equation could be large or small at a particular t, but it must be balanced in $f(\sigma(t-L))$ by a term comparable to $f(\sigma(t))$. This is particularly important since the result is independent of L.

Next we consider the system

$$\begin{aligned} x' &= Ax + \int_{t-L}^{t} b(s-t)f(\sigma(s))\,ds \\ \sigma' &= c^T x - rf(\sigma) \end{aligned} \tag{2.9.12}$$

where A, c, r, and f are as before. Here, $b : R \to R^n$ is continuous and we ask that

$$\sup_{t\geq 0} \int_0^t e^{-\beta(t-u)}K \int_0^u |e^{A(u-v)} \int_{v-L}^{v} b(s-v)|\,dv\,du \leq \alpha < 1. \tag{2.9.13}$$

Theorem 2.9.2. *Let (2.9.5), (2.9.6), and (2.9.13) hold. Then the zero solution of (2.9.12) is asymptotically stable.*

Proof. For the given initial condition (x_0, ψ) we have the unique solution $(x_1(t), \sigma_1(t))$ as in the proof of Theorem 2.9.1. Then

$$x(t) = x_0 e^{At} + \int_0^t e^{A(t-u)} \int_{u-L}^{u} b(s-u)f(\sigma(s))\,ds\,du$$

and

$$\sigma' = c^T \left[x_0 e^{At} + \int_0^t e^{A(t-u)} \int_{u-L}^{u} b(s-u)f(\sigma(s))dsdu \right] - rf(\sigma).$$

Again, we define

$$d(t) = \frac{rf(\sigma_1(t))}{\sigma_1(t)} \tag{2.9.14}$$

so that along that fixed solution, we have

$$\sigma' = -d(t)\sigma + c^T \left[x_0 e^{At} + \int_0^t e^{A(t-u)} \int_{u-L}^u b(s-u) f(\sigma(s))\, ds\, du \right] \tag{2.9.15}$$

so

$$\begin{aligned} \sigma(t) &= \sigma(0) e^{-\int_0^t d(s)\, ds} \\ &+ \int_0^t e^{-\int_u^t d(s)\, ds} c^T \left[x_0 e^{Au} \right. \\ &+ \left. \int_0^u e^{A(u-v)} \int_{v-L}^v b(s-v) f(\sigma(s))\, ds\, dv \right] du. \end{aligned}$$

We now construct the same set H and the corresponding mapping P as in the proof of Theorem 2.9.1. Then $P : H \to H$ for the same reasons as before. Also, P will be a contraction in case (2.9.13) holds.

A remark similar to Remark 2.9.1 can be made here.

Finally, we can carry out the former investigations for

$$\begin{aligned} x' &= Ax + \int_{-\infty}^t b(s-t) f(\sigma(s))\, ds \\ \sigma' &= c^T x - rf(\sigma) \end{aligned} \tag{2.9.16}$$

under the assumption that

$$\int_{-\infty}^0 |b(u)|\, du \tag{2.9.17}$$

exists. The only point of difficulty will be in the argument that $(P\phi)(t) \to 0$ where we examine

$$\int_0^t e^{-\int_u^t d(s)\, ds} c^T \int_0^u e^{A(u-v)} \int_{-\infty}^v b(s-v) f(\sigma(s))\, ds\, dv\, du.$$

For $-\infty < J < v$, we can write

$$\int_{-\infty}^v b(s-v) f(\sigma(s))\, ds = \int_{-\infty}^J b(s-v) f(\sigma(s))\, ds + \int_J^v b(s-v) f(\sigma(s))\, ds.$$

Now

$$\int_{-\infty}^{J} b(s-v)f(\sigma(s))\,ds = \int_{-\infty}^{J-v} b(w)f(\sigma(w+v))\,dw$$

and as $v \to \infty$, this term tends to zero. The term

$$\int_{J}^{v} b(s-v)f(\sigma(s))\,ds$$

can be made small taking J large since $\sigma(v) \to 0$. We leave it to the reader to state and prove the result.

2.10 The Sunflower Equation

This section will serve as an introduction to the kinds of problems we consider in the next chapter without physical motivation. We will see a natural sequence of transformations on a classical problem which lead us to fixed point mappings. These same transformations will be seen throughout the next chapter.

The sunflower is a large and beautiful plant which bends back and forth in a seemingly periodic manner. This has been studied since at least 1827 and there is a nice history given in the paper by Somolinos (1978) who derives the sunflower equation and proves the existence of periodic solutions both by bifurcation theory and a fixed point argument in a cone. Our interest here is not in periodicity, but rather in the fact that the equation evolves into the very form we wish to study. Nevertheless, we wish to briefly describe the behavior and the current explanation.

Bend a plant to one side and hold it there for a time. There is a growth hormone, auxin, which accumulates on the lower side of the plant due to gravity. Stop bending the plant. Nothing happens for, perhaps, twenty minutes (the time-delay, r). Then that lower side of the plant starts to grow until it bends the plant over in the the opposite direction. Gravity brings auxin to the new low side of the plant. After twenty minutes that side begins to grow until it bends the plant back to its original position. It oscillates back and forth. In some plants the ocillations die out. It is thought that in others the oscillations are self-exciting, gaining in amplitude until they reach a steady state, a periodic motion. This is one of the ideas which motivates the present study. If the motion starts small, will it stay small, or can it grow very large? This is a type of stability question. Suppose we bend the tip a small amount, hold it there, and then release it (so that $x'(0) = 0$). Will it approach a small periodic solution, or will it grow significantly to approach a large amplitude periodic solution? Our subsequent result can be interpreted in these terms.

The sunflower equation, as derived by Somolinos, is

$$x'' + (a/r)x' + (b/r)\sin x(t-r) = 0 \tag{2.10.1}$$

where x is the angle between the stem and the vertical, $a \geq b > 0$, and $a \geq 1$, while r is a positive constant denoting the reaction time of the process. A unique solution is specified by a value $x'(0)$ and a continuous function $\psi : [-r, 0] \to R$; here, we will reduce the computation by asking that $x'(0) = 0$, but let ψ be arbitrary and $\psi(0)$ small. This is consistent with bending the plant for a time and then simply releasing it. For a fixed ψ, write (2.10.1) as a system

$$\begin{aligned} x' &= y \\ y' &= -\frac{a}{r}y - \frac{b}{r}\sin x(t-r) \end{aligned} \tag{2.10.2}$$

and then use the variation of parameters formula on the second equation to write

$$x' = y(t) = -\int_0^t e^{-(a/r)(t-s)}(b/r)\sin x(s-r)\,ds. \tag{2.10.3}$$

Next, write

$$x' = -(b/a)\sin x(t-r) + \frac{d}{dt}\int_0^t e^{-(a/r)(t-s)}(b/a)\sin x(s-r)\,ds. \tag{2.10.4}$$

By adding and subtracting $(b/a)\sin x$ and introducing the neutral term we have

$$\begin{aligned} x' = -(b/a)x + (b/a)(x - \sin x) + (b/a)\frac{d}{dt}\int_{t-r}^t \sin x(u)\,du \\ + \frac{d}{dt}\int_0^t e^{-(a/r)(t-s)}(b/a)\sin x(s-r)\,ds. \end{aligned} \tag{2.10.5}$$

Notice that $x - \sin x$ is odd, increasing, has the sign of x, and is Lipschitz with constant $\alpha < 1$ on a certain interval $[-L, L]$. This is the continuing theme throughout the next chapter.

We now use the variation of parameters formula to write

$$x(t) = \psi(0)e^{-(b/a)t} + \int_0^t e^{-(b/a)(t-s)}[(b/a)(x(s) - \sin x(s))$$
$$+ (b/a)\frac{d}{ds}\int_{s-r}^{s} \sin x(u)\, du$$
$$+ (b/a)\frac{d}{ds}\int_0^s e^{-(a/r)(s-u)} \sin x(u-r)\, du]\, ds. \qquad (2.10.6)$$

Integrate those neutral terms by parts as we have done so many times before and obtain

$$x(t) = \psi(0)e^{-(b/a)t}$$
$$+ \int_0^t e^{-(b/a)(t-s)}(b/a)(x(s) - \sin x(s))\, ds$$
$$+ (b/a)\int_{t-r}^{t} \sin x(u) du - e^{-(b/a)t}(b/a)\int_{-r}^{0} \sin \psi(u)\, du$$
$$- \int_0^t e^{-(b/a)(t-s)}(b/a)^2 \int_{s-r}^{s} \sin x(u)\, du\, ds$$
$$+ (b/a)\int_0^t e^{-(a/r)(t-u)} \sin x(u-r)\, du$$
$$- \int_0^t e^{-(b/a)(t-s)}(b/a)^2 \int_0^s e^{-(a/r)(s-u)} \sin x(u-r)\, du\, ds. \qquad (2.10.7)$$

Theorem 2.10.1. *Let $0 < L \le \pi/2$ and suppose that $\psi : [-r, 0] \to R$ is continuous and satifies*

$$|\psi(0)| + 3\frac{br}{a} + 2\frac{br}{a^2} \le \sin L. \qquad (2.10.8)$$

Then the solution of (2.10.3) satisfies $|x(t, 0, \psi)| \le L$ for $t \ge 0$.

Proof. Define

$$\mathcal{M} = \{\phi : [-r, \infty) \to R \,|\, \phi_0 = \psi, \phi \in \mathcal{C}, |\phi(t)| \le L \text{ for } t \ge 0\}.$$

Define a mapping $P : \mathcal{M} \to \mathcal{M}$ using (2.10.7) as we have done before. Notice that if $\phi \in \mathcal{M}$ then from (2.10.7) we have

$$|(P\phi)(t)| \le |\psi(0)| + (L - \sin L) + \frac{2br}{a} + \frac{br}{a} + \frac{br}{a^2} + \frac{br}{a^2}$$
$$= |\psi(0)| + L - \sin L + 3\frac{br}{a} + 2\frac{br}{a^2}$$
$$\le L - \sin L + \sin L = L$$

by (2.10.8). Because the mapping is given by integrals and the integrands are Lipschitz we can define a weighted metric such that P is a contraction. This will complete the proof.

We leave as an exercise to prove that the same result holds when we let $x'(0)$ be nonzero. Somolinos (1978; p. 477) poses the problem of proving that there is a periodic solution when there is a proper type of periodic forcing function. The reader should be able to do that with the contraction mappings for a small forcing function.

Chapter 3

Borrowing Coefficients

3.0 Introduction: Schauder's Theorem

Throughout this chapter we will construct closed bounded sets $\mathcal{M}$ and a mapping $P : \mathcal{M} \to \mathcal{M}$. That construction will be accomplished without reference to a Lipschitz condition. Then we will assume a Lipschitz condition and introduce a weighted metric so that in the weighted space the mapping P will be a contraction. This type of work was done in Section 2.5 and the first part of it was also done in Section 1.7. In each of those places there is a choice which is explained more fully in the introduction to Chapter 4. If the functions fail to satisfy the required Lipschitz condition, it is possible to introduce a weighted norm, obtaining a Banach space called a $\mathcal{C}_g$ space, in which the set $\mathcal{M}$ can be refined to a compact set just in case we can prove equicontinuity; this, of course, goes beyond the Ascoli theorem since the domain is $[0, \infty)$. We then use Schauder's theorem, for example, to obtain a fixed point. In this arrangement, we virtually always must settle for stability instead of asymptotic stability. Moreover, we must remember that our conclusion is that for the given continuous initial function, there is a bounded solution. If we do not have the Lipschitz condition then we may not have uniqueness and the mapping may have more than one fixed point. Often, however, we can argue that any fixed point will be in the set $\mathcal{M}$ and, hence, be bounded.

3.1 Highly Nonlinear Delay Equations

Chapter 2 focused on very well-known classical problems and showed that fixed point theory can both improve and add new dimensions to the theory concerning those problems. By contrast, this chapter will focus on contrived problems in which we attempt to show that fixed point theory leads us to totally new kinds of relations which can substantially change how we think about relationships in differential equations. The results of this section appeared in Burton (2004e).

Sections 3.1 and 3.2 contrast sharply. In this section we study scalar delay equations of the form

$$x'(t) = -a(t)g_1(x(t-r)) - b(t)g_2(x(t-r))$$

where the g_i are odd increasing functions and $g_2(x)/g_1(x) \to 0$ as $|x| \to 0$. We show that stability properties can be proved by studying $x'(t) = -[a(t) + b(t)]g_1(x(t-r))$. In effect, we can borrow the coefficient of the higher order term and add it to the coefficient of the lower order term to stabilize the equation. The next section covers the same kinds of problems except that $\lim_{x\to 0} g_2(x)/g_1(x)$ exists, but is not zero.

Liapunov's theorem tells us that if the linear approximation of a differential equation is uniformly asymptotically stable then the higher order terms can be ignored. Here, we take the reverse view: Higher order terms can be used with lower order terms to stabilize an equation. This is not necessarily a surprise. The linear scalar equation

$$x'' = 0$$

is unstable, but if we add a nonlinear term like x^3 to obtain

$$x'' + x^3 = 0$$

then we have a stable equation. Moreover, if we begin with the stable linear equation

$$x'' + x = 0$$

and add the nonlinear term $(1 + x^2)x'$ to obtain

$$x'' + (1 + x^2)x' + x = 0$$

then we have a uniformly asymptotically stable equation. The problems we consider here will be of a very different nature.

For future use we begin with an equation very much like some we have studied earlier in the form of

$$x'(t) = -a(t)g(x(t-r))$$

where r is a positive constant, $a(t)$ can change sign, while g is nonlinear, $xg(x) > 0$ for $x \neq 0$, and g satisfies monotonicity and growth conditions.

Three things of interest are accomplished. First, we prove a fixed point theorem of contraction type which is suited to such stability problems; it is closely related to the proposition used in Section 2.5, but is suited exactly to our problems here. Next, we obtain a stability criterion for $x' = -a(t)g(x(t-r))$ when $a(t)$ can change sign, but on average is positive. Then we obtain results which allow us to "average" the coefficients of terms of different orders in order to get a stability result. We show that stability of the zero solution of

$$x'(t) = -a(t)x(t-r) - b(t)x^{2n+1}(t-r)$$

can be established from the stability of

$$x'(t) = -(a(t) + b(t))x(t-r);$$

the coefficient of the higher order term is simply borrowed and added to the coefficient of the lower order term. There is, of course, a "cost" for doing so. In the same way, stability of the zero solution of

$$x'(t) = -a(t)x^{2n+1}(t-r) - b(t)x^{2n+3}$$

can be established from examination of

$$x'(t) = -(a(t) + b(t))x^{2n+1}(t-r).$$

The aforementioned *cost* goes down as n increases. This does seem to be a type of result not previously seen in the theory of differential equations. In one of the later results we show that our first equation can still be stable when $a(t)$ changes sign and is, on average, negative.

The focus will be on polynomial problems in order to give clear insight into what can be proved. The same techniques work on far more general problems.

It is well known that if $f(t,x)$ satisfies a uniform Lipschitz condition in x in the supremum norm, then a new norm can be defined (a weighted norm) so that $\int_0^t f(s,x(s))ds$ is a contraction. But if $f(t,s,x)$ does not satisfy a uniform Lipschitz condition in x then it can be challenging to define a norm so that $\int_0^t f(t,s,x(s))ds$ is a contraction.

We will have a mapping with several terms which needs to be a contraction. The terms will be considered individually with each term being a contraction and having a contraction constant as small as we please. The following theorem shows how this is done and it will be used repeatedly in this chapter. We have used similar results twice in Chapter 2.

Theorem 3.1.1. *Let $L > 0$, $\psi(0)$ be a fixed number,*

$$\mathcal{M} = \{\phi : [0, \infty) \to R \mid \phi \in \mathcal{C}, \phi(0) = \psi(0), |\phi(t)| \leq L\},$$

and $f : [-L, L] \to R$ satisfy a Lipschitz condition with constant $K > 0$. Suppose also that $a : [0, \infty) \to R$ is continuous, $h : [0, \infty) \to R$ is continuous, and for $\phi \in \mathcal{M}$ define

$$(P\phi)(t) = h(t) + \int_0^t e^{-\int_s^t a(u+r)\,du} a(s+r) f(\phi(s))\, ds.$$

If $P : \mathcal{M} \to \mathcal{M}$ then for each $d > 1$ there is a metric ρ on $\mathcal{M}$ such that P is a contraction with constant $1/d$ and $(\mathcal{M}, \rho)$ is a complete metric space.

Proof. Let $(\mathcal{X}, |\cdot|_K)$ be the Banach space of continuous $\phi : [0, \infty) \to R$ for which

$$|\phi|_K := \sup_{t \geq 0} e^{-(dK+2)\int_0^t |a(s+r)|\,ds} |\phi(t)|$$

exists. If $\phi, \eta \in \mathcal{M}$ then

$$\begin{aligned}
&|P\phi - P\eta|_K \\
&\leq \sup_{t\geq 0} e^{-(dK+2)\int_0^t |a(s+r)|\,ds} \int_0^t e^{-\int_s^t a(u+r)\,du} |a(s+r)| \\
&\times |f(\phi(s)) - f(\eta(s))|\, ds \\
&\leq \sup_{t\geq 0} \int_0^t e^{-\int_s^t a(u+r)\,du} |a(s+r)| K |\phi(s) - \eta(s)| \\
&\times e^{-(dK+2)(\int_0^s |a(u+r)|du + \int_s^t |a(u+r)|\,du)}\, ds \\
&\leq |\phi - \eta|_K \sup_{t\geq 0} \int_0^t e^{-\int_s^t a(u+r)\,du} \\
&\times |a(s+r)| K e^{-(dK+2)\int_s^t |a(u+r)|\,du}\, ds \\
&\leq |\phi - \eta|_K \sup_{t\geq 0} \int_0^t e^{-dK\int_s^t |a(u+r)|\,du} K |a(s+r)|\, ds \\
&\leq |\phi - \eta|_K \frac{K}{dK} = (1/d)|\phi - \eta|_K.
\end{aligned}$$

Now $\mathcal{M}$ is a subset of the Banach space $\mathcal{X}$ and $\mathcal{M}$ is closed so $\mathcal{M}$ is complete. Thus, $P : \mathcal{M} \to \mathcal{M}$ has a unique fixed point. The proof is complete.

Let r be a positive constant, $a : [0, \infty) \to R$ be continuous, and consider the scalar equation

$$x'(t) = -a(t)g(x(t-r)) \tag{3.1.1}$$

with continuous initial function $\psi : [-r, 0] \to R$, where g is continuous, locally Lipschitz, and odd, while $x - g(x)$ is nondecreasing and $g(x)$ is increasing on an interval $[0, L]$ for some $L > 0$. This requirement may be reduced, in part, when g has a derivative by writing

$$a(t)g(x(t-r)) = (Da(t))(g(x(t-r))/D),$$

where $|\frac{d}{dx}g(x)| \leq D$ on $[0, L]$, and then renaming $a(t)$ as $Da(t)$.

Earlier we studied this equation for $n = 0$ (the linear case). We proved that if there is an $\alpha < 1$ such that

$$\int_{t-r}^{t} |a(u+r)|\, du + \int_0^t |a(s+r)| e^{-\int_s^t a(u+r)\, du} \Big| \int_{s-r}^{s} a(u+r)\, du \Big|\, ds \leq \alpha$$

and if $\int_0^t a(s)\, ds \to \infty$ as $t \to \infty$, then the zero solution of (3.1.1) is asymptotically stable.

To prepare for our next theorem write (3.1.1) as

$$x'(t) = -a(t+r)g(x(t)) + \frac{d}{dt} \int_{t-r}^{t} a(s+r)g(x(s))\, ds$$

and then as

$$x'(t) = -a(t+r)x(t) + a(t+r)[x(t) - g(x(t))] + \frac{d}{dt} \int_{t-r}^{t} a(s+r)g(x(s))\, ds.$$

For each $t_0 \geq 0$, Equation (3.1.1) requires a continuous initial function $\psi : [t_0 - r, t_0] \to R$ to specify a solution $x(t, t_0, \psi)$. In this problem the computations are the same for any $t_0 \geq 0$ so we take $t_0 = 0$. By the variation of parameters formula we have

$$\begin{aligned} x(t) =& \psi(0)e^{-\int_0^t a(s+r)\, ds} + \int_0^t e^{-\int_s^t a(u+r)\, du} a(s+r)[x(s) - g(x(s))]\, ds \\ &+ \int_0^t e^{-\int_s^t a(u+r)\, du} \frac{d}{ds} \int_{s-r}^{s} a(u+r)g(x(u))\, du\, ds. \end{aligned}$$

Integration by parts of the last term yields

$$
\begin{aligned}
x(t) &= \psi(0)e^{-\int_0^t a(s+r)\,ds} \\
&\quad - e^{-\int_0^t a(u+r)\,du} \int_{-r}^{0} a(u+r)g(\psi(u))\,du \\
&\quad + \int_0^t e^{-\int_s^t a(u+r)\,du} a(s+r)[x(s) - g(x(s))]\,ds \\
&\quad + \int_{t-r}^{t} a(u+r)g(x(u))\,du \\
&\quad - \int_0^t e^{-\int_s^t a(u+r)du} a(s+r) \int_{s-r}^{s} a(u+r)g(x(u))\,du\,ds. \qquad (3.1.2)
\end{aligned}
$$

Note that if $0 < L_1 < L$, then the conditions on g given with (3.1.1) hold on $[-L_1, L_1]$. Also, note that if $\phi : [-r, \infty) \to R$ with $\phi_0 = \psi$, if ϕ is continuous and $|\phi(t)| \leq L$, then for $t \geq 0$ we have

$$
|\phi(t) - g(\phi(t))| \leq L - g(L)
$$

since $x - g(x)$ is odd and nondecreasing on $(0, L)$. The symbol ϕ_0 denotes the segment of ϕ on $[-r, 0]$.

For any continuous ψ on $[-r, 0]$ with $|\psi(t)| < L$ we take

$$
\mathcal{M} = \{\phi : [-r, \infty) \to R \mid \phi_0 = \psi, \phi \in \mathcal{C}, |\phi(t)| \leq L\}.
$$

The size of ψ will be further restricted later.

Theorem 3.1.2. *Let g be odd, increasing on $[0, L]$, satisfy a Lipschitz condition, and let $x - g(x)$ be nondecreasing on $[0, L]$. Suppose also that for each $L_1 \in (0, L]$ we have*

$$
\begin{aligned}
&|L_1 - g(L_1)| \sup_{t \geq 0} \int_0^t e^{-\int_s^t a(u+r)du} |a(s+r)|\,ds \\
&+ g(L_1) \sup_{t \geq 0} \int_0^t e^{-\int_s^t a(u+r)\,du} |a(s+r)| \int_{s-r}^{s} |a(u+r)|\,du\,ds \\
&+ g(L_1) \sup_{t \geq 0} \int_{t-r}^{t} |a(u+r)|\,du < L_1 \qquad (3.1.3)
\end{aligned}
$$

and there exists $J > 0$ such that

$$
-\int_0^t a(s+r)\,ds \leq J \quad \text{for} \quad t \geq 0. \qquad (3.1.4)
$$

Then the zero solution of (3.1.1) is stable.

Proof. We will first define a mapping $P : \mathcal{M} \to \mathcal{M}$ using (3.1.2) so that for $\phi \in \mathcal{M}$ we have

$$(P\phi)(t) = \psi(t), \quad -r \le t \le 0,$$

and for $t \ge 0$

$$\begin{aligned}(P\phi)(t) &= \psi(0)e^{-\int_0^t a(s+r)\,ds} - e^{-\int_0^t a(u+r)\,du}\int_{-r}^0 a(u+r)g(\psi(u))\,du \\ &+ \int_0^t e^{-\int_s^t a(u+r)\,du}a(s+r)[\phi(s) - g(\phi(s))]\,ds \\ &- \int_0^t e^{-\int_s^t a(u+r)du}a(s+r)\int_{s-r}^s a(u+r)g(\phi(u))\,du\,ds \\ &+ \int_{t-r}^t a(u+r)g(\phi(u))\,du. \qquad (3.1.5)\end{aligned}$$

By (3.1.3) there is an $\alpha < 1$ such that if $\phi \in \mathcal{M}$ then

$$\begin{aligned}&|(P\phi)(t)| \\ &\le \|\psi\|e^J + e^J\|g(\psi)\|\int_{-r}^0 |a(s+r)|\,ds \\ &+ |L - g(L)|\sup_{t\ge 0}\int_0^t e^{-\int_s^t a(u+r)\,du}|a(s+r)|\,ds \\ &+ g(L)\int_{t-r}^t |a(u+r)|\,du \\ &+ g(L)\int_0^t e^{-\int_s^t a(u+r)\,du}|a(s+r)|\int_{s-r}^s |a(u+r)|\,du\,ds \\ &\le e^J[\|\psi\| + \|g(\psi)\|\int_{-r}^0 |a(s+r)|\,ds] + \alpha L.\end{aligned}$$

Choose $\delta > 0$ so that $\|\psi\| < \delta$ and K the Lipschitz constant for g on $[0, L]$ implies that

$$e^J[\delta + K\delta\int_{-r}^0 |a(s+r)|\,ds] < (1-\alpha)L.$$

Then $|(P\phi)(t)| \le L$ so we can show that $P : \mathcal{M} \to \mathcal{M}$. Since the mapping is given by integrals and the functions are Lipschitz we will be able to show that P is a contraction. (At this point, if (3.1.3) holds only for L, itself, then we have a boundedness result. Moreover, if the Lipschitz condition fails, we could now give a stability proof using Schauder's theorem.) For a given $\varepsilon > 0$, $\varepsilon < L$, substitute ε for L and obtain the usual stability proof since a fixed point of P will be a unique solution and lie in $\mathcal{M}$.

We now want to change the metric so that we will have a contraction. For $\phi, \eta \in \mathcal{M}$ we have

$$\begin{aligned}
&|(P\phi)(t) - (P\eta)(t)| \\
&\leq \int_0^t e^{-\int_s^t a(u+r)\,du} |a(s+r)||\phi(s) - g(\phi(s)) - \eta(s) + g(\eta(s))|\,ds \\
&+ \int_0^t e^{-\int_s^t a(u+r)\,du} |a(s+r)| \int_{s-r}^s |a(u+r)||g(\phi(u)) - g(\eta(u))|\,du\,ds \\
&+ \int_{t-r}^t |a(u+r)||g(\phi(u)) - g(\eta(u))|\,du. \qquad (3.1.6)
\end{aligned}$$

Now $g(x)$ and $x - g(x)$ both satisfy a Lipschitz condition with the same constant K, so we proceed as in the proof of Theorem 3.1.1 and take the metric on $\mathcal{M}$ as that induced by the norm

$$|\phi|_K := \sup_{t \geq 0} e^{-(dK+2)\int_0^t |a(u+r)|\,du} |\phi(t)|$$

where we will find that $d > 3$ will suffice. As shown in Theorem 3.1.1, the first term on the right-hand-side of (3.1.6) has a contraction constant $1/d$. The second term satisfies

$$\begin{aligned}
&\sup_{t \geq 0} e^{-(dK+2)\int_0^t |a(u+r)|\,du} \int_{t-r}^t |a(u+r)||g(\phi(u)) - g(\eta(u))|\,du \\
&\leq \sup_{t \geq 0} \int_{t-r}^t |a(u+r)|K|\phi(u) - \eta(u)|e^{-(dK+2)\int_0^u |a(s+r)|\,ds} \\
&\times e^{-(dK+2)\int_u^t |a(s+r)|\,ds}\,du \\
&\leq \sup_{t \geq 0} \int_{t-r}^t |a(s+r)|Ke^{-(dK+2)\int_s^t |a(u+r)|\,du}\,ds |\phi - \eta|_K \\
&\leq (K/dK)|\phi - \eta|_K \\
&= (1/d)|\phi - \eta|_K.
\end{aligned}$$

Multiply the third term by $e^{-(dk+2)\int_s^t |a(u+r)|\,du}$ obtaining

$$\begin{aligned}
&\int_0^t e^{-\int_s^t a(u+r)\,du}|a(s+r)|\int_{s-r}^s |a(u+r)|K|\phi(u)-\eta(u))| \\
&\times e^{-(dK+2)(\int_0^u |a(v+r)|\,dv)-(dK)\int_u^t |a(v+r)|\,dv)-2\int_s^t |a(v+r)|\,dv}\,du\,ds \\
&\le |\phi-\eta|_K \sup_{t\ge 0}\int_{t-r}^t |a(u+r)|Ke^{-dK\int_u^t |a(v+r)|\,dv}\,du \\
&\le (1/d)|\phi-\eta|_K.
\end{aligned}$$

We take $d > 3$ for a contraction. As in the proof of Theorem 3.1.1 there is a unique fixed point. This completes the proof.

It is a fairly simple matter to use exactly the same proof as that of Theorem 3.1.2 to show that if (3.1.1) satisfies the conditions of Theorem 3.1.2 and if (3.1.1) is perturbed with a higher order term then the conclusion of Theorem 3.1.2 still holds for the perturbed system. But if the linear part does not satisfy Theorem 3.1.2 then we note an interesting fact: a higher order term can stabilize the equation in the sense that by borrowing from the coefficient of a higher order term we can fulfill conditions parallel to those of Theorem 3.1.2.

Remark 3.1.1. *We could continue here with*

$$x'(t) = -a(t)x(t-r) - b(t)g(x(t-r))$$

where $x - g(x)$ is odd, positive and increasing on $(0, L)$. But the impact of the next two results will depend crucially on the reader being able to see the exact value of two constants. See Remark 3.1.6 for a comparison of results.

Thus, we consider

$$x'(t) = -a(t)x(t-r) - b(t)x^3(t-r) \tag{3.1.7}$$

where we assume that

$$c(t) := b(t) + a(t) \ge 0, \tag{3.1.8}$$

and that there is a constant $\alpha < 1$ with

$$\sup_{t\ge 0} 2\int_{t-r}^t c(u+r)\,du + \frac{2}{3}\int_0^t e^{-\int_s^t c(u+r)\,du}|b(s)|\,ds \le \alpha. \tag{3.1.9}$$

Example 3.1.1. If $1 < k < 3/2$ then

$$x'(t) = -(1 - k\sin t)x(t-r)$$

will not satisfy (3.1.3), but when

$$2r + (2/3)k < 1$$

then

$$x'(t) = -(1 - k\sin t)x(t-r) - (k\sin t)x^3(t-r)$$

will satisfy (3.1.9) and the conditions of our next theorem will hold.

Problem. It would be very interesting to see a Liapunov functional constructed for this last equation showing stability. Such construction might well lead Liapunov theory into a fruitful new phase. Indeed, it seems challenging even for $r = 0$ since the condition then includes $k < 3/2$.

Remark 3.1.2. *Theorem 3.1.2 averaged the values of $a(t)$ to produce stability. Our next result will produce stability by averaging the coefficients of terms having different powers.*

Theorem 3.1.3. *If (3.1.8) and (3.1.9) hold then there is a $\delta > 0$ such that if ψ is a continuous initial function on $[-r, 0]$ with $\|\psi\| < \delta$ then the solution of (3.1.7) satisfies $|x(t, 0, \psi)| < 1/\sqrt{3}$ for all $t \geq 0$.*

Proof. Write (3.1.7) as

$$\begin{aligned}
x' &= -a(t)x(t-r) - b(t)x(t-r) + b(t)[x(t-r) - x^3(t-r)] \\
&= -c(t)x(t-r) + b(t)[x(t-r) - x^3(t-r)] \\
&= -c(t+r)x(t) + \frac{d}{dt}\int_{t-r}^{t} c(s+r)x(s)\,ds \\
&+ b(t)[x(t-r) - x^3(t-r)]
\end{aligned}$$

so that by the variation of parameters formula, followed by integration by parts we have

$$\begin{aligned}
x(t) &= \psi(0)e^{-\int_0^t c(s+r)\,ds} \\
&\quad - e^{-\int_0^t c(u+r)\,du} \int_{-r}^{0} c(u+r)\psi(u)\,du \\
&\quad + \int_0^t e^{-\int_s^t c(u+r)\,du} b(s)[x(s-r) - x^3(s-r)]\,ds \\
&\quad + \int_{t-r}^{t} c(u+r)x(u)\,du \\
&\quad - \int_0^t e^{-\int_s^t c(u+r)\,du} c(s+r) \int_{s-r}^{s} c(u+r)x(u)\,du\,ds.
\end{aligned}$$

Now $f(x) = x - x^3$ has a maximum of $2/3\sqrt{3}$ at $1/\sqrt{3}$. As $x - x^3$ increases on $(0, 1/\sqrt{3})$ we could work on any shorter interval. We will show here that if ψ is small enough and if $|\phi(t)| \leq 1/\sqrt{3}$ then $|(P\phi)(t)| \leq 1/\sqrt{3}$. We use the equation for x which we just obtained to define P as we did in (3.1.5) and have

$$\begin{aligned}
|(P\phi)(t)| \leq \|\psi\| &+ \|\psi\| \int_{-r}^{0} c(u+r)\,du \\
&+ \int_0^t e^{-\int_s^t c(u+r)\,du} |b(s)|(2\sqrt{3}/9)\,ds \\
&+ 2\sup_{t\geq 0} \int_{t-r}^{t} c(u+r)\,du(1/\sqrt{3}) \\
&\leq 1/\sqrt{3}
\end{aligned}$$

provided that (3.1.9) holds and $\|\psi\| < \delta$ where $\delta + \delta \int_{-r}^{0} c(u+r)\,du < 1 - \alpha$. Our mapping set for any such fixed ψ is

$$\mathcal{M} = \{\phi : [-r, \infty) \to R \,|\, \phi_0 = \psi, |\phi(t)| \leq 1/\sqrt{3}\} \tag{3.1.10}$$

and $P : \mathcal{M} \to \mathcal{M}$.

The contraction argument parallel to Theorem 3.1.1 uses the weight for the norm as

$$e^{-2d\int_0^t [|b(s)| + c(s+r)]\,ds}$$

since for $\phi, \eta \in \mathcal{M}$ we have

$$|(P\phi)(t)-(P\eta)(t)| \\ \leq \int_0^t e^{-\int_s^t c(u+r)\,du}|b(s)||\phi(s)-\eta(s)|\,ds \\ +2\int_{t-r}^t c(u+r)|\phi(s)-\eta(s)|\,ds.$$

Here, we obtain an extension of the result in the last theorem by starting with a nonlinear term which does not necessarily satisfy the conditions of Theorem 3.1.2. Yet, when there is a suitable higher order term present, it may be used to stabilize the equation. Moreover, the coefficient corresponding to the 2/3 in (3.1.9) is reduced to a small fraction of that 2/3 as the order increases. This means that the penalty for merging the coefficients reduces as the order increases.

Consider the equation

$$x'(t) = -a(t)x^{2n+1}(t-r) - b(t)x^{2n+3}(t-r) \tag{3.1.11}$$

where n is a positive integer. Let

$$c(t) := a(t) + b(t) \geq 0.$$

Remark 3.1.3. *As remarked before, we could work with*

$$x'(t) = -a(t)g_1(x(t-r)) - b(t)g_2(x(t-r))$$

where $g_2(x)/g_1(x) \to 0$ as $|x| \to 0$, g_i odd, $g_1(x) - g_2(x)$ positive and increasing on $(0, L)$. But the impact of the next result depends on the reader seeing the coefficient of the first integral in Theorem 3.1.4. See Remark 3.1.6 for a summary.

Remark 3.1.4. *Here, we enlarge on the process begun in the proof of Theorem 3.12. The order of these steps is critical, as is explained between (3.1.25) and (3.1.27). The following steps are taken in the proof of the next theorem.*

(i) Add and subtract the term $b(t)x^{2n+1}(t-r)$ to yield the term $-c(t)x^{2n+1}(t-r)$.

(ii) Two steps are now taken to enable us to use the variation of parameters formula:

(a) First, write

$$-c(t)x^{2n+1}(t-r) = -c(t+r)x^{2n+1}(t) + \frac{d}{dt}\int_{t-r}^{t} c(s+r)x^{2n+1}(s)\,ds.$$

If we try exact linearization here, we lose the exact properties of $c(t)$ which are crucially needed in comparing $b(t)$ and $c(t)$ in the averaging statement of our theorem below.

(b) Finally, write

$$-c(t+r)x^{2n+1}(t) = -c(t+r)x(t) + c(t+r)[x(t) - x^{2n+1}(t)].$$

Theorem 3.1.4. *Let $a(t) + b(t) = c(t) \geq 0$, $L^2 = \frac{2n+1}{2n+3}$, and let*

$$(1-L^2)\sup_{t\geq 0}\int_0^t e^{-\int_s^t c(u+r)\,du}|b(s)|\,ds + 2\sup_{t\geq 0}\int_{t-r}^{t} c(s+r)ds < 1.$$

Then there is a $\delta > 0$ such that if the continuous initial function ψ for (3.1.11) satisfies $\|\psi\| < \delta$ then $|x(t,0,\psi)| < L$.

Proof. Write (3.1.11) as

$$\begin{aligned}x'(t) &= \frac{d}{dt}\int_{t-r}^{t} c(s+r)x^{2n+1}(s)\,ds \\ &\quad + b(t)[x^{2n+1}(t-r) - x^{2n+3}(t-r)] \\ &\quad - c(t+r)x(t) + c(t+r)[x(t) - x^{2n+1}(t)]. \qquad (3.1.12)\end{aligned}$$

As we have done before, use the variation of parameters formula, integrate by parts, and obtain

$$\begin{aligned}x(t) &= e^{-\int_0^t c(s+r)\,ds}\psi(0) - e^{\int_0^t c(u+r)du}\int_{-r}^{0} c(u+r)\,du\psi^{2n+1}(u)\,du \\ &\quad + \int_0^t e^{-\int_s^t c(u+r)\,du}c(s+r)[x(s) - x^{2n+1}(s)]\,ds \\ &\quad + \int_0^t e^{-\int_s^t c(u+r)\,du}b(s)[x^{2n+1}(s-r) - x^{2n+3}(s-r)]\,ds \\ &\quad + \int_{t-r}^{t} c(u+r)x^{2n+1}(u)\,du \\ &\quad - \int_0^t e^{-\int_s^t c(u+r)\,du}c(s+r)\int_{s-r}^{s} c(u+r)x^{2n+1}(u)\,duds. \qquad (3.1.13)\end{aligned}$$

Now

$$f(x) = x^{2n+1} - x^{2n+3} \tag{3.1.14}$$

has a local maximum at

$$x = \sqrt{\frac{2n+1}{2n+3}} = L \tag{3.1.15}$$

and is increasing on $(0, L)$. Moreover,

$$f(L) = \left(\frac{2n+1}{2n+3}\right)^{\frac{2n+1}{2}} - \left(\frac{2n+1}{2n+3}\right)^{\frac{2n+3}{2}}. \tag{3.1.16}$$

Use (3.1.13) to define a mapping P as in (3.1.15) of a set

$$\mathcal{M} = \{\phi : [-r, \infty) \to R \mid \phi_0 = \psi, |\phi(t)| \leq L\} \tag{3.1.17}$$

into itself where ψ is a sufficiently small initial function. That mapping in $\mathcal{M}$ will be possible if

$$L(1 - L^{2n}) + f(L) \sup_{t\geq 0} \int_0^t e^{-\int_s^t c(u+r)\,du} |b(s)|\, ds$$
$$+ 2L^{2n+1} \sup_{t\geq 0} \int_{t-r}^t c(s+r) ds < L.$$

As $f(L)/L^{2n+1} = 1 - L^2$ this will reduce to

$$(1-L^2) \sup_{t\geq 0} \int_0^t e^{-\int_s^t c(u+r)\,du} |b(s)|\, ds + 2 \sup_{t\geq 0} \int_{t-r}^t c(s+r)\, ds < 1. \tag{3.1.18}$$

The remainder of the proof is exactly as before.

Remark 3.1.5. *The critical value for applications in (3.1.18) is*

$$1 - L^2 = 1 - \frac{2n+1}{2n+3} = \frac{2}{2n+3} \tag{3.1.19}$$

and we note that this tends to zero as $n \to \infty$.

Example 3.1.2. Let

$$a(t) = 1 - 2\sin t, \qquad b(t) = 2\sin t, \qquad c(t) = 1.$$

Then

$$\int_0^t e^{-\int_s^t 1du}|b(s)|\,ds \le 2$$

so to satisfy (3.1.18) we need

$$2\frac{2}{2n+3} + 2r < 1$$

or

$$r < \frac{2n-1}{4n+6}.$$

The idea used here will also work for the nonlinear equation

$$x'(t) = -a(t)g(x(t-r)) \tag{3.1.20}$$

We can remove a portion of $a(t)$ which does not fit our theorem, provided that the removed portion has a sufficiently small average.

Theorem 3.1.5. *Let the conditions with (3.1.1) hold for (3.1.20) and suppose that*

$$a(t) = c(t) - b(t) \tag{3.1.21}$$

where $c(t) \ge 0$ and continuous, while

$$\sup_{t\ge 0}\int_0^t e^{-\int_s^t c(u+r)\,du}|b(s)|\,ds + 2\sup_{t\ge 0}\int_{t-r}^t c(u+r)\,du < 1. \tag{3.1.22}$$

Then the zero solution of (3.1.20) is stable.

Proof. We give the proof for $t_0 = 0$. Write the equation as

$$\begin{aligned}
x'(t) &= -c(t)g(x(t-r)) + b(t)g(x(t-r)) \\
&= -c(t+r)g(x(t)) + \frac{d}{dt}\int_{t-r}^t c(s+r)g(x(s))\,ds + b(t)g(x(t-r)) \\
&= -c(t+r)x(t) + c(t+r)[x(t) - g(x(t))] \\
&\quad + \frac{d}{dt}\int_{t-r}^t c(s+r)g(x(s))\,ds + b(t)g(x(t-r)).
\end{aligned}$$

Use the variation of parameters formula with an initial function ψ and integrate the neutral term by parts as we have done before and obtain

$$\begin{aligned}
x(t) &= \psi(0)e^{-\int_0^t c(s+r)\,ds} \\
&\quad - e^{-\int_0^t c(u+r)\,du}\int_{-r}^0 c(u+r)g(\psi(u))\,du \\
&\quad + \int_0^t e^{-\int_s^t c(u+r)du}c(s+r)[x(s)-g(x(s))]\,ds \\
&\quad + \int_0^t e^{-\int_s^t c(u+r)\,du}b(s)g(x(s-r))\,ds \\
&\quad + \int_{t-r}^t c(u+r)g(x(u))\,du \\
&\quad - \int_0^t e^{-\int_s^t c(u+r)du}c(s+r)\int_{s-r}^s c(u+r)g(x(u))\,du\,ds.
\end{aligned}$$

Now $f(x) = x - g(x)$ has a maximum on $[0, L]$ at L. Take

$$\mathcal{M} = \{\phi : [-r,\infty) \to R \mid \phi_0 = \psi, |\phi(t)| \le L, \phi \in \mathcal{C}\}.$$

Define a mapping P on $\mathcal{M}$ using the last equation in x, as before. We have

$$\begin{aligned}
|(P\phi)(t)| &\le \|\psi\| + g(\|\psi\|)\sup_{t\ge 0}\int_{t-r}^t c(u+r)\,du \\
&\quad + L - g(L) + g(L)\sup_{t\ge 0}\int_0^t e^{-\int_s^t c(u+r)\,du}|b(s)|\,ds \\
&\quad + 2g(L)\int_{t-r}^t c(u+r)\,du.
\end{aligned}$$

In order to say that $P : \mathcal{M} \to \mathcal{M}$ we need

$$\begin{aligned}
&L - g(L) + g(L)\sup_{t\ge 0}\int_0^t e^{-\int_s^t c(u+r)\,du}|b(s)|\,ds \\
&+ 2g(L)\sup_{t\ge 0}\int_{t-r}^t c(u+r)\,du < L.
\end{aligned}$$

When we subtract L from each side and divide by $g(L)$ we arrive at (3.1.22).

Theorem 3.1.2 was a result in which the second integral in (3.1.22) yielded stability by averaging the values of c. Theorem 3.1.3 showed how the fixed point method averaged the coefficients of terms of different powers. We now see that the integral in (3.1.22) can average an oscillating large term to make its effect smaller than a small constant term.

Example 3.1.3. Consider the scalar equation

$$x'(t) = -(1 - k\cos^2 pt)g(x(t-r)) \tag{3.1.23}$$

where $1 < k < 2$, g satisfies the conditions with (3.1.1), and p is a large positive integer. Conditions of Theorem 3.1.2 would not be satisfied.

To show the behavior of solutions, referring to Theorem 3.1.5, we take

$$c(t) = 1, \qquad b(t) = k\cos^2 pt,$$

so that (3.1.22) asks that

$$\sup_{t\geq 0}\int_0^t e^{-(t-s)}k\cos^2 ps\,ds + 2r < 1.$$

We will show that for p large enough and r small enough, then (3.1.22) is satisfied. Now

$$k\cos^2 pt = (k/2)(1+\cos 2pt)$$

and

$$\begin{aligned}
(k/2)&\int_0^t e^{-(t-s)}(1+\cos 2ps)\,ds \\
&\leq (k/2) + (k/2)e^{-t}\int_0^t e^s\cos 2ps\,ds \\
&= \frac{k}{2} + \frac{ke^{-t}}{2(1+4p^2)}\left[e^s(\cos 2ps + 2p\sin 2ps)\right|_0^t \\
&= \frac{k}{2} + \frac{ke^{-t}}{2(1+4p^2)}[e^t(\cos 2pt + 2p\sin 2pt) - 1] \\
&= \frac{k}{2} + \frac{k}{2(1+4p^2)}[\cos 2pt + 2p\sin 2pt - e^{-t}] \\
&\leq \frac{k}{2} + \frac{k(1+2p)}{2(1+4p^2)} \\
&\to \frac{k}{2}
\end{aligned}$$

as $p\to\infty$. Thus, for p large enough and r small enough, then (3.1.22) is satisfied.

Remark 3.1.6. *Compare (3.1.22) with (3.1.9) and (3.1.18). That crucial coefficient of the integral containing $|b(s)|$ is 1 in (3.1.22), 2/3 in (3.1.9), and $(1-L^2)$ in (3.1.18). The higher order term made a significant contribution to stability.*

It is quickly verified that the technique here works with any number of terms. In the equation

$$x'(t) = -a(t)x^{2n+1}(t-r) - b(t)x^{2n+3}(t-r) - p(t)x^{2n+5}(t-r) \quad (3.1.24)$$

we simply get an additional term

$$\int_0^t e^{-\int_s^t c(u+r)\,du} p(s)[x^{2n+1}(s-r) - x^{2n+5}(s-r)]\,ds \quad (3.1.25)$$

and $c(t) = a(t) + b(t) + p(t)$. The only step which requires caution is that the term $-c(t)x^{2n+1}(t-r)$ must be converted to the neutral term

$$-c(t+r)x^{2n+1}(t) + \frac{d}{dt}\int_{t-r}^t c(s+r)x^{2n+1}(s)\,ds; \quad (3.1.26)$$

do not add in the linear term and then convert it to the neutral term. If that order is inverted, then one obtains integrals of the form

$$\int_0^t e^{-\int_s^t c(u+r)du} c(s)...\,ds \quad (3.1.27)$$

which are not readily estimated.

This process also shows that the terms need not enter as consecutive odd exponents in x; if the coefficient in the next higher order term is insufficient to show stability, we can continue to still higher order terms.

This will acknowledge that much of the work in this section was first published in *Fixed Point Theory*, as detailed in Burton (2004e), which is published by the House of the Book of Science.

3.2 Borrowing and Addition of Terms

In applying Liapunov's theorem to the stability study of $x' = x - 2\sin x$ we understand that we can replace $\sin x$ by x and study stability of $x' = -x$. The reason reduces to the fact that $\lim_{x\to 0} \frac{\sin x}{x} = 1$. In another vein, it is an easy application of Liapunov's direct method to see that in the study of stability of $x' = -(1 - 100\sin t)x^3 - 100\sin t\sin^3 x$ we may simply replace $\sin^3 x$ by x^3 and study the stability of $x' = -x^3$. The reason reduces to the fact that $\lim_{x\to 0} \frac{\sin^3 x}{x^3} = 1$. In this section we use contraction mappings to study stability of three prominent classes of functional differential equations involving two functionals, $g(x_t)$ and $G(x_t)$, having the property that $\lim_{x\to 0} \frac{G(x)}{g(x)} = 1$. We show that we can replace G with g and study the stability of the resulting equation. These results appeared in Burton (2004c).

We begin with the equation

$$x'(t) = -a(t)g(x(t-r)) - b(t)G(x(t-r)) \tag{3.2.1}$$

with a continuous initial function

$$\psi : [-r, 0] \to R. \tag{3.2.2}$$

Several of the steps we take are unnecessary if either $b(t) = 0$, if $r = 0$, or if $g(x) = x$. But those cases can be readily distinguished from our final mapping. Let

$$c(t) := a(t) + b(t). \tag{3.2.3}$$

Theorem 3.2.1. *A solution $x(t, 0, \psi)$ of (3.2.1) can be expressed as*

$$\begin{aligned} x(t) &= \psi(0)e^{-\int_0^t c(s+r)\,ds} \\ &\quad - e^{-\int_0^t c(u+r)\,du} \int_{-r}^{0} c(u+r)g(\psi(u))\,du \\ &\quad + \int_0^t e^{-\int_s^t c(u+r)\,du} c(s+r)[x(s) - g(x(s))]\,ds \\ &\quad + \int_0^t e^{-\int_s^t c(u+r)\,du} b(s)[g(x(s-r) - G(x(s-r))]\,ds \\ &\quad + \int_{t-r}^{t} c(u+r)g(x(u))\,du \\ &\quad - \int_0^t e^{-\int_s^t c(u+r)\,du} c(s+r) \int_{s-r}^{s} c(u+r)g(x(u))\,du\,ds. \end{aligned} \tag{3.2.4}$$

Proof. We write

$$\begin{aligned}x'(t) &= -a(t)g(x(t-r)) - b(t)g(x(t-r)) \\ &\quad + b(t)[g(x(t-r)) - G(x(t-r))] \\ &= -c(t+r)g(x(t)) + \frac{d}{dt}\int_{t-r}^{t} c(s+r)g(x(s))\,ds \\ &\quad + b(t)[g(x(t-r)) - G(x(t-r))] \\ &= -c(t+r)x(t) + c(t+r)[x(t) - g(x(t))] \\ &\quad + b(t)[g(x(t-r)) - G(x(t-r))] \\ &\quad + \frac{d}{dt}\int_{t-r}^{t} c(s+r)g(x(s))\,ds.\end{aligned}$$

By the variation of parameters formula

$$\begin{aligned}x(t) &= \psi(0)e^{-\int_0^t c(s+r)\,ds} \\ &\quad + \int_0^t e^{-\int_s^t c(u+r)\,du}c(s+r)[x(s) - g(x(s))]\,ds \\ &\quad + \int_0^t e^{-\int_s^t c(u+r)\,du}b(s)[g(x(s-r)) - G(x(s-r))]\,ds \\ &\quad + \int_0^t e^{-\int_s^t c(u+r)\,du}\frac{d}{ds}\int_{s-r}^{s} c(u+r)g(x(u))\,du.\end{aligned} \tag{3.2.5}$$

If we integrate the last term by parts we obtain the desired form.

We will now look at the case for $\lim_{x\to 0}\frac{G(x)}{g(x)} = 1$. In this discussion we will have expressions of the form

$$a(t,s)g(x) + b(t,s)G(x) \tag{3.2.6}$$

with an $L > 0$ so that $g, G : [-L, L] \to R$ are continuous, a, b are continuous for $0 \leq s \leq t < \infty$,

$$g \quad \text{and} \quad G \quad \text{are odd and satisfy a Lipschitz condition,} \tag{3.2.7}$$

and

$$xg(x) > 0 \quad \text{and} \quad xG(x) > 0 \quad \text{if} \quad x \neq 0. \tag{3.2.8}$$

Condition (3.2.8) is often necessary for stability in the problems considered. But now we come to three conditions which seem unusual. We want to show that they can be substantially reduced. We will ask that

$$0 \leq x - g(x) \quad \text{and} \quad x - g(x) \quad \text{is nondecreasing on} \quad [0, L]. \tag{3.2.9}$$

Note 1. If $|\frac{d}{dx}g(x)| \leq D$ on $[0, L]$, for some $D > 0$, then write (3.2.6) as

$$[Da(t,s)][g(x)/D] + b(t,s)G(x)$$

and we then see that $0 \leq x - g(x)/D$ and $x - g(x)/D$ is nondecreasing. Thus, grouping terms accomplishes (3.2.9). Next, we need

$$\lim_{x \to 0} \frac{G(x)}{g(x)} = 1. \tag{3.2.10}$$

Note 2. If $\lim_{x\to 0} \frac{DG(x)}{g(x)} = \xi \neq 0$, then rewrite (3.2.6) as

$$Da(t,s)g(x)/D + \xi b(t,s)G(x)/\xi$$

and rename $G(x)/\xi$ as G. Thus, our two critical conditions (3.2.9) and (3.2.10) can both be avoided if $\xi \neq 0$ and if $\frac{d}{dx}g$ exists.

Finally, we need conditions we can not so easily avoid:

$$\begin{aligned} &g,\, G \quad \text{odd and Lipschitz,} \\ &\qquad g(x),\ |g(x) - G(x)| \quad \text{nondecreasing on} \quad [0, L]. \end{aligned} \tag{3.2.11}$$

The term $|g(x) - G(x)|$ can be treated as in our Note 1.

Consider the equation

$$x'(t) = -a(t)g(x(t-r)) - b(t)G(x(t-r)). \tag{3.2.12}$$

Theorem 3.2.2. *Suppose that $c(t) := a(t) + b(t) \geq 0$ and that (3.2.8) through (3.2.11) hold. If, in addition,*

$$\sup_{t \geq 0} \int_{t-r}^{t} c(s+r)\, ds < 1/2, \tag{3.2.13}$$

and

$$\int_0^t e^{-\int_s^t c(u+r)\, du} |b(s)|\, ds \quad \text{is bounded} \tag{3.2.14}$$

then the zero solution of (3.2.12) is stable.

Proof. By (3.2.13) there is a $\beta < 1$ with

$$2 \sup_{t \geq 0} \int_{t-r}^{t} c(u+r)\, du \leq \beta. \tag{3.2.15}$$

Take

$$\alpha = \frac{1+\beta}{2} \tag{3.2.16}$$

and fix $L_1 > 0$ such that $0 < L < L_1$ implies that

$$\begin{aligned} \frac{|g(L) - G(L)|}{g(L)} &\sup_{t \geq 0} \int_0^t e^{-\int_s^t c(u+s)du} |b(s)|\, ds \\ &+ 2 \sup_{t \geq 0} \int_{t-r}^{t} c(u+r)\, du \leq \alpha. \end{aligned} \tag{3.2.17}$$

Find $\delta > 0$ such that $\|\psi\| < \delta$ implies that

$$\|\psi\| + \|g(\psi)\| \int_{-r}^{0} c(u+r)\, du \leq g(L)(1-\alpha). \tag{3.2.18}$$

Define

$$\mathcal{M} = \{\phi : [-r, \infty) \to R \,|\, \phi_0 = \psi, \phi \in \mathcal{C}, |\phi(t)| \leq L\}$$

and use (3.2.4) to define P as we have done before. For $\phi \in \mathcal{M}$ we have

$$\begin{aligned} |(P\phi)(t)| \leq \|\psi\| + \|g(\psi)\| &\int_{-r}^{0} c(u+r)\, du + L - g(L) \\ &+ |g(L) - G(L)| \sup_{t \geq 0} \int_0^t e^{-\int_s^t c(u+r)\, du} |b(s)|\, ds \\ &+ 2g(L) \sup_{t \geq 0} \int_{t-r}^{t} c(s+r)\, ds \\ &\leq g(L)(1-\alpha) + L - g(L) + \alpha g(L) \leq L. \end{aligned}$$

We have shown that $P : \mathcal{M} \to \mathcal{M}$ and we can change the metric to one with an exponential weight so that P is a contraction with fixed point ϕ. As the argument works for a given $\varepsilon = L < L_1$ this will prove stability at $t_0 = 0$. The proof for a general t_0 is completely parallel.

Example 3.2.1. Let $0 < r < 1/2$, $g(x) = x$, $G(x) = (\operatorname{sgn} x)\ln(1+x)$, $a(t) = 1 - 2\sin t$, $b(t) = 2\sin t$. Then $c(t) = 1$, (3.2.10) and (3.2.14) are satisfied, while $\int_{t-r}^{t} c(s+r)\,ds = r < 1/2$ satisfies (3.2.15). Thus,

$$x'(t) = -(1-2\sin t)x(t-r) - 2\sin t(\operatorname{sgn} x(t-r))\ln(1+|x(t-r)|) \quad (3.2.19)$$

is stable for

$$r < 1/2. \quad (3.2.20)$$

Condition (3.2.10) is critical. Yet, without it we can still borrow the $b(t)$ and add it to $a(t)$, but there is a cost. Consider

$$x'(t) = -a(t)g(x(t-r)) - b(t)G(x(t-r)) \quad (3.2.21)$$

with a, b continuous and with a continuous initial function

$$\psi : [-r, 0] \to R. \quad (3.2.22)$$

Using Theorem 3.2.1 we define the mapping equation $(P\phi)(t) = \psi(t)$ for $-r \le t \le 0$ and

$$\begin{aligned}(P\phi)(t) &= \psi(0)e^{-\int_0^t c(s+r)\,ds} \\ &\quad - e^{-\int_0^t c(u+r)\,du}\int_{-r}^{0} c(u+r)g(\psi(u))\,du \\ &\quad + \int_0^t e^{-\int_s^t c(u+r)\,du}c(s+r)[\phi(s) - g(\phi(s))]\,ds \\ &\quad + \int_0^t e^{-\int_s^t c(u+r)\,du}b(s)[g(\phi(s-r)) - G(\phi(s-r))]\,ds \\ &\quad - \int_0^t e^{-\int_s^t c(u+r)\,du}c(s+r)\int_{s-r}^{s} c(u+r)g(\phi(u))\,du\,ds \\ &\quad + \int_{t-r}^{t} c(u+r)g(\phi(u))\,du. \end{aligned} \quad (3.2.23)$$

Theorem 3.2.3. *Let (3.2.8), (3.2.9) and (3.2.11) hold. Define*

$$H(L) := \sup_{0 \le x \le L} |g(x) - G(x)| \quad (3.2.24)$$

and

$$c(t) := a(t) + b(t) \ge 0. \quad (3.2.25)$$

Assume that

$$\frac{H(L)}{g(L)}\sup_{t\geq 0}\int_0^t e^{-\int_s^t c(u+r)\,du}|b(s)|\,ds$$
$$+\,2\sup_{t\geq 0}\int_{t-r}^t c(u+r)\,du<1. \tag{3.2.26}$$

Under these conditions there is a $\delta>0$ such that if ψ is a continuous initial function with $\|\psi\|<\delta$ then the unique solution $x(t,0,\psi)$ of (3.2.21) satisfies $|x(t,0,\psi)|<L$ for $t\geq 0$.

Proof. Define

$$\mathcal{M}=\{\phi:[-r,\infty)\to R\,|\,\phi\in\mathcal{C},\phi_0=\psi,|\phi(t)|\leq L\}.$$

Now $\phi\in\mathcal{M}$ so $|\phi(t)|\leq L$ and by our assumptions we will get

$$|(P\phi)(t)|\leq\|\psi\|+\|g(\psi)\|\int_{-r}^0 c(u+r)\,du$$
$$+\,L-g(L)+H(L)\sup_{t\geq 0}\int_0^t e^{-\int_s^t c(u+r)du}|b(s)|\,ds$$
$$+\,2g(L)\sup_{t\geq 0}\int_{t-r}^t c(u+r)\,du\leq L$$

upon application of (3.2.26), provided that $\|\psi\|<L_1$ and L_1 is small enough. This will yield $P:\mathcal{M}\to\mathcal{M}$. Because $P:\mathcal{M}\to\mathcal{M}$, because the functions are Lipschitz, and because the mapping is represented by integrals, it is possible to define an exponentially weighted metric under which P is a contraction. The unique fixed point is a solution of (3.2.21).

Remark 3.2.1. *If we compare (3.2.26) with (3.2.14) we see that the first term in (3.2.26)*

$$\frac{H(L)}{g(L)}\sup_{t\geq 0}\int_0^t e^{-\int_s^t c(u+r)\,du}|b(s)|\,ds \tag{3.2.27}$$

may be called the "cost" of borrowing $b(t)$ and adding it to $a(t)$. When (3.2.10) holds, there is no cost.

We now study the case of multiple delays. Consider the scalar equation

$$x'(t) = -a(t)x(t-r) - b(t)x(t-h) - d(t)x(t/2) \tag{3.2.28}$$

where a, b, d are continuous with $0 \leq h \leq r$, where

$$c(t) := a(t+r) + b(t+h) + 2d(2t), \tag{3.2.29}$$

and where

$$-\int_0^t c(s)ds \quad \text{is bounded above.} \tag{3.2.30}$$

Theorem 3.2.4. *If (3.2.30) holds and if there is an $\alpha < 1$ with*

$$\begin{aligned}
&\int_{t-r}^t |a(u+r)|\,du + \int_{t-h}^t |b(u+h)|\,du + \int_{t/2}^t 2|d(2u)|\,du \\
&+ \int_0^t e^{-\int_s^t c(u)\,du}|c(s)|\Big[\int_{s-r}^s |a(u+r)|\,du \\
&+ \int_{s-h}^s |b(u+h)|\,du + \int_{s/2}^s 2|d(2u)|\,du\Big]\,ds \\
&\leq \alpha
\end{aligned} \tag{3.2.31}$$

then for each continuous initial function $\psi : [-r, 0] \to R$ the unique solution $x(t, 0, \psi)$ of (3.2.28) is bounded; the zero solution of (3.2.28) is stable. If, in addition,

$$\int_0^t c(u)du \to \infty \quad \text{as} \quad t \to \infty, \tag{3.2.32}$$

then the zero solution of (3.2.28) is asymptotically stable.

Proof. We can write (3.2.28) as

$$\begin{aligned}
x'(t) &= -a(t+r)x(t) - b(t+h)x(t) - 2d(2t)x(t) \\
&\quad + \frac{d}{dt}\bigg(\int_{t-r}^t a(s+r)x(s)\,ds \\
&\quad + \int_{t-h}^t b(s+h)x(s)\,ds + \int_{t/2}^t 2d(2s)x(s)\,ds\bigg)
\end{aligned}$$

and for a continuous initial function $\psi : [-r, 0] \to R$ we have

$$x(t) = \psi(0)e^{-\int_0^t c(s)\,ds} + \int_0^t e^{-\int_s^t c(u)\,du} \frac{d}{ds}\bigg(\int_{s-r}^s a(u+r)x(u)\,du + \int_{s-h}^s b(u+h)x(u)\,du + \int_{s/2}^s 2d(2u)x(u)\,du\bigg).$$

Upon integration by parts we obtain

$$\begin{aligned} x(t) &= \psi(0)e^{-\int_0^t c(s)\,ds} \\ &- e^{-\int_0^t c(u)\,du}\bigg(\int_{-r}^0 a(u+r)\psi(u)\,du + \int_{-h}^0 b(u+h)\psi(u)du\bigg) \\ &+ \int_{t-r}^t a(u+r)x(u)\,du + \int_{t-h}^t b(u+h)x(u)\,du \\ &+ \int_{t/2}^t 2d(2u)x(u)\,du \\ &- \int_0^t e^{-\int_s^t c(u)\,du}c(s)\bigg(\int_{s-r}^s a(u+r)x(u)\,du + \int_{s-h}^s b(u+h)x(u)\,du \\ &+ \int_{s/2}^s 2d(2u)x(u)\,du\bigg)\,ds. \end{aligned} \tag{3.2.33}$$

We can use this equation to define a mapping P as we have done before. The conditions of the theorem will show that P will map bounded continuous functions into bounded continuous functions. Moreover, P is a contraction because of (3.2.31). This can be used to show the stability. With the addition of (3.2.32), the mapping set can be taken as the complete metric (supremum metric) space of bounded continuous functions (agreeing with the initial function on the initial interval) which tend to zero as $t \to \infty$. The classical proof that the convolution of an L^1 function with a function which tends to zero does itself tend to zero can be used to prove that $P\phi$ is in the set whenever ϕ is.

This example is linear, but it works equally well for nonlinear problems. There are simply more terms in the calculations. We can consider

$$x'(t) = -a(t)g(x(t-r)) - b(t)G(x(t-h)) - d(t)Q(x(t/2)), \tag{3.2.34}$$

obtain ode terms by the introduction of the neutral integrals, obtain linear terms by adding and subtracting, and proceed just as we did in Theorem 3.2.1.

Here are the details. We ask that $|x - g(x)|, |x - G(x)|, |x - Q(x)|$ are all nondecreasing on an interval $[0, L]$ and that g, G, Q are all odd and have the sign of x on $[-L, L]$. Write (3.2.34) as

$$\begin{aligned} x'(t) = -a(t)x(t-r) &+ a(t)[x(t-r) - g(x(t-r))] \\ &- b(t)x(t-h) + b(t)[x(t-h) - G(x(t-h))] \\ &- d(t)x(t/2) + d(t)[x(t/2) - Q(x(t/2))]. \end{aligned}$$

Thus, in the variation of parameters formula (3.2.33) add the term

$$\begin{aligned} \int_0^t e^{-\int_s^t c(u)\,du} \Big(& a(s)[x(s-r) - g(x(s-r))] \\ &+ b(s)[x(s-h) - G(x(s-h))] \\ &+ d(s)[x(s/2) - Q(x(s/2))] \Big)\, ds. \end{aligned} \tag{3.2.35}$$

Define the mapping set as we did in the proof of Theorem 3.2.4 and define the mapping P using (3.2.33), augmented by (3.2.35). Thus, for $|\phi(t)| \leq L$ we will have

$$\begin{aligned} &|(P\phi)(t)| \leq \psi(0)e^{-\int_0^t c(s)\,ds} \\ &- e^{-\int_0^t c(u)\,du}\left(\int_{-r}^0 a(u+r)\psi(u)\,du + \int_{-h}^0 b(u+h)\psi(u)\,du\right) \\ &+ L\int_{t-r}^t |a(u+r)|\,du + L\int_{t-h}^t |b(u+h)|\,du + L\int_{t/2}^t 2|d(2u)|\,du \\ &+ \int_0^t e^{-\int_s^t c(u)\,du}|c(s)|\left(L\int_{s-r}^s |a(u+r)|\,du\right. \\ &\left. + L\int_{s-h}^s |b(u+h)|\,du + L\int_{s/2}^s 2|d(2u)|\,du\right) ds \\ &+ |L - g(L)|\int_0^t e^{-\int_s^t c(u)du}|a(s)|\,ds \\ &+ |L - G(L)|\int_0^t e^{-\int_s^t c(u)\,du}|b(s)|\,ds \\ &+ |L - Q(L)|\int_0^t e^{-\int_s^t c(u)du}|d(s)|\,ds. \end{aligned} \tag{3.2.36}$$

The condition for stability is that this quantity be bounded by L; that will mean that our usual set $\mathcal{M}$ will be mapped into itself by P. We then need to ask that g, G, Q satisfy a local Lipschitz condition so that the metric can be changed to make P a contraction. Three basic results will then follow from (3.2.36). The last three terms in (3.2.36) represent the cost of changing the nonlinear terms to linear terms; Theorem 3.2.7 shows how to make the cost go to zero.

Theorem 3.2.5. *Suppose there is an $L > 0$ such that g, G, Q are locally Lipschitz, odd, and have the sign of x on $[-L, L]$, that $|x - g(x)|, |x - G(x)|, |x - Q(x)|$ are nondecreasing on $[0, L]$, and that there is an $\alpha < 1$ with*

$$\begin{aligned}
&L\int_{t-r}^{t} |a(u+r)|\,du + L\int_{t-h}^{t} |b(u+h)|\,du + L\int_{t/2}^{t} 2|d(2u)|\,du \\
&+ L\int_0^t e^{-\int_s^t c(u)\,du}|c(s)|\left(\int_{s-r}^{s} |a(u+r)|\,du\right. \\
&\left.+ \int_{s-h}^{s} |b(u+h)|\,du + \int_{s/2}^{s} 2|d(2u)|\,du\right) ds \\
&+ |L - g(L)|\int_0^t e^{-\int_s^t c(u)\,du}|a(s)|\,ds \\
&+ |L - G(L)|\int_0^t e^{-\int_s^t c(u)\,du}|b(s)|\,ds \\
&+ |L - Q(L)|\int_0^t e^{-\int_s^t c(u)du}|d(s)|\,ds \le \alpha L. \qquad (3.2.37)
\end{aligned}$$

If (3.2.30) holds then the zero solution of (3.2.34) is stable.

Theorem 3.2.6. *If $a(t), b(t), d(t)$ are all non-negative, then (3.2.37) in Theorem 3.2.5 can be replaced by*

$$\begin{aligned}
&2L\int_{t-r}^{t} a(u+r)\,du + 2L\int_{t-h}^{t} b(u+h)\,du + 2L\int_{t/2}^{t} 2d(2u)\,du \\
&+ |L - g(L)|\int_0^t e^{-\int_s^t c(u)du}a(s)\,ds + |L - G(L)|\int_0^t e^{-\int_s^t c(u)du}b(s)\,ds \\
&+ |L - Q(L)|\int_0^t e^{-\int_s^t c(u)du}d(s)\,ds \le \alpha L. \qquad (3.2.38)
\end{aligned}$$

Theorem 3.2.7. *If $a(t), b(t), d(t)$ are all non-negative and if*

$$\lim_{x\to 0}\frac{g(x)}{x} = \lim_{x\to 0}\frac{G(x)}{x} = \lim_{x\to 0}\frac{Q(x)}{x} = 1 \qquad (3.2.39)$$

then (3.2.37) in Theorem 3.2.6 can be replaced by

$$2L\int_{t-r}^{t} a(u+r)\,du + 2L\int_{t-h}^{t} b(u+h)\,du + 2L\int_{t/2}^{t} 2d(2u)\,du \le \alpha L. \tag{3.2.40}$$

There are numerous other combinations. Not all of the limits in (3.2.39) need be 1, but (3.2.40) is then modified. In Theorems 3.2.6 and 3.2.7 we do not need to ask that all of the coefficients be positive, but that will demand other changes.

We will now study an averaged delay equation. Consider the scalar equation

$$x'(t) = -\int_{t-r}^{t} a(t,s)g(x(s))\,ds - \int_{t-r}^{t} b(t,s)G(x(s))\,ds \tag{3.2.41}$$

where r is a positive constant and $a,\, b : [0,\infty)\times[0,\infty) \to R$ are continuous. Write (3.2.41) as

$$x'(t) = -\int_{t-r}^{t} [a(t,s)+b(t,s)]g(x(s))\,ds + \int_{t-r}^{t} b(t,s)[g(x(s)) - G(x(s))]\,ds.$$

Define

$$A(t,s) := \int_{t-s}^{r} [a(u+s,s)+b(u+s,s)]\,du \tag{3.2.42}$$

and ask that

$$A(t,t) = \int_{0}^{r} [a(u+t,t)+b(u+t,t)]\,du \ge 0 \tag{3.2.43}$$

with

$$\beta := 2\sup_{t\ge 0}\int_{t-r}^{t} |A(t,u)|\,du < 1 \tag{3.2.44}$$

and

$$\int_{0}^{t} e^{-\int_s^t A(u,u)\,du}\int_{s-r}^{s} |b(s,u)|\,du\,ds \quad \text{is bounded.} \tag{3.2.45}$$

Theorem 3.2.8. *Let (3.2.8)-(3.2.11) and (3.2.43)-(3.2.45) hold. Then the zero solution of (3.2.41) is stable.*

Proof. We first define some constants. For

$$\alpha = \frac{1+\beta}{2}$$

fix $L > 0$ so that

$$\frac{|g(L) - G(L)|}{g(L)} \sup_{t \geq 0} \int_0^t e^{-\int_s^t A(u,u)\,du} \int_{s-r}^s |b(s,u)|\,du\,ds + \beta < \alpha.$$

For this L, find $\delta > 0$ so that

$$\delta + g(\delta) \int_{-r}^0 |A(0,u)|\,du < g(L)(1-\alpha).$$

Then let $\|\psi\| < \delta$.

Now write (3.2.41) as

$$\begin{aligned} x'(t) &= -A(t,t)g(x(t)) + \frac{d}{dt}\int_{t-r}^t A(t,s)g(x(s))\,ds \\ &\quad + \int_{t-r}^t b(t,s)[g(x(s)) - G(x(s))]\,ds \\ &= -A(t,t)x(t) + A(t,t)[x - g(x)] \\ &\quad + \int_{t-r}^t b(t,s)[g(x(s)) - G(x(s))]\,ds \\ &\quad + \frac{d}{dt}\int_{t-r}^t A(t,s)g(x(s))\,ds. \end{aligned}$$

To specify a solution we need a continuous initial function $\psi : [-r, 0] \to [-L, L]$. By the variation of parameters formula we obtain

$$\begin{aligned} x(t) =& \psi(0)e^{-\int_0^t A(s,s)\,ds} - e^{-\int_0^t A(u,u)\,du} \int_{-r}^0 A(0,u)g(\psi(u))\,du \\ &+ \int_0^t e^{-\int_s^t A(u,u)\,du} A(s,s)[x(s) - g(x(s))]\,ds \\ &+ \int_0^t e^{-\int_s^t A(u,u)\,du} \int_{s-r}^s b(s,u)[g(x(u)) - G(x(u))]\,du\,ds \\ &+ \int_{t-r}^t A(t,u)g(x(u))\,du \\ &- \int_0^t e^{-\int_s^t A(u,u)\,du} A(s,s) \int_{s-r}^s A(s,u)g(x(u))\,du\,ds. \end{aligned}$$

Next, define a mapping set

$$\mathcal{M} = \{\phi : [-r, \infty) \to R \mid \phi_0 = \psi, \phi \in \mathcal{C}, |\phi(t)| \le L\}.$$

Then use the above formula for $x(t)$ to define a mapping $P : \mathcal{M} \to \mathcal{M}$ by $\phi \in \mathcal{M}$ implies $(P\phi)(t) = \psi(t)$ on $[-r, 0]$ and for $t > 0$ define

$$\begin{aligned}
(P\phi)(t) &= \psi(0)e^{-\int_0^t A(s,s)\,ds} - e^{-\int_0^t A(u,u)\,du} \int_{-r}^{0} A(0,u)g(\psi(u))\,du \\
&+ \int_0^t e^{-\int_s^t A(u,u)\,du} A(s,s)[\phi(s) - g(\phi(s))]\,ds \\
&+ \int_0^t e^{-\int_s^t A(u,u)\,du} \int_{s-r}^{s} b(s,u)[g(\phi(u)) - G(\phi(u))]\,du \\
&+ \int_{t-r}^{t} A(t,u)g(\phi(u))\,du \\
&- \int_0^t e^{-\int_s^t A(u,u)\,du} A(s,s) \int_{s-r}^{s} A(s,u)g(\phi(u))\,du\,ds.
\end{aligned}$$

Clearly, if $\phi \in \mathcal{M}$ then $P\phi$ is continuous. Also,

$$\begin{aligned}
|(P\phi)(t)| &\le \|\psi\| + \|g(\psi)\| \int_{-r}^{0} |A(0,u)|du + L - g(L) \\
&+ |g(L) - G(L)| \sup_{t \ge 0} \int_0^t e^{-\int_s^t A(u,u)\,du} \int_{s-r}^{s} |b(s,u)|\,du\,ds \\
&+ 2g(L) \sup_{t \ge 0} \int_{t-r}^{t} |A(t,u)|\,du \\
&\le \delta + g(\delta) \int_{-r}^{0} |A(0,u)|du + L - g(L) + g(L)\alpha \\
&\le g(L)(1-\alpha) + L - g(L) + g(L)\alpha \\
&= L.
\end{aligned}$$

It now follows that $P : \mathcal{M} \to \mathcal{M}$. Define a metric with an exponential weight which makes P a contraction with fixed point, a solution of (3.2.41). For a given ε sufficiently small, substitute ε for L and have a stability relation.

Remark 3.2.2. *Equations (3.2.1) and (3.2.41) can be combined and a stability result proved in the same way.*

We now study a Volterra equation. Consider the scalar equation

$$x'(t) = -\int_0^t a(t,s)g(x(s))\,ds - \int_0^t b(t,s)G(x(s))\,ds \tag{3.2.46}$$

where $a, b : [0,\infty) \times [0,\infty) \to R$ is continuous. In addition, suppose that there is a function $D(t,s)$ with

$$\frac{\partial D(t,s)}{\partial t} = -[a(t,s) + b(t,s)] \tag{3.2.47}$$

and

$$D(t,t) \geq 0. \tag{3.2.48}$$

For example, it may be possible to select

$$D(t,s) = \int_t^\infty [a(v,s) + b(v,s)]\,dv \tag{3.2.49}$$

if the integral exists.

Next, suppose that

$$\int_0^t e^{-\int_s^t D(u,u)\,du} \int_0^s |b(s,u)|\,du\,ds \quad \text{is bounded,} \tag{3.2.50}$$

and there is a $\beta < 1$ with

$$2\int_0^t |D(t,u)|du < \beta. \tag{3.2.51}$$

If (3.2.49) holds then we could write (3.2.51) as

$$2\int_0^t \left| \int_t^\infty [a(v,u) + b(v,u)]\,dv \right| du < \beta. \tag{3.2.52}$$

Theorem 3.2.9. *Let (3.2.8)-(3.2.11) hold. If (3.2.47), (3.2.48), (3.2.50), and (3.2.51) hold, then the zero solution of (3.2.46) is stable.*

Proof. We will need some constants defined in order to show that we have a proper mapping. For

$$\alpha = \frac{1+\beta}{2}$$

fix $L > 0$ so that

$$\frac{|g(L) - G(L)|}{g(L)} \int_0^t e^{-\int_s^t D(u,u)\,du} \int_0^s |b(s,u)|\,du\,ds + \beta < \alpha.$$

For this L, let

$$|x_0| < g(L)(1-\alpha).$$

Next, write (3.2.46) as

$$\begin{aligned}
x'(t) = &- \int_0^t [a(t,s) + b(t,s)]g(x(s))\,ds \\
&+ \int_0^t b(t,s)[g(x(s)) - G(x(s))]\,ds \\
= &-D(t,t)g(x(t)) + \frac{d}{dt}\int_0^t D(t,s)g(x(s))\,ds \\
&+ \int_0^t b(t,s)[g(x(s)) - G(x(s))]\,ds \\
= &-D(t,t)x(t) + D(t,t)[x(t) - g(x(t))] \\
&+ \frac{d}{dt}\int_0^t D(t,s)g(x(s))\,ds \\
&+ \int_0^t b(t,s)[g(x(s)) - G(x(s))]\,ds.
\end{aligned}$$

By the variation of parameters formula and integration by parts we obtain

$$\begin{aligned}
x(t) = &x_0 e^{-\int_0^t D(s,s)\,ds} \\
&+ \int_0^t e^{-\int_s^t D(u,u)\,du} D(s,s)[x(s) - g(x(s))]\,ds \\
&+ \int_0^t e^{-\int_s^t D(u,u)\,du} \int_0^s b(s,u)[g(x(u)) - G(x(u))]\,du\,ds \\
&+ \int_0^t D(t,u)g(x(u))\,du \\
&- \int_0^t e^{-\int_s^t D(u,u)du} D(s,s) \int_0^s D(s,u)g(x(u))\,du\,ds.
\end{aligned}$$

Define a set

$$\mathcal{M} = \{\phi : [0,\infty) \to R \mid \phi(0) = x_0, \phi \in \mathcal{C}, |\phi(t)| \leq L\}$$

and then use our equation for x to define a mapping $P : \mathcal{M} \to \mathcal{M}$ by $\phi \in \mathcal{M}$ implies that

$$\begin{aligned}(P\phi)(t) &= x_0 e^{-\int_0^t D(s,s)\,ds} \\ &\quad + \int_0^t e^{-\int_s^t D(u,u)\,du} D(s,s)[\phi(s) - g(\phi(s))]\,ds \\ &\quad + \int_0^t e^{-\int_s^t D(u,u)\,du} \int_0^s b(s,u)[g(\phi(u)) - G(\phi(u))]\,du\,ds \\ &\quad + \int_0^t D(t,u)g(\phi(u))\,du \\ &\quad - \int_0^t e^{-\int_s^t D(u,u)\,du} D(s,s) \int_0^s D(s,u)g(\phi(u))\,du\,ds.\end{aligned}$$

To see that P does map $\mathcal{M} \to \mathcal{M}$, note that if $\phi \in \mathcal{M}$ then $P\phi$ is continuous and $(P\phi)(0) = x_0$. Moreover, by the constants defined above it follows that

$$|(P\phi)(t)| \leq g(L)(1-\alpha) + L - g(L) + \alpha g(L) = L.$$

This shows that $P : \mathcal{M} \to \mathcal{M}$. It is now possible to define a metric with exponential weight showing that P is a contraction with unique fixed point $\phi \in \mathcal{M}$. That fixed point is a solution of (3.2.46). To see that we have stability, for a given $\varepsilon > 0$ with $\varepsilon < L$ for which our constants were chosen, substitute ε for L and take $\delta < g(L)(1-\alpha)$.

This will acknowledge that much of the work in this section was first published in *Dynamic Systems and Applications*, as detailed in Burton (2004c), which is published by Dynamic Publishers.

Chapter 4

Schauder's Theorem: A Choice

4.0 Introduction

In each result of the last chapter there was constructed a closed and bounded set $\mathcal{M}$ and a mapping $P : \mathcal{M} \to \mathcal{M}$. Only after that construction was the assumption used that the functions satisfied a Lipschitz condition. There was then constructed a weighted metric so that P became a contraction. The contraction mapping theorem is so much easier to use than Schauder's theorem because the former operates in a complete metric space, while the latter operates in a compact set. But if the functions do not satisfy a Lipschitz condition then one can turn to Schauder's theorem and there are great similarities in the processes of arriving at fixed points. In the following pages we will give several examples of how to show that mappings are equicontinuous. Here, Theorem 1.2.2 is central and we repeat it for reference.

Theorem 1.2.2. *Let $R^+ = [0, \infty)$ and let $q : R^+ \to R^+$ be a continuous function such that $q(t) \to 0$ as $t \to \infty$. If $\{\phi_k(t)\}$ is an equicontinuous sequence of R^d-valued functions on R^+ with $|\phi_k(t)| \leq q(t)$ for $t \in R^+$, then there is a subsequence that converges uniformly on R^+ to a continuous function $\phi(t)$ with $|\phi(t)| \leq q(t)$ for $t \in R^+$, where $|\cdot|$ denotes the Euclidean norm on R^d.*

The fact that we have the function q as an upper bound on the set is what makes it all work. But there is a way around this by using a weighted norm, as we did so often in the last chapter. We can change the closed bounded (in the supremum norm) set into a compact set if we can prove equicontinuity, but we will only conclude that the solution is a bounded function. The enabling result is found in Burton (1985; p. 169).

Theorem 4.0.1. *Let $g : [0, \infty) \to [1, \infty)$ be a continuous strictly increasing function with $g(0) = 1$ and $g(r) \to \infty$ as $r \to \infty$. Let $(\mathcal{S}, |\cdot|_g)$ be the space of continuous functions $f : [0, \infty) \to R^n$ for which*

$$\sup_{0 \le t < \infty} |f(t)/g(t)| =: |f|_g$$

exists. Then $(\mathcal{S}, |\cdot|_g)$ is a Banach space (called a $\mathcal{C}_g$-space). Moreover, for positive constants M and K the set

$$L := \{f \in \mathcal{S} \mid |f(t)| \le M \text{ on } [0, \infty), |f(u) - f(v)| \le K|u - v|\}$$

is compact.

Using this result, in all of our problems in which we mapped a closed bounded set into itself, if we do not have a Lipschitz condition and if we can prove equicontinuity, then we can use a weighted norm, obtain a fixed point, and conclude that there is a bounded solution.

We now study several problems in which we prove equicontinuity. Recall that we mentioned early on the advantage of contraction mappings over Schauder's and other fixed point theorems in that contractions called for a complete metric space, while Schauder's result asks for compactness. Our main purpose in this chapter is to illustrate technique. The problems studied can be done in more than one way. But here we show certain ways of working with Schauder's theorem. In particular, we attempt to give a great amount of detail concerning various kinds of stabilities.

While there are many problems in which we simply do not have a Lipschitz condition, this chapter serves as a motivation to try to work with contractions whenever possible. In that connection, there are very interesting problems concerning the possibility of proving contractions without a Lipschitz condition in a way somewhat parallel to proving a contraction when the Lipschitz constant is large, so long as we are working inside an integral. When we write $g(t, x) - g(t, y)$, that expression is almost certainly an integral

$$\int_y^x h(t, s)\, ds$$

where g is an anti-derivative of h. Thus, if our work is in the integrand of an integral with respect to t, as virtually all of ours is, we may have a double integral represented by $(P\phi)(t) - (P\eta)(t)$ in our contraction step. Several things should now be investigated. For example, can we possibly interchange the order of integration?

4.1 A Second Order Equation

So many of the equations we have considered are scalar equations and for them the variation of parameters formula is quite simple. It is known that if a linear equation has a coefficient matrix which is the sum of a scalar matrix and a skew symmetric matrix, then the variation of parameters formula is also quite simple. Thus, investigators strive to put linear systems in that form, plus a small perturbation. Transformations of that sort for both ordinary and partial differential equations are seen in Burton and Hatvani (1993). Our first result here is taken from Burton and Furumochi (2001b) in which we derive a transformation of the type just described. That transformation has proved to be very useful in several contexts. Morosanu and Vladimirescu (2005a,b) have used it with considerable success on the same problem, but with different methods.

Consider the scalar equation

$$x'' + 2f(t)x' + x + g(t,x) = 0, \quad t \in R^+, \tag{4.1.1}$$

where $R^+ := [0,\infty)$, $f(t)$ and $g(t,x)$ are continuous, $f(t) > 0$, $f(t) \to 0$ as $t \to \infty$, $\int_0^t f(s)ds \to \infty$ as $t \to \infty$, and

$$|f'(t) + f^2(t)| \le Kf(t),\, t \in R^+,\, K < 1, \tag{4.1.2}$$

$$|g(t,x)| \le Mf(t)|x|^\alpha,\, t \in R^+, \tag{4.1.3$_a$}$$

$$|g(t,x) - g(t,y)| \le L(\delta)f(t)|x-y|, \; t \in R^+, \; |x|, \; |y| \le \delta. \tag{4.1.3$_b$}$$

Here, K, M and $\alpha > 1$ are constant, and $L(\delta)$ is continuous and increasing. The linear part of equations such as (4.1.1) has been extensively studied, as may be seen in Hatvani (1996), for example. Condition (4.1.2) will allow f to be arbitrarily near zero for arbitrarily long periods of time. This is in contrast to our conditions in Section 2.4. There are intricate conditions which must be developed before we can state our main result, Theorem 4.1.1. The reader may wish, at this point, to look ahead at that result before continuing with the development here. Much of the work concerns construction of the function q of Theorem 1.2.2; this seems to be a main theme in our work with Schauder's theorem.

Change (4.1.1) to a system

$$\begin{aligned} x' &= y - f(t)x \\ y' &= (f'(t) + f^2(t) - 1)x - f(t)y - g(t,x) \end{aligned}$$

and write it as

$$z' = \begin{pmatrix} -f(t) & 1 \\ -1 & -f(t) \end{pmatrix} z + \begin{pmatrix} 0 & 0 \\ f'(t) + f^2(t) & 0 \end{pmatrix} z + \begin{pmatrix} 0 \\ -g(t,x) \end{pmatrix},$$

or

$$z' = A(t)z + B(t)z + F(t,z). \tag{4.1.4}$$

Notice that

$$A(t)\int_0^t A(s)ds = \left(\int_0^t A(s)ds\right)A(t).$$

Hence, the principal matrix solution of $z' = A(t)z$ is

$$\exp\left(\int_0^t A(s)ds\right) = \exp\left(-\int_0^t f(s)ds\right)\begin{pmatrix} \cos t & \sin t \\ -\sin t & \cos t \end{pmatrix}.$$

If $U(t) = (u_{ij}(t))$, then we define the norm by the maximum of the row sums of $(|u_{ij}(t)|)$ and we denote that norm by $\|U(t)\|$. We then see that the norms of the matrices in (4.1.4) satisfy

$$\left\|\exp\left(\int_s^t A(u)\,du\right)\right\| \leq \sqrt{2}\exp\left(-\int_s^t f(s)\,ds\right), \quad t \geq s \geq 0 \tag{4.1.5}$$

and

$$\|B(t)\| = |f'(t) + f^2(t)| \leq Kf(t),\, t \in R^+.$$

Now the solution $z(t)$ of (4.1.4) with $z(0) = z_0$ is

$$z(t) = e^{\int_0^t A(s)\,ds}\left(z_0 + \int_0^t e^{-\int_0^s A(u)du}(B(s)z(s) + F(s,z(s)))\,ds\right).$$

We will use Schauder's first theorem to prove that for each small z_0, a solution through z_0 tends to zero as $t \to \infty$. Recall that Schauder's first theorem was found in Chapter 1. We repeat it here for ready reference.

Theorem 1.2.5 Schauder's first fixed point theorem. *Let $\mathcal{M}$ be a nonempty compact convex subset of a Banach space and let $P : \mathcal{M} \to \mathcal{M}$ be continuous. Then P has a fixed point in $\mathcal{M}$.*

Here, we will take $\mathcal{C}$ to be the Banach space of bounded and continuous functions $\phi : R^+ \to R^2$ with the supremum norm, $\|\phi\|$ (which will cause no confusion with the matrix norm given above).

Let a be a number with $0 < a < ((1-K)/M)^\beta$, where $\beta := 1/(\alpha - 1)$, and let $|z_0| \le a$. Define

$$S_0 := \{\phi : R^+ \to R^2 \mid \phi(0) = z_0, |\phi(t)| \le q(t) \text{ on } R^+, \phi \in \mathcal{C}\},$$

where $|\cdot|$ denotes the Euclidean norm on R^2 and $q(t)$ is the unique solution of the initial value problem

$$x' = f(t)(K - 1 + Mx^{\alpha-1})x, \; x(0) = a.$$

Define a map P on S_0 by

$$(P\phi)(t) := e^{\int_0^t A(s)\,ds}\left(z_0 + \int_0^t e^{-\int_0^s A(u)\,du}(B(s)\phi(s) + F(s, \phi(s)))\,ds\right),$$

and maps P_i for $i = 1, 2, 3$ by

$$(P_1\phi)(t) := e^{\int_0^t A(s)\,ds} z_0,$$

$$(P_2\phi)(t) := e^{\int_0^t A(s)\,ds} \int_0^t e^{-\int_0^s A(u)\,du} B(s)\phi(s)\,ds,$$

and

$$(P_3\phi)(t) := e^{\int_0^t A(s)\,ds} \int_0^t e^{-\int_0^s A(u)\,du} F(s, \phi(s))\,ds.$$

Note that $(P_1\phi)(0) = z_0$ and $(P_2\phi)(0) = (P_3\phi)(0) = 0$.

Lemma 4.1.1. *If $\phi \in S_0$ then $|(P\phi)(t)| \le q(t), t \in R^+$.*

Proof. Let $\phi(t) = \begin{pmatrix} x(t) \\ y(t) \end{pmatrix}$. Then, for $t \in R^+$ we have

$$\begin{aligned}
|(P_1\phi)(t)| &= |e^{\int_0^t A(s)\,ds}\phi(0)| = \left| e^{-\int_0^t f(s)\,ds} \begin{pmatrix} x(0)\cos t + y(0)\sin t \\ -x(0)\sin t + y(0)\cos t \end{pmatrix} \right| \\
&= e^{-\int_0^t f(s)\,ds}|\phi(0)| \\
&\le a e^{-\int_0^t f(s)\,ds}.
\end{aligned}$$

Next, for $t \in R^+$ we obtain

$$\begin{aligned}
|(P_2\phi)(t)| =& \left| \int_0^t e^{\int_s^t A(u)\,du} B(s)\phi(s)\,ds \right| \\
=& \left| \int_0^t e^{\int_s^t A(u)\,du} \begin{pmatrix} 0 \\ 1 \end{pmatrix} (f'(s) + f^2(s))x(s)\,ds \right| \\
=& \left| \int_0^t e^{-\int_s^t f(u)\,du} \begin{pmatrix} \sin(t-s) \\ \cos(t-s) \end{pmatrix} (f'(s) + f^2(s))x(s)\,ds \right| \\
\leq& K \int_0^t e^{-\int_s^t f(u)\,du} f(s)|x(s)|\,ds.
\end{aligned}$$

Similarly, for $t \in R^+$ we have

$$\begin{aligned}
|(P_3\phi)(t)| =& \left| \int_0^t e^{\int_s^t A(u)du} F(s, \phi(s))\,ds \right| \\
=& \left| \int_0^t e^{-\int_s^t f(u)\,du} \begin{pmatrix} \sin(t-s) \\ \cos(t-s) \end{pmatrix} g(s, x(s))\,ds \right| \\
\leq& M \left| \int_0^t e^{-\int_s^t f(u)\,du} f(s)|x(s)|^\alpha\,ds \right|.
\end{aligned}$$

Recall that $x(t)$ is the first component of $\phi(t)$ and in the definition of S_0 we have $|\phi(t)| \leq q(t)$ so $|x(t)| \leq |\phi(t)| \leq q(t)$. This will now be used.

For $t \in R^+$ we obtain

$$\begin{aligned}
|(P\phi)(t)| \leq& ae^{-\int_0^t f(s)\,ds} + K \int_0^t e^{-\int_s^t f(u)du} f(s)|x(s)|\,ds \\
&+ M \int_0^t e^{-\int_s^t f(u)\,du} f(s)|x(s)|^\alpha\,ds \\
\leq& ae^{-\int_0^t f(s)\,ds} + K \int_0^t e^{-\int_s^t f(u)du} f(s)q(s)\,ds \\
&+ M \int_0^t e^{-\int_s^t f(u)\,du} f(s)q^\alpha(s)\,ds \\
=&: r(t).
\end{aligned}$$

Since we have $q'(t) = -f(t)q(t) + f(t)q(t)(K + Mq^{\alpha-1}(t))$, we obtain

$$\begin{aligned}
r(t) =& ae^{-\int_0^t f(s)\,ds} + \int_0^t e^{-\int_s^t f(u)\,du}(Kf(s)q(s) + Mf(s)q^\alpha(s))\,ds \\
=& ae^{-\int_0^t f(s)\,ds} + \int_0^t e^{-\int_s^t f(u)\,du}(q'(s) + f(s)q(s))\,ds.
\end{aligned}$$

If we integrate $\int_0^t e^{-\int_s^t f(u)du} q'(s)\,ds$ by parts we get $r(t) = q(t)$ on R^+, which gives the desired inequality.

Lemma 4.1.2. *There is a continuous increasing function $\delta = \delta(\varepsilon)$: $(0, 2a) \to (0, \infty)$ with*

$$|q(t_0) - q(t_1)| \leq \varepsilon \text{ if } 0 \leq t_0 < t_1 < t_0 + \delta \tag{4.1.6}$$

and

$$|(P\phi)(t_0) - (P\phi)(t_1)| \leq \varepsilon \text{ if } \phi \in S_0 \text{ and } 0 \leq t_0 < t_1 < t_0 + \delta. \tag{4.1.7}$$

Proof. First, it is easy to see that for any ε with $0 < \varepsilon < 2a$ there is a $\delta_1 > 0$ such that

$$|q(t_0) - q(t_1)| \leq \varepsilon \text{ if } 0 \leq t_0 < t_1 < t_0 + \delta_1. \tag{4.1.8}$$

Next, for any $\phi \in S_0$ we have

$$(P_1\phi)'(t) = A(t)e^{\int_0^t A(s)\,ds} z_0.$$

Let G be a number with $0 < f(t) \leq G$ on R^+. Then from (4.1.5) with $s = 0$ we obtain

$$\begin{aligned} |(P_1\phi)'(t)| \leq& \|A(t)\| \|e^{\int_0^t A(s)\,ds}\| |z_0| \\ \leq& \sqrt{2}(1+G)e^{-\int_0^t f(s)\,ds} a \\ \leq& \sqrt{2}(1+G)a. \end{aligned}$$

Thus, for any ε with $0 < \varepsilon < 2a$ there is a $\delta_2 > 0$ such that

$$|(P_1\phi)(t_0) - (P_1\phi)(t_1)| \leq \varepsilon \text{ if } 0 \leq t_0 < t_1 < t_0 + \delta_2. \tag{4.1.9}$$

Now let $T > 1$ be a number such that

$$q(t) \leq \varepsilon/6 \text{ if } t \geq T - 1.$$

Then, since we have $|(P_i\phi)(t)| \leq q(t), (i = 2, 3)$, it is easy to see that

$$|(P_i\phi)(t_0) - (P_i\phi)(t_1)| \leq \varepsilon/3 \text{ if } T - 1 \leq t_0 < t_1. \tag{4.1.10$_i$}$$

For any t_0 and t_1 with $0 \le t_0 < t_1 \le T$ we have

$$\begin{aligned}
&|(P_2\phi)(t_0) - (P_2\phi)(t_1)| \\
&\quad = \left| \int_0^{t_0} e^{\int_s^{t_0} A(u)\,du} B(s)\phi(s)\,ds - \int_0^{t_1} e^{\int_s^{t_1} A(u)\,du} B(s)\phi(s)\,ds \right| \\
&\quad \le \left| \int_0^{t_0} (e^{\int_s^{t_0} A(u)\,du} - e^{\int_s^{t_1} A(u)\,du}) B(s)\phi(s)\,ds \right| \\
&\quad + \left| \int_{t_0}^{t_1} e^{\int_s^{t_1} A(u)\,du} B(s)\phi(s)\,ds \right| \\
&\quad \le \int_0^{t_0} \|e^{\int_s^{t_0} A(u)\,du} - e^{\int_s^{t_1} A(u)\,du}\| \|B(s)\| |\phi(s)|\,ds \\
&\quad + \int_{t_0}^{t_1} \|e^{\int_s^{t_1} A(u)\,du}\| \|B(s)\| |\phi(s)|\,ds \\
&\quad \le Ka \int_0^{t_0} \|e^{\int_s^{t_0} A(u)\,du} - e^{\int_s^{t_1} A(u)\,du}\| f(s)\,ds \\
&\quad + \sqrt{2}Ka \int_{t_0}^{t_1} e^{-\int_s^{t_1} f(u)\,du} f(s)\,ds \\
&\quad \le GKa \int_0^{t_0} \|e^{\int_s^{t_0} A(u)\,du} - e^{\int_s^{t_1} A(u)\,du}\|\,ds + \sqrt{2}GKa|t_0 - t_1|.
\end{aligned}$$

For any η with $0 < \eta < T$, let

$$\begin{aligned}
d(\eta) := \sup\{&\|e^{\int_s^{t_0} A(u)\,du} \\
&- e^{\int_s^{t_1} A(u)\,du}\| \,|\, 0 \le s \le t_0 < t_1 \le T \text{ and } t_1 \le t_0 + \eta\}.
\end{aligned}$$

It is clear that $d(\eta) \to 0^+$ as $\eta \to 0^+$. Let δ_3 be a number such that $0 < \delta_3 < 1$, and that $d(\delta_3) \le \varepsilon/(6GKTa)$ and $\delta_3 \le \varepsilon/(6\sqrt{2}GKa)$. Then we have

$$|(P_2\phi)(t_0) - (P_2\phi)(t_1)| \le \varepsilon/3 \text{ if } 0 \le t_0 < t_1 \le T \text{ and } t_1 < t_0 + \delta_3,$$

which, together with $(4.1.10_2)$, yields

$$|(P_2\phi)(t_0) - (P_2\phi)(t_1)| \le \varepsilon/3 \text{ if } 0 \le t_0 < t_1 < t_0 + \delta_3. \tag{4.1.11}$$

Similarly, for any t_0 and t_1 with $0 \le t_0 < t_1 \le T$ we have

$$\begin{aligned}
&|(P_3\phi)(t_0) - (P_3\phi)(t_1)| \\
&= |\int_0^{t_0} e^{\int_s^{t_0} A(u)\,du} F(s,\phi(s))\,ds - \int_0^{t_1} e^{\int_s^{t_1} A(u)\,du} F(s,\phi(s))\,ds| \\
&\le M \int_0^{t_0} \|e^{\int_s^{t_0} A(u)\,du} - e^{\int_s^{t_1} A(u)\,du}\| f(s)|x(s)|^\alpha\,ds \\
&+ M \int_{t_0}^{t_1} \|e^{\int_s^{t_1} A(u)\,du}\| f(s)|x(s)|^\alpha\,ds \\
&\le GMa^\alpha \int_0^{t_0} \|e^{\int_s^{t_0} A(u)\,du} - e^{\int_s^{t_1} A(u)\,du}\|\,ds \\
&+ \sqrt{2}GMa^\alpha \int_{t_0}^{t_1} e^{-\int_s^{t_1} f(u)\,du}\,ds \\
&\le GMa^\alpha \int_0^{t_0} \|e^{\int_s^{t_0} A(u)\,du} - e^{\int_s^{t_1} A(u)\,du}\|ds + \sqrt{2}GMa^\alpha|t_0 - t_1|.
\end{aligned}$$

Finally, let δ_4 be a number such that $0 < \delta_4 < 1$, and that

$$d(\delta_4) \le \epsilon/(6GMTa^\alpha) \text{ and } \delta_4 \le \varepsilon/(6\sqrt{2}GMa^\alpha).$$

Then we obtain

$$|(P_3\phi)(t_0) - (P_3\phi)(t_1)| \le \varepsilon/3 \text{ if } 0 \le t_0 < t_1 \le T \text{ and } t_1 \le t_0 + \delta_4,$$

which together with $(4.1.10)_3$, yields

$$|(P_3\phi)(t_0) - (P_3\phi)(t_1)| \le \varepsilon/3 \text{ if } 0 \le t_0 < t_1 < t_0 + \delta_4. \tag{4.1.12}$$

Thus, from (4.1.8), (4.1.9), (4.1.11) and (4.1.12), for $\delta_5 := \min\{\delta_i : 1 \le i \le 4\}$, we have (4.1.6) and (4.1.7) with $\delta = \delta_5$. Since we may assume that $\delta_5(\varepsilon)$ is nondecreasing, we can easily conclude that there is a continuous increasing function $\delta : (0, 2a) \to (0, \infty)$ which satisfies (4.1.6) and (4.1.7).

At this point we remind the reader of Theorem 1.2.2 which was repeated for reference in Section 4.0.

Remark 4.1.1. *This result is a generalization of the Ascoli-Arzela lemma in the following sense. Let $\{\phi_k(t)\}$ be a uniformly bounded and equicontinuous sequence of R^d-valued functions on $[0,T]$, and let $|\phi_k(t)| \le Q$ on $[0,T]$. Define $q(t) := Qe^{T-t}$ for $t \in R^+$, and let ϕ_k be an extension to R of ϕ_k obtained by defining $\phi_k(t) := \phi_k(T)e^{T-t}$ for $t > T$. Then, q and $\{\phi_k(t)\}$ satisfy the assumptions of Theorem 1.2.2, and hence, the original sequence $\{\phi_k(t)\}$ contains a subsequence that converges uniformly on $[0,T]$ to a continuous function.*

Let S be a set of functions $\phi \in S_0$ such that for the function δ in Lemma 4.1.2,

$$|\phi(t_0) - \phi(t_1)| \leq \epsilon \text{ if } 0 \leq t_0 < t_1 < t_0 + \delta.$$

Lemma 4.1.3. *The set S is a compact convex nonempty subset of C, and the map $P : S \to S$ is continuous.*

Proof. Since the function $\phi(t) := (q(t)/a)z_0$ is contained in S, S is nonempty. Clearly S is a convex subset of C. Moreover, Theorem 1.2.2 implies that S is compact.

Next, we prove that the map $P : S \to S$ is continuous. For any $\phi \in S$, let $\xi := P\phi$. Clearly we have $\xi(0) = z_0$ and $\xi \in C$. Next, from Lemma 4.1.1 we obtain

$$|\xi(t)| \leq q(t), \ t \in R^+.$$

Thus, we have $\xi \in S_0$, which together with Lemma 4.1.2, implies that $\xi \in S$. Hence, P maps S into S. We need to prove that P is continuous. For any $\phi_i \in S$ $(i = 1, 2)$ and $t \in R^+$ we have

$$\begin{aligned}
&|(P\phi_1)(t) - (P\phi_2)(t)| \\
&\quad \leq \left| \int_0^t e^{\int_s^t A(u)du} (B(s)(\phi_1(s) - \phi_2(s)) \right. \\
&\quad \left. + (F(s, \phi_1(s)) - F(s, \phi_2(s)))\, ds \right| \\
&\quad \leq \sqrt{2} \int_0^t e^{-\int_s^t f(u)\,du} |Kf(s) + L(a)f(s)||\phi_1(s) - \phi_2(s)|\, ds \\
&\quad \leq \sqrt{2}(K + L(a)\|\phi_1 - \phi_2\| \int_0^t e^{-\int_s^t f(u)\,du} f(s)\, ds) \\
&\quad \leq \sqrt{2}(K + L(a)\|\phi_1 - \phi_2\|),
\end{aligned}$$

which implies that P is continuous.

In view of Schauder's first theorem we have the following result.

Theorem 4.1.1. *Under assumptions (4.1.2), (4.1.3$_a$), and (4.1.3$_b$), let a be a number with $0 < a < ((1 - K)/M)^\beta$, where $\beta := 1/(\alpha - 1)$, and let $|z_0| \leq a$. Then the solution $z(t, z_0)$ of (4.1.4) satisfies*

$$|z(t, z_0)| \leq q(t).$$

We conclude with an example.

Example 4.1.1. Consider the equation

$$x'' + [2/(t+1)]x' + x + x^2/(t+1) = 0, \; t \in R^+. \tag{4.1.14}$$

(Thus, $f(t) = 1/(t+1)$ and $g(t,x) = x^2/(t+1)$.) Clearly, $f(t) \to 0$ as $t \to \infty$, $\int_0^\infty f(s)ds = \infty$. Moreover, it is easy to see that (4.1.2) holds with $K = 0$, that (3_a) holds with $M = 1$ and $\alpha = 2$, and that (3_b) holds with $L(\delta) = 2\delta$. Change (4.1.14) to a system $z' = G(t,z)$ as before. Let a be a number with $0 < a < 1$, and let $|z_0| \leq a$. By our result, the solution $z(t, z_0)$ satisfies

$$|z(t, z_0)| \leq a/[1 + (1-a)t]$$

This will acknowledge that much of the work in this section was first published in *Funkcialaj Ekvacioj*, as detailed in Burton and Furumochi (2001b), which is published by Japana Matematik Societo.

4.2 Some Cubic Equations

In this section we consider several equations of the general type

$$x'(t) = -a(t)x^3(t) + b(t)x^3(t - r(t))$$

and use Schauder's fixed point theorem to obtain stability, uniform stability, asymptotic stability, and uniform asymptotic stability. Distributed delays are among the types considered. This work may be found in Burton and Furumochi (2005). We have considered this type of equation in both Chapters 2 and 3 with much different methods. This illustrates a choice of methods. More importantly, from the details with the simple function x^3 the investigator can see how to extend the results to include a $g(x)$ which is not locally Lipschitz.

We will be working in a space with weighted norm. Let r_0 be a fixed nonnegative constant and let $h : [-r_0, \infty) \to [1, \infty)$ be any strictly increasing and continuous function with $h(-r_0) = 1$, $h(t) \to \infty$ as $t \to \infty$. For any $t_0 \in R^+ := [0, \infty)$, let C_{t_0} be the space of continuous functions $\phi : [t_0 - r_0, \infty) \to R := (-\infty, \infty)$ with

$$\|\phi\|_h := \sup\left\{ \frac{|\phi(t)|}{h(t-t_0)} : t \geq t_0 - r_0 \right\} < \infty.$$

Then, clearly $\|\cdot\|_h$ is a norm on C_{t_0}, and $(C_{t_0}, \|\cdot\|_h)$ is a Banach space.

First we state a lemma without proof which may be found in Burton (1985; p. 169]).

Lemma 4.2.1. *If the set $\{\phi_k(t)\}$ of R-valued functions on $[t_0 - r_0, \infty)$ is uniformly bounded and equi-continuous, then there is a bounded and continuous function ϕ and a subsequence $\{\phi_{k_j}(t)\}$ such that $\|\phi_{k_j} - \phi\|_h \to 0$ as $j \to \infty$.*

Consider the scalar nonlinear equation

$$x'(t) = -a(t)x^3(t) + b(t)x^3(t - r(t)), \qquad t \in R^+ \tag{4.2.1}$$

where $a, r : R^+ \to R^+$ and $b : R^+ \to R$ are continuous. Let α be any fixed number with $0 < \alpha \leq 1/\sqrt{3}$. We assume that there are constants $r_0 \geq 0$ and $\gamma > 0$ so that

$$t - r(t) \geq -r_0, \tag{4.2.2}$$

$$\sigma = \sigma(t_0) := \sup_{t \geq t_0} \int_{t_0}^{t} (\gamma^3|b(s)| - a(s))\, ds < \infty \text{ for any } t_0 \in R^+, \tag{4.2.3}$$

and

$$\sup_{t \geq t_0 \geq 0} \left(\frac{\frac{1}{2\delta^2} + \int_{t_0}^{t} (a(s) - \gamma^3|b(s)|)\, ds}{\frac{1}{2\delta^2} + \int_{t_0}^{\tau(t)} (a(s) - \gamma^3|b(s)|)\, ds} \right)^{1/2} \leq \gamma \text{ for any } \delta \in (0, \eta], \tag{4.2.4}$$

where $\tau = \tau(t) := \max(t_0, t - r(t))$, and η is a number defined by

$$\eta = \eta(t_0) := \left(\frac{1}{\alpha^2} + 2\sigma(t_0) \right)^{-1/2}. \tag{4.2.5}$$

Corresponding to Equation (4.2.1), consider the scalar nonlinear equation

$$q' = (\gamma^3|b(t)| - a(t))q^3, \qquad t \in R^+. \tag{4.2.6}$$

Let $q : [t_0 - r_0, \infty) \to R$ be a continuous function such that $q(t) = \eta$ on $[t_0 - r_0, t_0]$, and $q(t)$ is the unique solution of the initial value problem

$$q' = (\gamma^3|b(t)| - a(t))q^3, \qquad q(t_0) = \eta, \qquad t \geq t_0.$$

Then $q(t)$ can be expressed as

$$\begin{aligned} q(t) &= \eta e^{-\int_{t_0}^{t} a(s)\, ds} + \int_{t_0}^{t} e^{-\int_s^t a(u) du} a(s)(q(s) - q^3(s))\, ds \\ &\quad + \gamma^3 \int_{t_0}^{t} e^{-\int_s^t a(u)\, du} |b(s)| q^3(s)\, ds \\ &= \left(\frac{1}{\eta^2} + 2 \int_{t_0}^{t} (a(s) - \gamma^3|b(s)|)\, ds \right)^{-1/2}, \qquad t \geq t_0, \end{aligned} \tag{4.2.7}$$

which together with (4.2.3) and (4.2.5), implies

$$0 < q(t) \leq \alpha, \qquad t \geq t_0. \tag{4.2.8}$$

Theorem 4.2.1. *Suppose that (4.2.2)-(4.2.4) hold. Then we have:*

(i) The zero solution of Equation (4.2.1) is stable.

(ii) If $\sigma^* := \sup\{\sigma(t) : t \in R^+\} < \infty$, then the zero solution of Equation (4.2.1) is uniformly stable.

(iii) If

$$\int_{t_0}^{t} (a(s) - \gamma^3 |b(s)|)\, ds \to \infty \qquad \text{as } t \to \infty, \tag{4.2.9}$$

then the zero solution of Equation (4.2.1) is asymptotically stable.

(iv) In addition to $\sigma^* < \infty$, if

$$\int_{t_0}^{t} (a(s) - \gamma^3 |b(s)|)\, ds \to \infty \qquad \text{uniformly for } t_0 \in R^+ \text{as } t \to \infty, \tag{4.2.10}$$

then the zero solution of Equation (4.2.1) is uniformly asymptotically stable.

Proof. (i) It is easy to see that the zero solution of Equation (4.2.6) is stable. Thus, for any $\varepsilon \in (0, \alpha]$ and $t_0 \in R^+$, there is a $\delta = \delta(\varepsilon, t_0)$ such that $0 < \delta \leq \eta$, and that for any $t_0 \in R^+$ and q_0 with $|q_0| \leq \delta$, we have $|q(t, t_0, q_0)| < \varepsilon$ for all $t \geq t_0$. For the t_0, let $(C_{t_0}, \|\cdot\|_h)$ be the Banach space of continuous functions $\phi : [t_0 - r_0, \infty) \to R$ with the norm $\|\cdot\|_h$. For a continuous function $\psi : [-r_0, 0] \to R$ with $\sup_{-r_0 \leq \theta \leq 0} |\psi(\theta)| \leq \delta$, let S be the set of continuous functions $\phi : [t_0 - r_0, \infty) \to R$ such that $\phi(t) = \psi(t - t_0)$ for $t_0 - r_0 \leq t \leq t_0$, $|\phi(t)| \leq q(t)$ for $t \geq t_0$, and $|\phi(t_1) - \phi(t_2)| \leq L|t_1 - t_2|$ for $t_1, t_2 \in R^+$ with $t_0 \leq \tau_1 \leq t_1, t_2 \leq \tau_2$, where $q(t)$ is defined by (4.2.7) with $\eta = \delta$, and where $L : R^+ \times R^+ \to R^+$ is a function defined by

$$L(\tau_1, \tau_2) := \max\{2a(t)\alpha + (a(t) + \gamma^3|b(t)|)\alpha^3 : \tau_1 \leq t \leq \tau_2\}.$$

Using (4.2.8), we obtain

$$|q'(t)| \leq (a(t) + \gamma^3|b(t)|)\alpha^3, \quad t \geq t_0.$$

Thus the function $\xi(t)$ defined by

$$\xi(t) := \begin{cases} \psi(t - t_0), & t_0 - r_0 \leq t \leq t_0 \\ \frac{\psi(0) q(t)}{\delta}, & t \geq t_0 \end{cases}$$

is an element of S, and from Lemma 4.2.1, S is a compact convex nonempty subset of C_{t_0}. Define a mapping P for $\phi \in S$ by

$$(P\phi)(t) := \begin{cases} \psi(t - t_0), & t_0 - r_0 \le t \le t_0 \\ \psi(0)e^{-\int_{t_0}^t a(s)\,ds} + \int_{t_0}^t e^{-\int_s^t a(u)\,du} a(s)(\phi(s) - \phi^3(s))\,ds \\ + \int_{t_0}^t e^{-\int_s^t a(u)\,du} b(s)\phi^3(s - r(s))\,ds, & t \ge t_0. \end{cases}$$

Then we have $(P\phi)(t) = \psi(t - t_0)$ for $t_0 - r_0 \le t \le t_0$, and from (4.2.4) and (4.2.7) with $\eta = \delta$ we obtain

$$\begin{aligned} |(P\phi)(t)| &\le \delta e^{-\int_{t_0}^t a(s)\,ds} + \int_{t_0}^t e^{-\int_s^t a(u)du} a(s)(q(s) - q^3(s))\,ds \\ &+ \int_{t_0}^t e^{-\int_s^t a(u)\,du} |b(s)| q^3(s - r(s))\,ds \\ &\le \delta e^{-\int_{t_0}^t a(s)\,ds} + \int_{t_0}^t e^{-\int_s^t a(u)\,du} a(s)(q(s) - q^3(s))\,ds \\ &+ \gamma^3 \int_{t_0}^t e^{-\int_s^t a(u)\,du} |b(s)| q^3(s)\,ds \\ &= q(t), \quad t \ge t_0. \end{aligned}$$

Moreover, it is easy to see that

$$(P\phi)'(t) = -a(t)(P\phi)(t) + a(t)(\phi(t) - \phi^3(t)) + b(t)\phi^3(t - r(t)), \quad t > t_0,$$

which implies

$$\begin{aligned} |(P\phi)'(t)| &\le a(t)q(t) + a(t)(q(t) - q^3(t)) + |b(t)| q^3(t - r(t)) \\ &\le 2a(t)\alpha + (a(t) + \gamma^3 |b(t)|)\alpha^3, \qquad t > t_0, \end{aligned}$$

and hence, P maps S into S. Clearly P is continuous. Thus, by Schauder's first theorem, P has a fixed point ϕ in S and that is the solution $x(t, t_0, \phi)$ of Equation (4.2.1) which satisfies

$$|x(t, t_0, \psi)| \le q(t) = q(t, t_0, \delta) < \varepsilon, \qquad t \ge t_0,$$

and hence, the zero solution of Equation (4.2.1) is stable.

(ii) If $\sigma^* < \infty$, then the zero solution of Equation (4.2.6) is uniformly stable. Since the uniform stability of the zero solution of Equation (4.2.1) can be similarly proved as in the proof of (i), we omit the details.

(iii) Assumption (4.2.9) implies that $q(t) \to 0$ as $t \to \infty$, and hence, the zero solution of Equation (4.2.6) is asymptotically stable. The proof of asymptotic stability of the zero solution of Equation (4.2.1) is similar to that of (i).

(iv) Assumptions $\sigma^* < \infty$ and (4.2.10) imply that the zero solution of Equation (4.2.6) is uniformly asymptotically stable. The proof of uniform asymptotic stability of the zero solution f (4.2.1) is similar to that of (i).

Example 4.2.1. Define functions $a, b, r : R^+ \to R^+$ by $a(t) := 2 + |t \sin t|, b(t) := \max(1, 1 + 2t \sin t)/27$ and $r(t) := 1/(t+1)$, and let $\alpha = 1/\sqrt{3}$. Then it can be seen that (4.2.2) with $r_0 = 1$, (4.2.3) and (4.2.4) with $\gamma = 3$ hold, and $\sigma^* = \infty$. Thus, concerning the stabilities of the zero solution of the equation

$$x'(t) = -a(t)x^3(t) + b(t)x^3(t - r(t)), \qquad t \in R^+, \tag{4.2.11}$$

Theorem 4.2.1 does not assure uniform stability, but assures stability.

Remark 4.2.1. *Notice the relations between a and b in (4.2.6) – (4.2.10). Simple examples are constructed by letting a and b have a common factor which will determine the character of the divergence of the integrals in (4.2.9) and (4.2.10). Thus, these examples are contrived to show the required behavior and to minimize the calculations.*

Example 4.2.2. Define a function $c : R^+ \to R^+$ by

$$c(t) := \begin{cases} n - n^2|t - 2n\pi|, & |t - 2n\pi| \leq \frac{1}{n} \\ 0, & \text{otherwise}, \end{cases}$$

where n denotes positive integers, and define a function $b : R^+ \to R^+$ by $b(t) := \max(c(t), 1 + \cos t)$. Let $a(t) = 9b(t)$, $r(t) = 1$ and $\alpha = 1/\sqrt{3}$. Then it is easily seen that (4.2.2) with $r_0 = 1$, (4.2.3) and (4.2.4) with $\gamma = 2$ hold, and $\eta = 1/\sqrt{3}$. Moreover, $\sigma^* < \infty$ and (4.2.10) hold. Thus, by Theorem 4.2.1, the zero solution of the equation

$$x'(t) = -9b(t)x^3(t) + b(t)x^3(t - 1), \qquad t \in R^+ \tag{4.2.12}$$

is uniformly asymptotically stable.

Example 4.2.3. Let $a(t) = 9/(t+2)$, $b(t) = 1/(t+2)$, $r(t) = 1$ and $\alpha = 1/\sqrt{3}$. Then (4.2.2) with $r_0 = 1$, (4.2.3) and (4.2.4) with $\gamma = 2$ hold, and $\eta = 1/\sqrt{3}$. Moreover, (4.2.10) does not hold, but $\sigma^* < \infty$ and (4.2.9) hold. Thus, concerning the stabilities of the zero solution of the equation

$$x'(t) = -\frac{9}{t+2}x^3(t) + \frac{1}{t+2}x^3(t - 1), \qquad t \in R^+, \tag{4.2.13}$$

Theorem 4.2.1 does not assure uniform asymptotic stability, but assures uniform stability and asymptotic stability.

Next consider the scalar nonlinear integrodifferential equation

$$x'(t) = -a(t)x^3(t) + \int_{t-r(t))}^{t} b(t,s)x^3(s)\,ds, \qquad t \in R^+, \tag{4.2.14}$$

where $a, r : R^+ \to R^+$ and $b : R^+ \times R \to R$ are continuous. Let α be any fixed number with $0 < \alpha \leq 1/\sqrt{3}$. We assume that (4.2.2) holds, and that there is a constant $\gamma > 0$ so that

$$\begin{aligned}\sigma = \sigma(t_0) :&= \sup_{t \geq t_0} \int_{t_0}^{t} (\gamma^3 \int_{s-r(s)}^{s} |b(s,u)| du - a(s)) ds \\ &< \infty \text{ for any } t_0 \in R^+\end{aligned} \tag{4.2.15}$$

and

$$\begin{aligned}&\sup_{t \geq t_0 \geq 0} \left\{ \sup_{\tau \leq v \leq t} \left(\frac{\frac{1}{2\delta^2} + \int_{t_0}^{t} (a(s) - \gamma^3 \int_{s-r(s)}^{s} |b(s,u)|\,du)\,ds}{\frac{1}{2\delta^2} + \int_{t_0}^{v} (a(s) - \gamma^3 \int_{s-r(s)}^{s} |b(s,u)|\,du)\,ds} \right)^{1/2} \right\} \\ &\leq \gamma \text{ for any } \delta \in (0, \eta],\end{aligned} \tag{4.2.16}$$

where $\tau = \tau(t) := \max(t_0, t - r(t))$, and η is defined by (4.2.5).

Corresponding to Equation (4.2.14), consider the scalar nonlinear equation

$$q' = \left(\gamma^3 \int_{t-r(t)}^{t} |b(t,s)|\,ds - a(t) \right) q^3, \qquad t \in R^+.$$

Let $q : [t_0 - r_0, \infty) \to R^+$ be a continuous function such that $q(t) = \eta$ on $[t_0 - r_0, t_0]$, and that $q(t)$ is the unique solution of the initial value problem

$$q' = \left(\gamma^3 \int_{t-r(t)}^{t} |b(t,s)|\,ds - a(t) \right) q^3, \qquad q(t_0) = \eta, \qquad t \geq t_0.$$

Then $q(t)$ can be expressed as

$$q(t) = \left(\frac{1}{\eta^2} + 2 \int_{t_0}^{t} (a(s) - \gamma^3 \int_{s-r(s)}^{s} |b(s,u)|\,du)\,ds \right)^{-1/2}, \qquad t \geq t_0,$$

which together with (4.2.5) and (4.2.15), implies (4.2.8).

Theorem 4.2.2. *Suppose that (4.2.2), (4.2.15) and (4.2.16) hold. Then:*

(i) The zero solution of Equation (4.2.14) is stable.

(ii) If $\sigma^* := \sup\{\sigma(t) : t \in R^+\} < \infty$, then the zero solution of Equation (4.2.14) is uniformly stable.

(iii) If

$$\int_{t_0}^{t}\left(a(s)-\gamma^3\int_{s-r(s)}^{s}|b(s,u)|\,du\right)ds \to \infty \qquad \text{as } t\to\infty, \qquad (4.2.17)$$

then the zero solution of Equation (4.2.14) is asymptotically stable.

(iv) In addition to $\sigma^* < \infty$, if

$$\int_{t_0}^{t}\left(a(s)-\gamma^3\int_{s-r(s)}^{s}|b(s,u)|\,du\right)ds \to \infty$$
$$\text{uniformly for } t_0 \in R^+ \text{ as } t\to\infty, \qquad (4.2.18)$$

then the zero solution of Equation (4.2.14) is uniformly asymptotically stable.

This theorem can be proved by using the set S in the proof of Theorem 4.2.1 for the above function $q(t)$, a function $L = L(\tau_1, \tau_2)$ with

$$2a(t)\alpha + \left(a(t)+\gamma^3\int_{t-r(t)}^{t}|b(t,s)|\,ds\right)\alpha^3 \le L \text{ for } \tau_1 \le t \le \tau_2,$$

and by defining a mapping P for $\phi \in S$ by

$$(P\phi)(t) := \begin{cases} \psi(t-t_0), & t_0 - r_0 \le t \le t_0 \\ \psi(0)e^{-\int_{t_0}^t a(u)\,du} + \int_{t_0}^t e^{-\int_s^t a(u)du}a(s)(\phi(s)-\phi^3(s))\,ds \\ +\int_{t_0}^t e^{-\int_s^t a(u)\,du}\int_{s-r(s)}^{s} b(s,u)\phi^3(u)\,du\,ds, & t > t_0. \end{cases}$$

We omit the details of the proof.

Example 4.2.4. Let $a(t) = 9(1+\sin t)$, $b(t,s) = 1+\sin t$, $r(t) = 1$ and $\alpha = 1/\sqrt{3}$. Then (4.2.2) with $r_0 = 1$, (4.2.15) and (4.2.16) with $\gamma = 2$ hold, and $\eta = 1/\sqrt{3}$. Moreover, $\sigma^* < \infty$ and (4.2.18) hold. Thus, by Theorem 4.2.2, the zero solution of the equation

$$x'(t) = -9(1+\sin t)x^3(t) + (1+\sin t)\int_{t-1}^{t} x^3(s)\,ds, \qquad t \in R^+$$

is uniformly asymptotically stable.

Example 4.2.5. Let $a(t) = 9/(t+2)$, $b(t,s) = 1/(t+2)$, $r(t) = 1$ and $\alpha = 1/\sqrt{3}$. Then (4.2.2) with $r_0 = 1$, (4.2.15) and (4.2.16) with $\gamma = 2$ hold, and $\eta = 1/\sqrt{3}$. Moreover, (4.2.18) does not hold, but $\sigma^* < \infty$ and

(4.2.17) do hold. Thus, concerning the stabilities of the zero solution of the equation

$$x'(t) = -\frac{9}{t+2}x^3(t) + \frac{1}{t+2}\int_{t-1}^{t} x^3(s)\,ds, \qquad t \in R^+,$$

Theorem 4.2.2 does not assure uniform asymptotic stability, but assures uniform stability and asymptotic stability.

Next, consider the scalar nonlinear equation

$$x'(t) = -a(t)x^3(t) + b(t)x(t-r(t))x^2(t), \qquad t \in R^+, \tag{4.2.19}$$

where $a, r : R^+ \to R^+$ and $b : R^+ \to R$ are continuous. Let α be any fixed number with $0 < \alpha \leq 1/\sqrt{3}$. We assume that (4.2.2) holds, and that there is a constant $\gamma > 0$ so that

$$\sigma = \sigma(t_0) := \sup_{t \geq t_0} \int_{t_0}^{t} (\gamma|b(s)| - a(s))ds < \infty \text{ for any } t_0 \in R^+ \tag{4.2.20}$$

and

$$\sup_{t \geq t_0 \geq 0} \left(\frac{\frac{1}{2\delta^2} + \int_{t_0}^{t} (a(s) - \gamma|b(s)|)\,ds}{\frac{1}{2\delta^2} + \int_{t_0}^{\tau} (a(s) - \gamma|(b(s)|)\,ds} \right)^{1/2} \leq \gamma \text{ for any } \delta \in (0, \eta], \tag{4.2.21}$$

where $\tau = \tau(t) := \max(t_0, t - r(t))$, and η is define by (4.2.5).

Corresponding to Equation (4.2.19), consider the scalar nonlinear equation

$$q' = (\gamma|b(t)| - a(t))q^3, \qquad t \in R^+.$$

Let $q : [t_0 - r_0, \infty) \to R$ be a continuous function such that $q(t) = \eta$ on $[t_0 - r_0, t_0]$, and that $q(t)$ is the unique solution of the initial value problem

$$q' = (\gamma|b(t)| - a(t))q^3, \qquad q(t_0) = \eta, \qquad t \geq t_0.$$

Then $q(t)$ can be expressed as

$$q(t) = \left(\frac{1}{\eta^2} + 2\int_{t_0}^{t} (a(s) - \gamma|b(s)|)\,ds \right)^{-1/2}, \qquad t \geq t_0,$$

which together with (4.2.5) and (4.2.20), implies (4.2.8).

Theorem 4.2.3. *Suppose that (4.2.2), (4.2.20) and (4.2.21) hold. Then:*

(i) The zero solution of Equation (4.2.19) is stable.

(ii) If $\sigma^* := \sup\{\sigma(t) : t \in R^+\} < \infty$, then the zero solution of Equation (4.2.19) is uniformly stable.

(iii) If

$$\int_{t_0}^{t} (a(s) - \gamma|b(s)|)\, ds \to \infty \quad \text{as } t \to \infty, \tag{42.22}$$

then the zero solution of Equation (4.2.19) is asymptotically stable.

(iv) In addition to $\sigma^* < \infty$, if

$$\int_{t_0}^{t} (a(s) - \gamma|b(s)|)\, ds \to \infty \quad \text{uniformly for } t_0 \in R^+ \text{ as } t \to \infty, \tag{4.2.23}$$

then the zero solution of Equation (4.2.19) is uniformly asymptotically stable.

This theorem can be proved by using the set S in the proof of Theorem 4.2.1 for the above function $q(t)$, a function $L = L(\tau_1, \tau_2)$ with

$$2a(t)\alpha + (a(t) + \gamma|b(t)|)\alpha^3 \leq L, \qquad \tau_1 \leq t \leq \tau_2,$$

and by defining a mapping P for $\phi \in S$ by

$$(P\phi)(t) := \begin{cases} \psi(t - t_0), & t_0 - r_0 \leq t \leq t_0, \\ \psi(0)e^{-\int_{t_0}^{t} a(s)\, ds} + \int_{t_0}^{t} e^{-\int_s^t a(u)\, du} a(s)(\phi(s) - \phi^3(s))\, ds & \\ + \int_{t_0}^{t} e^{-\int_s^t a(u)\, du} b(s)\phi(s - r(s))\phi^2(s)\, ds, & t > t_0. \end{cases}$$

We omit the details of the proof.

Remark 4.2.2. *As mentioned in Remark 4.2.1, the common factor in a and b dictates the divergence of the integrals and allows for fairly simple computations; it is certainly not necessary to have the common factor.*

Example 4.2.6. Let $a(t) = 3(1 + \sin t)$, $b(t) = 1 + \sin t$, $r(t) = 1$ and $\alpha = 1/\sqrt{3}$. Then (4.2.2) with $r_0 = 1$, (4.2.20) and (4.2.21) with $\gamma = 2$ hold, and $\eta = 1/\sqrt{3}$. Moreover, $\sigma^* < \infty$ and (4.2.23) hold. Thus, by Theorem 4.2.3, the zero solution of the equation

$$x'(t) = -3(1 + \sin t)x^3(t) + (1 + \sin t)x(t - 1)x^2(t), \qquad t \in R^+$$

is uniformly asymptotically stable.

Example 4.2.7. Let $a(t) = 3/(t+2)$, $b(t) = 1/(t+2)$, $r(t) = 1$ and $\alpha = 1/\sqrt{3}$. Then (4.2.2) with $r_0 = 1$, (4.2.20) and (4.2.21) with $\gamma = 2$ hold, and $\eta = 1/\sqrt{3}$. Moreover, (4.2.23) does not hold, but $\sigma^* < \infty$ and (4.2.22) hold. Thus, concerning the stabilities of the zero solution of the equation

$$x'(t) = -\frac{3}{t+2}x^3(t) + \frac{1}{t+2}x(t-1)x^2(t), \qquad t \in R^+,$$

Theorem 4.2.3 does not assure uniform asymptotic stability, but assures uniform stability and asymptotic stability.

Finally, for completeness of our discussion of Equation (4.2.1), consider the scalar nonlinear equation

$$x'(t) = -a(t)x^3(t) + b(t)x^2(t - r(t))x(t), \qquad t \in R^+, \tag{4.2.24}$$

where $a, r : R^+ \to R^+$ and $b : R^+ \to R$ are continuous. Let α be any fixed number with $0 < \alpha \leq 1/\sqrt{3}$. We assume that (4.2.2) holds, and that there is a constant $\gamma > 0$ so that

$$\sigma = \sigma(t_0) := \sup_{t \geq t_0} \int_{t_0}^{t} (\gamma^2|b(s)| - a(s))ds < \infty \text{ for any } t_0 \in R^+ \tag{4.2.25}$$

and

$$\sup_{t \geq t_0 \geq 0} \left(\frac{\frac{1}{2\delta^2} + \int_{t_0}^{t} (a(s) - \gamma^2|b(s)|)\, ds}{\frac{1}{2\delta^2} + \int_{t_0}^{\tau} (a(s) - \gamma^2|b(s)|)\, ds} \right)^{1/2} \leq \gamma \text{ for any } \delta \in (0, \eta], \tag{4.2.26}$$

where $\tau = \tau(t) := \max(t_0, t - r(t))$, and η is defined by (4.2.5).

Corresponding to Equation (4.2.24), consider the scalar nonlinear equation

$$q'(t) = (\gamma^2|b(t)| - a(t))q^3, \qquad t \in R^+.$$

Let $q : [t_0 - r_0, \infty) \to R$ be a continuous function such that $q(t) = \eta$ on $[t_0 - r_0, t_0]$, and that $q(t)$ is the unique solution of the initial value problem

$$q' = (\gamma^2|b(t)| - a(t))q^3, \qquad q(t_0) = \eta, \qquad t \geq t_0.$$

Then $q(t)$ can be expressed as

$$q(t) = \left(\frac{1}{\eta^2} + 2\int_{t_0}^{t} (a(s) - \gamma^2|b(s)|)\, ds \right)^{-1/2}, \qquad t \geq t_0,$$

which together with (4.2.5) and (4.2.25), implies (4.2.8).

Theorem 4.2.4. *Suppose that (4.2.2), (4.2.25) and (4.2.26) hold. Then:*

(i) The zero solution of Equation (4.2.24) is stable.

(ii) If $\sigma^* := \sup\{\sigma(t) : t \in R^+\} < \infty$, then the zero solution of Equation (4.2.24) is uniformly stable.

(iii) If

$$\int_{t_0}^{t} (a(s) - \gamma^2 |b(s)|)\, ds \to \infty \qquad \text{as } t \to \infty, \tag{4.2.27}$$

then the zero solution of Equation (4.2.24) is asymptotically stable.

(iv) In addition to $\sigma^* < \infty$, if

$$\int_{t_0}^{t} (a(s) - \gamma^2 |b(s)|)\, ds \to \infty \qquad \text{uniformly for } t_0 \in R^+ \text{ as } t \to \infty, \tag{4.2.28}$$

then the zero solution of Equation (4.2.24) is uniformly asymptotically stable.

This theorem can be proved by using the set S in the proof of Theorem 4.2.1 for the above function $q(t)$, a function $L = L(\tau_1, \tau_2)$ with

$$2a(t)\alpha + (a(t) + \gamma^2|b(t)|)\alpha^3 \le L, \qquad \tau_1 \le t \le \tau_2,$$

and by defining a mapping P for $\phi \in S$ by

$$(P\phi)(t) := \begin{cases} \psi(t - t_0), & t_0 - r_0 \le t \le t_0, \\ \psi(0)e^{-\int_{t_0}^t a(s)\,ds} + \int_{t_0}^t e^{-\int_s^t a(u)du} a(s)(\phi(s) - \phi^3(s))\, ds \\ + \int_{t_0}^t e^{-\int_s^t a(u)du} b(s)\phi^2(s - r(s))\phi(s)\, ds, & t > t_0. \end{cases}$$

Example 4.2.8. Let $a(t) = 5(1 + \sin t)$, $b(t) = 1 + \sin t$, $r(t) = 1$ and $\alpha = 1/\sqrt{3}$. Then (4.2.2) with $r_0 = 1$, (4.2.25) and (4.2.26) with $\gamma = 2$ hold, and $\eta = 1/\sqrt{3}$. Moreover, $\sigma^* < \infty$ and (4.2.28) hold. Thus, by Theorem 4.2.4, the zero solution of the equation

$$x'(t) = -5(1 + \sin t)x^3(t) + (1 + \sin t)x^2(t - 1)x(t), \qquad t \in R^+$$

is uniformly asymptotically stable.

Example 4.2.9. Let $a(t) = 5/(t + 2)$, $b(t) = 1/(t + 2)$, $r(t) = 1$ and $\alpha = 1/\sqrt{3}$. Then (4.2.2) with $r_0 = 1$, (4.2.25) and (4.2.26) with $\gamma = 2$ hold, and $\eta = 1/\sqrt{3}$. Moreover, (4.2.28) does not hold, but $\sigma^* < \infty$ and (4.2.27) do hold. Thus, concerning the stabilities of the zero solution of the equation

$$x'(t) = -\frac{5}{t + 2}x^3(t) + \frac{1}{t + 2}x^2(t - 1)x(t), \qquad t \in R^+,$$

Theorem 4.2.4 does not assure uniform asymptotic stability, but assures uniform stability and asymptotic stability.

Tetsuo Furumochi has played a leading role in the study of stability by Schauder's theorem. We refer the reader to Furumochi (2004) for some recent extensions of his earlier work. Additional results by Schauder's theorem are found in Burton and Furumochi (2002a) for the equations

$$x'(t) = -a(t)x(t) - b(t)g(x(t - r(t))),$$

$$x'(t) = -a(t)x(t) + \int_{t-r(t)}^{t} b(t,s)g(x(s))\,ds,$$

and

$$x'' + ax' + bx(t - r) = 0.$$

This will acknowledge that most of the work in this section was first published in *Nonlinear Studies*, as detailed in Burton and Furumochi (2005), which is published by I and S Publishers.

Chapter 5

Boundedness, Periodicity, and Stability

5.0 Introduction

Virtually all of our results have aimed at writing an equation as

$$x' = a(t)x + f(t, x_t) \tag{5.0.1}$$

and then using the variation of parameters formula to write

$$x(t) = e^{\int_0^t a(s)\,ds}x(0) + \int_0^t e^{\int_s^t a(u)\,du} f(s, x_s)\,ds.$$

From this we define a mapping

$$(P\phi)(t) = e^{-\int_0^t a(s)\,ds}x(0) + \int_0^t e^{\int_0^t a(u)\,du} f(s, \phi_s)\,ds \tag{5.0.2}$$

which we show will map some set of bounded functions into itself. The functional f may be very complicated and it may be of neutral type, while a may be a matrix, but symbolically this is the total situation.

Once we have gotten this far, it is but a simple step to go so much farther. Suppose there is a bounded forcing function so that (5.0.1) is really

$$x' = a(t)x + f(t, x_t) + p(t) \tag{5.0.3}$$

and (5.0.2) becomes

$$(P\phi)(t) = e^{-\int_0^t a(s)\,ds}x(0) + \int_0^t e^{\int_0^t a(u)\,du}[f(s, \phi_s) + p(s)]\,ds. \tag{5.0.4}$$

Almost always this will still map bounded functions into bounded functions.

Notice that if P was a contraction without p, it still is: *The function p simply drops out of the contraction argument.* Thus, there should be a fixed point, a bounded solution. In many cases we can argue that if certain periodicity assumptions hold then the fixed point is at least asymptotically periodic. Schaefer's fixed point theorem is also useful in such problems.

We had begun briefly discussing these ideas in Section 1.5 where we performed the integration for the mapping P by integrating from $-\infty$ to t That discussion will be continued in these sections.

5.1 The Bernoulli Equation Revisited

The following material was taken from Burton and Furumochi (2002a).

We discussed the Bernoulli equation in several sections in Chapter 2 and we continue with that discussion here. That equation is the source of many ideas. Let $a : [0, \infty) \to (0, \infty)$ and $b : [0, \infty) \to R$ be continuous, $r_1 \geq 0$ be constant, $r_2 : [0, \infty) \to [0, \gamma]$ for $\gamma > 0$ with r_2 continuous, and consider the scalar equation

$$x'(t) = -a(t)x(t - r_1) + b(t)x^{1/3}(t - r_2(t)). \tag{5.1.1}$$

Suppose there are constants $0 < \beta < 1$ and $K > 0$ satisfying

$$\sup_{t \geq 0} |b(t)/a(t + r_1)| \leq \beta \tag{5.1.2}$$

and if $|t_2 - t_1| \leq 1$ then

$$\left|\int_{t_1}^{t_2} a(u + r_1)\, du\right| \leq K|t_2 - t_1|, \tag{5.1.3}$$

while for $t \geq 0$ we have

$$2 \sup_{t \geq 0} \int_{t-r_1}^{t} a(u + r_1)\, du + \beta < 1. \tag{5.1.4}$$

Remark 5.1.1. *Notice that (5.1.2) and an upcoming (5.1.2*) are pointwise conditions, as are some of the conditions which we see with Schauder arguments. We also saw pointwise conditions in the sunflower equation. Virtually all the other conditions used in this book are averages.*

From existence theory we can conclude that for each continuous initial function $\psi : [-r, 0] \to R$ (where $r \geq \max[r_1, r_2]$) there is a continuous solution $x(t, 0, \psi)$ on an interval $[0, T)$ for some $T > 0$ and $x(t, 0, \psi) = \psi(t)$ on $[-r, 0]$.

When the r_i are positive, then this becomes a formidable problem. But if $r_1 = r_2 = 0$, then (5.1.1) is an elementary Bernoulli equation which can be explicitly integrated for exact solutions. Several interesting types of behavior can be distinguished in that elementary case and these lead us to conjectures when $r_i > 0$. Cases of interest are:

i) The zero solution of

$$x' = -2x + x^{1/3}$$

is unstable, but all solutions are bounded.

ii) The zero solution of

$$x' = \frac{-2t}{t^2+1}x + \frac{x^{1/3}}{t^2+1}$$

is asymptotically stable; although the unstable term $x^{1/3}$ dominates the stable term $-x$, the coefficient $\frac{t}{t^2+1}$ strongly dominates the coefficient $\frac{1}{t^2+1}$ and that plays an important role in compactness arguments given later.

iii) The forced equation

$$x' = -2x + x^{1/3} + \sin t$$

has a periodic solution.

We will use fixed point theory to obtain counterparts of these properties where $r_i > 0$. In particular, for moderate size r_i, fixed point theory can treat the formidable case almost as easily as integration can treat the elementary case. The first result needed is Krasnoselskii's theorem from Section 1.2, restated here for convenience.

Theorem 1.2.7 Krasnoselskii. *Let $\mathcal{M}$ be a closed convex non-empty subset of a Banach space $(\mathcal{S}, \|\cdot\|)$. Suppose that A and B map $\mathcal{M}$ into $\mathcal{S}$ such that*

(i) $Ax + By \in \mathcal{M} (\forall x, y \in \mathcal{M})$,

(ii) *A is continuous and $A\mathcal{M}$ is contained in a compact set,*

(iii) *B is a contraction with constant $\alpha < 1$.*

Then there is a $y \in \mathcal{M}$ with $Ay + By = y$.

It may be noted that for $r_i = 0$ Equation (5.1.1) is a Bernoulli equation and the zero solution may not be unique to the right if $b > 0$; hence, stability is impossible. Moreover, our method of proof does not distinguish $b > 0$ from $b < 0$. The most we can hope for is bounded solutions. Not only do we obtain boundedness, but the proof and construction is very interesting.

Theorem 5.1.1. *If (5.1.2)-(5.1.4) hold and if ψ is a given continuous initial function which is sufficiently small, then there is a solution $x(t,0,\psi)$ of (5.1.1) on $[0,\infty)$ with $|x(t,0,\psi)| < 1$.*

Proof. Let $\psi : [-r,0] \to R$ be a given continuous initial function with $|\psi(t)| \le \Psi$ where Ψ is a positive constant to be determined, $\Psi < 1$. Let $h : [-r,\infty) \to [1,\infty)$ be any strictly increasing and continuous function with $h(-r) = 1$, $h(s) \to \infty$ as $s \to \infty$, together with a constant $\alpha < 1$, such that

$$2\int_{t-r_1}^{t} [a(s+r_1)h(s)/h(t)]\,ds \le \alpha. \tag{5.1.5}$$

Indeed, from (5.1.4) we have for any such function h that

$$2\int_{t-r_1}^{t} [a(s+r_1)h(s)/h(t)]\,ds \le 2\int_{t-r_1}^{t} [a(s+r_1)h(t)/h(t)]\,ds < 1.$$

Let $(\mathcal{S}, |\cdot|_h)$ be the Banach space of continuous $\phi : [-r,\infty) \to R$ with

$$|\phi|_h := \sup_{t\ge -r} |\phi(t)/h(t)| < \infty$$

and let $(\mathcal{M}, |\cdot|_h)$ be the complete metric space of $\phi \in \mathcal{S}$ such that

$$|\phi(t)| \le 1 \text{ for } -r \le t < \infty \text{ and } \phi(t) = \psi(t) \text{ on } [-r,0].$$

Write (5.1.1) as

$$x'(t) = -a(t+r_1)x(t) + (d/dt)\int_{t-r_1}^{t} a(s+r_1)x(s)\,ds + b(t)x^{1/3}(t-r_2(t))$$

so that by the variation of parameters formula and integration by parts we have

$$\begin{aligned} x(t) &= x_0 e^{-\int_0^t a(s+r_1)\,ds} \\ &\qquad + \int_0^t e^{-\int_s^t a(u+r_1)\,du}\left[\frac{d}{ds}\int_{s-r_1}^{s} a(u+r_1)x(u)\,du \right. \\ &\qquad \left. + b(s)x^{1/3}(s-r_2(s))\right] ds \end{aligned}$$

or

$$
\begin{aligned}
x(t) =& x_0 e^{-\int_0^t a(s+r_1)\,ds} + \int_{t-r_1}^t a(u+r_1)x(u)\,du \\
& - e^{-\int_0^t a(u+r_1)\,du} \int_{-r_1}^0 a(u+r_1)x(u)\,du \\
& - \int_0^t a(s+r_1)e^{-\int_s^t a(u+r_1)\,du} \int_{s-r_1}^s a(u+r_1)x(u)\,du\,ds \\
& + \int_0^t b(s)e^{-\int_s^t a(u+r_1)\,du} x^{1/3}(s-r_2(s))\,ds \qquad (5.1.6)
\end{aligned}
$$

where $x(t) = \psi(t)$ on $[-r, 0]$ and $\psi(0) = x_0$.

Define mappings $A, B : \mathcal{M} \to \mathcal{M}$ by $\phi \in \mathcal{M}$ implies that

$$
(A\phi)(t) = \int_0^t b(s)e^{-\int_s^t a(u+r_1)\,du} \phi^{1/3}(s-r_2(s))\,ds \qquad (5.1.7)
$$

and

$$
\begin{aligned}
(B\phi)(t) =& x_0 e^{-\int_0^t a(s+r_1)\,ds} + \int_{t-r_1}^t a(u+r_1)\phi(u)\,du \\
& - e^{-\int_0^t a(u+r_1)\,du} \int_{-r_1}^0 a(u+r_1)\psi(u)\,du \\
& - \int_0^t a(s+r_1)e^{-\int_s^t a(u+r_1)\,du} \int_{s-r_1}^s a(u+r_1)\phi(u)\,du\,ds. \\
& \qquad (5.1.8)
\end{aligned}
$$

We now show that $\phi, \eta \in \mathcal{M}$ implies that $A\phi + B\eta \in \mathcal{M}$. Let $\|\cdot\|$ be the supremum norm on $[-r, \infty)$ of $\phi \in \mathcal{S}$ if ϕ is bounded. Note that if $\phi, \eta \in \mathcal{M}$ then

$$
\begin{aligned}
&|(A\phi)(t) + (B\eta)(t)| \\
&\leq \|\psi\|e^{-\int_0^t a(s+r_1)\,ds} + \|\eta\| \int_{t-r_1}^t a(u+r_1)\,du \\
&+ e^{-\int_0^t a(u+r_1)\,du} \|\psi\| \int_{-r_1}^0 a(u+r_1)\,du \\
&+ \|\eta\| \sup_{t\geq 0} \int_{t-r_1}^t a(u+r_1)\,du + \|\phi\|^{1/3}\beta \\
&\leq \|\psi\|e^{-\int_0^t a(s+r_1)\,ds}\left(1 + \int_{-r_1}^0 a(u+r_1)\,du\right)
\end{aligned}
$$

$$+ 2 \sup_{t \geq 0} \int_{t-r_1}^{t} a(u+r_1)\,du + \beta$$
$$< 1$$

provided that $\|\psi\|$ is sufficiently small.

Next, we show that $A\mathcal{M}$ is equicontinuous. If $\phi \in \mathcal{M}$ and if $0 \leq t_1 < t_2$ with $t_2 - t_1 < \min[1, r_1]$ then

$$\begin{aligned}
&|(A\phi)(t_2) - (A\phi)(t_1)| \\
&\leq \left| \int_0^{t_2} b(s) e^{-\int_s^{t_2} a(u+r_1)\,du} \phi^{1/3}(s - r_2(s))\,ds \right. \\
&\quad \left. - \int_0^{t_1} b(s) e^{-\int_s^{t_1} a(u+r_1)\,du} \phi^{1/3}(s - r_2(s))\,ds \right| \\
&= \left| \int_{t_1}^{t_2} b(s) e^{-\int_s^{t_2} a(u+r_1)\,du} \phi^{1/3}(s - r_2(s))\,ds \right. \\
&\quad \left. + \int_0^{t_1} b(s) \left[e^{-\int_s^{t_2} a(u+r_1)\,du} - e^{-\int_s^{t_1} a(u+r_1)\,du} \right] \phi^{1/3}(s - r_2)\,ds \right| \\
&\leq \beta \int_{t_1}^{t_2} a(s+r_1)\,ds \\
&\quad + \int_0^{t_1} |b(s)| \left| e^{-\int_0^{t_2} a(u+r_1)\,du} - e^{-\int_0^{t_1} a(u+r_1)\,du} \right| e^{\int_0^s a(u+r_1)\,du}\,ds \\
&\leq \beta K |t_2 - t_1| + \beta | e^{-\int_0^{t_2} a(u+r_1)\,du} \\
&\quad - e^{-\int_0^{t_1} a(u+r_1)\,du} | [e^{\int_0^{t_1} a(u+r_1)\,du} - 1] \\
&\leq \beta K |t_2 - t_1| + \beta | e^{-\int_{t_1}^{t_2} a(u+r_1)\,du} - 1 | \\
&\leq \beta K |t_2 - t_1| + \beta \left| \int_{t_1}^{t_2} a(u+r_1)\,du \right| \\
&\leq 2\beta K |t_2 - t_1|,
\end{aligned}$$

by (5.1.3). In the above calculation, the error made by replacing $e^{-\int_{t_1}^{t_2} a(u+r_1)\,du}$ by 1 is less than the next term in its series, $|\int_{t_1}^{t_2} a(u+r_1)\,du|$. In the space $(\mathcal{S}, |\cdot|_h)$, the set $A\mathcal{M}$ resides in a compact set. To see this, one may consult Burton (1985; p. 169, Example 3.1.6(c)).

Now we show that there is an $\alpha < 1$ so that $B : \mathcal{M} \to \mathcal{M}$ and

$$|B\phi_1 - B\phi_2|_h \leq \alpha |\phi_1 - \phi_2|_h.$$

We have

$$
\begin{aligned}
&|(B\phi_1)(t) - (B\phi_2)(t)|/h(t) \\
&\leq \int_{t-r_1}^{t} a(u+r_1)|\phi_1(u) - \phi_2(u)|/h(t)\, du \\
&+ (1/h(t)) \int_0^t a(s+r_1)e^{-\int_s^t a(u+r_1)\, du} \\
&\times \int_{s-r_1}^{s} a(u+r_1)|\phi_1(u) - \phi_2(u)|\, du\, ds \\
&\leq \sup_{t\geq 0} \int_{t-r_1}^{t} [a(s+r_1)h(s)/h(t)]\, ds |\phi_1 - \phi_2|_h \\
&+ |\phi_1 - \phi_2|_h \sup_{t\geq 0} \int_{t-r_1}^{t} a(u+r_1)[h(u)/h(t)] du \\
&\leq 2|\phi_1 - \phi_2|_h \sup_{t\geq 0} \int_{t-r_1}^{t} a(u+r_1)[h(u)/h(t)]\, du \\
&\leq \alpha|\phi_1 - \phi_2|_h
\end{aligned}
$$

for some $\alpha < 1$ by (5.1.4).

Finally, we need to show A continuous.

Let $\varepsilon > 0$ be given and let $\phi \in \mathcal{M}$. We must find $\delta > 0$ such that $[\eta \in \mathcal{M}, |\phi - \eta|_h < \delta]$ implies that $|A\phi - A\eta|_h < \varepsilon\beta$.

Now $x^{1/3}$ is uniformly continuous on $[-1, 1]$ so for $\varepsilon > 0$ and a fixed $T > 0$ with $4/h(T) < \varepsilon$ there is a $\delta > 0$ such that $|x_1 - x_2| < \delta h(T)$ implies $|x_1^{1/3} - x_2^{1/3}| < \varepsilon/2$. Thus for $|\phi(t) - \eta(t)| < \delta h(t)$ and for $t > T$ we have

$$
\begin{aligned}
&|(A\phi)(t) - (A\eta)(t)|/h(t) \\
&\leq (1/h(t)) \int_0^t b(s)e^{-\int_s^t a(u+r_1)\, du}|\phi^{1/3}(s-r_2) - \eta^{1/3}(s - r_2(s))|\, ds \\
&\leq (1/h(t))\left[\int_0^T |b(s)|e^{-\int_s^t a(u+r_1)\, du}|\phi^{1/3}(s-r_2) - \eta^{1/3}(s-r_2(s))|\, ds \right. \\
&\left. + \int_T^t 2|b(s)|e^{-\int_s^t a(u+r_1)\, du}\, ds\right] \\
&\leq [(\beta\varepsilon)/(2h(t))] \int_0^T a(s+r_1)e^{-\int_s^t a(u+r_1)\, du}\, ds + (2\beta/h(T)) \\
&\leq (\beta\varepsilon/2) + (2\beta/h(T)) \\
&< \varepsilon\beta.
\end{aligned}
$$

The conditions of Krasnoselskii's theorem are satisfied and there is a fixed point.

For reference we restate Schaefer's theorem.

Theorem 1.2.10 Schaefer. *Let $(\mathcal{B}, \|\cdot\|)$ be a normed space, H a continuous mapping of $\mathcal{B}$ into $\mathcal{B}$ which is compact on each bounded subset X of $\mathcal{B}$. Then either*

(i) the equation $x = \lambda Hx$ has a solution for $\lambda = 1$, or

(ii) the set of all such solutions x, for $0 < \lambda < 1$, is unbounded.

Consider Equation (5.1.1) again with a perturbation

$$x'(t) = -a(t)x(t - r_1) + b(t)x^{1/3}(t - r_2(t)) + p(t) \tag{5.1.9}$$

in which $p : R \to R$ is continuous and there is a $T > 0$ with

$$r_2(t+T) = r_2(t), \quad p(t+T) = p(t), \quad a(t+T) = a(t), \\ \text{and } b(t+T) = b(t) \tag{5.1.10}$$

for all $t \in R$. We will need relaxed forms of (5.1.2)-(5.1.4). Assume there are positive constants β and K with

$$\sup_{t\geq 0} |b(t)/a(t+r_1)| \leq \beta \tag{5.1.11}$$

and if $|t_2 - t_1| \leq 1$, then

$$\left|\int_{t_1}^{t_2} a(u + r_1)\, du\right| \leq K|t_2 - t_1|, \tag{5.1.12}$$

while for $t \geq 0$ we have

$$2 \sup_{t\geq 0} \int_{t-r_1}^{t} a(u + r_1)\, du < 1. \tag{5.1.13}$$

Theorem 5.1.2. *If (5.1.10) - (5.1.13) hold, then (5.1.9) has a T-periodic solution.*

Proof. Write (5.1.9) as

$$x' = -a(t+r_1)x(t) + \frac{d}{dt}\int_{t-r_1}^{t} a(s+r_1)x(s)\,ds + b(t)x^{1/3}(t-r_2(t)) + p(t)$$

or

$$\begin{aligned}&(xe^{\int_0^t a(s+r_1)\,ds})' \\ &= e^{\int_0^t a(s+r_1)\,ds}\left[\frac{d}{dt}\int_{t-r_1}^{t} a(s+r_1)x(s)\,ds + b(t)x^{1/3}(t-r_2(t)) + p(t)\right].\end{aligned}$$

Note that since the function a is periodic and $a(t) > 0$ we have $\int_0^t a(s)\,ds \uparrow \infty$ as $t \to \infty$. If we consider only solutions of (5.1.9) bounded on R (if any), we can integrate from $-\infty$ to t and obtain (after integration by parts):

$$\begin{aligned}x(t) &= \int_{t-r_1}^{t} a(u+r_1)x(u)\,du \\ &- \int_{-\infty}^{t} e^{-\int_s^t a(u+r_1)\,du} a(s+r_1)\int_{s-r_1}^{s} a(u+r_1)x(u)\,du \\ &+ \int_{-\infty}^{t} e^{-\int_s^t a(u+r_1)du}[b(s)x^{1/3}(s-r_2(s)) + p(s)]ds.\end{aligned}$$

Now, if $(P_T, \|\cdot\|)$ is the Banach space of continuous T-periodic functions with the supremum norm, then we can define an operator $H : P_T \to P_T$ by $\phi \in P_T$ implies that

$$\begin{aligned}(H\phi)(t) = \lambda\Bigg[&\int_{t-r_1}^{t} a(u+r_1)\phi(u)\,du \\ &- \int_{-\infty}^{t} e^{-\int_s^t a(u+r_1)\,du} a(s+r_1)\int_{s-r_1}^{s} a(u+r_1)\phi(u)\,du \\ &+ \int_{-\infty}^{t} e^{-\int_s^t a(u+r_1)\,du}[b(s)\phi^{1/3}(s-r_2) + p(s)]\,ds\Bigg]\end{aligned} \tag{5.1.14}$$

where $0 < \lambda \leq 1$.

We now obtain an *a priori* bound on all fixed points of H. If $x \in P_T$ is a fixed point, then using $\|\cdot\|$ to denote the supremum norm we have $Q > 0$ with

$$\|x\| \leq \sup_{t\in R}\int_{t-r_1}^{t} a(u+r_1)\,du[2\|x\|] + \|x\|^{1/3}\sup_{t\in R}\frac{|b(t)|}{a(t+r_1)} + Q.$$

Taking into account (5.1.11)-(5.1.13), this will establish an *a priori* bound if

$$2\sup\int_{t-r_1}^{t} a(u+r_1)\,du \le \alpha < 1.$$

An argument just like that in the proof of Theorem 5.1.1 will show that H maps bounded sets into equicontinuous sets. Thus, H maps bounded sets into compact subsets of our space. By Schaefer's theorem there is a fixed point and it solves (5.1.9). This completes the proof.

This section is motivated by the Bernoulli equation

$$x'(t) = -(2/t)x(t) - (1/t^2)x^3(t).$$

Even without solving the equation, the investigator would conjecture that the asymptotically stable part

$$x' = -(2/t)x$$

would dominate the term $(1/t^2)x^3$ since x dominates x^3 for small x and $1/t^2$ is $L^1[1,\infty)$.

But that entire rationale vanishes when we consider

$$x'(t) = -a(t)x(t-r_1) + b(t)x^{1/3}(t-r_2(t)) \tag{5.1.15}$$

again where the exact conditions of Theorem 5.1.1 hold except (5.1.2) becomes: There is a $\beta < 1$ with

$$\sup_{t\ge 0}|b(t)/a(t+r_1)| \le \beta \text{ and } b(t)/a(t+r_1)\to 0 \text{ as } t\to\infty. \tag{5.1.2*}$$

Not only does $x^{1/3}$ dominate x for small x, but there is a delay in the linear part which makes it a nontrivial stability problem and $b(t)$ is not assumed to be $L^1[0,\infty)$.

Thus, the investigator recognizes that it is a challenge to prove asymptotic stability. Our point here is that fixed point theory is such a powerful tool that this becomes a simple problem.

Theorem 5.1.3. *Let (5.1.2*), (5.1.3), and (5.1.4) hold and assume that $\int_0^\infty a(s)ds = \infty$. If ψ is a given initial function which is sufficiently small, then (5.1.15) has a solution $x(t,0,\psi)\to 0$ as $t\to\infty$.*

Proof. All of the calculations of the proof of Theorem 5.1.1 hold with $h(t) = 1$ when $|\cdot|_h$ is replaced by the supremum norm $\|\cdot\|$.

Define $g : [0,\infty) \to (0,1]$ by

$$g(t) = \int_0^t a(s+r_1)e^{-\int_s^t a(u+r_1)\,du}[b(s)/a(s+r_1)]\,ds. \tag{5.1.16}$$

We can easily modify the classical proof that the convolution of an L^1-function [substitute $a(s+r_1)e^{-\int_s^t a(u+r_1)}\,du$] with a function tending to zero $[b(s)/a(s+r_1)]$ tends to zero to show that $g(t) \to 0$ as $t \to \infty$.

Add to $\mathcal{M}$ the condition that $\phi \in \mathcal{M}$ implies that $\phi(t) \to 0$ as $t \to \infty$. With A defined by (6.3) we see that for $\phi \in \mathcal{M}$ then $|(A\phi)(t)| \le g(t)$. Clearly $(B\phi)(t) \to 0$ as $t \to \infty$.

Since $A\mathcal{M}$ has been shown to be equicontinuous, A maps $\mathcal{M}$ into a compact subset of $\mathcal{M}$. By Krasnoselskii's theorem there is a $y \in \mathcal{M}$ with $Ay + By = y$. As $y \in \mathcal{M}$, $y(t) \to 0$ as $t \to \infty$. This completes the proof.

This will acknowledge that much of the material from this section was first published in *Dynamic Systems and Applications*, as detailed in Burton and Furumochi (2002a), which is published by Dynamic Publishers.

5.2 Becker's Resolvent and Perron's Theorem

The purpose of this section is to present a condition under which we can offer simple, unified, and concise proofs of the three most important properties of the resolvent of a linear Volterra equation. These results rest on Becker's form of the resolvent and very simple contraction mapping arguments. The work in this section can be found in Burton (2005c).

The resolvent is a function, $R(t,s)$, with which we can express the solution of

$$y'(t) = A(t)y(t) + \int_0^t C(t,s)y(s)ds + p(t), \quad y(0) = y_0,$$

as

$$y(t) = R(t,0)y_0 + \int_0^t R(t,s)p(s)\,ds.$$

If we can show that $\int_0^t |R(t,s)|\,ds$ is bounded, that $|R(t,s)|$ is bounded, and that $R(t,s) \to 0$ as $t \to \infty$ for fixed s, then we can show boundedness of solutions for bounded p, various stability properties, and a formula for periodic solutions.

Establishing these properties can be challenging and usually requires a variety of *ad hoc* conditions and techniques. The reader will find treatment of the resolvent, together with many references, in Burton (1985, 1983) as well as in the recent papers of Eloe, Islam, and Zhang (2000), Hino and Murakami (1996), and Zhang (2004).

While the main focus is on linear equations, we are also interested in what the results will say about equations of the form

$$x'(t) = A(t)h(x(t)) + \int_0^t C(t,s)g(x(s))\,ds \tag{5.2.1}$$

in which A, C, h, and g are continuous real valued functions. The equation is almost linear in the following sense. It is assumed that there is a $K > 0$ so that for all $x \in R$ we have

$$|h(x) - x| \le K, \qquad |g(x) - x| \le K. \tag{5.2.2}$$

This equation is motivated by a forced Liénard equation with distributed delay

$$x'' + f(x)x' + \int_0^t C(t,s)g(x(s))\,ds = e(t)$$

which we discuss at the end of this section. The conditions in (5.2.2) can allow $f(0) < 0$, giving rise to negative damping and relaxation oscillations. Thus, our term "almost linear" refers only to behavior for very large values of the dependent variable.

We will study stability properties of (5.2.1) through information obtained from the forced linear equation

$$y'(t) = A(t)y(t) + \int_0^t C(t,s)y(s)\,ds + p(t), \quad y(0) = y_0 \tag{5.2.3}$$

Here, we focus on the resolvent in Becker's (1979; p. 11) form and Becker, Burton, and Krisztin (1988) which holds for R, A, C being matrices:

$$\frac{\partial R(t,\tau)}{\partial t} = A(t)R(t,\tau) + \int_\tau^t C(t,u)R(u,\tau)\,du, \qquad R(\tau,\tau) = I. \tag{5.2.4}$$

As $' = \frac{d}{dt}$, it will cause no confusion if we write (5.2.4) also as

$$R'(t,\tau) = A(t)R(t,\tau) + \int_\tau^t C(t,u)R(u,\tau)\,du, \qquad R(\tau,\tau) = I.$$

For the significance of using Becker's resolvent, see the remark after the proof of Theorem 5.2.1.

Under conditions to be given, we will show that solutions of (5.2.3) are bounded for every bounded continuous $p(t)$. If we then write the solution of (5.2.3) as

$$y(t) = R(t,0)y_0 + \int_0^t R(t,s)p(s)\,ds$$

and apply Perron's (1930) theorem (or Burton (1985; p. 114)), we can conclude that there is a positive number W with

$$\int_0^t |R(t,s)|ds \leq W \tag{5.2.5}$$

for all $t \geq 0$. We can also apply the aforementioned stability criterion to (5.2.4) and find a fixed positive number Z such that

$$|R(t,s)| \leq Z, \qquad 0 \leq s \leq t < \infty \tag{5.2.6}$$

and that for fixed s

$$|R(t,s)| \to 0 \tag{5.2.7}$$

as $t \to \infty$.

After establishing these properties of R, we write (5.2.1) as

$$\begin{aligned} x'(t) =& A(t)x(t) + \int_0^t C(t,s)x(s)\,ds \\ & - A(t)[x(t) - h(x(t))] - \int_0^t C(t,s)[x(s) - g(x(s))]\,ds \end{aligned} \tag{5.2.8}$$

and express its solution as

$$\begin{aligned} x(t) =& R(t,0)x_0 + \int_0^t R(t,s)[A(s)(x(s) - h(x(s)))]\,ds \\ & + \int_0^t R(t,s)\int_0^s C(s,u)[x(u) - g(x(u))]\,du\,ds. \end{aligned} \tag{5.2.9}$$

We then ask that A and the integral of C be bounded and conclude from (5.2.9) that all solutions of (5.2.1) are bounded. The result says: In (5.2.1) replace $g(x)$ and $h(x)$ by x to obtain

$$y'(t) = A(t)y(t) + \int_0^t C(t,s)y(s)\,ds. \tag{5.2.1a}$$

If $|g(x) - x|$ and $|h(x) - x|$ are bounded and if the stability condition holds then the difference of the solutions of (5.2.1) and (5.2.1a) are bounded.

By way of application of the theory, we point out that the result on boundedness of solutions of (5.2.3) allows us to write down the formula for a periodic solution of the limiting equation for (5.2.3) under periodicity assumptions on the coefficients. We also briefly introduce the reader to stability using the technique and, as mentioned above, discuss the Liénard equation with distributed delay.

We ask that $A : [0, \infty) \to R$ and $C : [0, \infty) \times [0, \infty) \to R$ be continuous. Then for each $y_0 \in R$ there is one and only one solution $y(t, 0, y_0)$ of (5.2.3) satisfying $y(0, 0, y_0) = y_0$ and is defined for $t \geq 0$. See Burton (1985; p. 175) for details on existence and uniqueness. Later, we will be interested in more general solutions.

Invert (5.2.3), transforming it into (5.2.16) below, by asking that

$$G(t,s) := -\int_t^{\infty} C(u,s)du \tag{5.2.10}$$

exist and be continuous for $0 \leq s \leq t < \infty$ so that we can write (5.2.3) as

$$y'(t) = [A(t) - G(t,t)]y(t) + \frac{d}{dt}\int_0^t G(t,s)y(s)\,ds + p(t)$$

or for $Q(t) := A(t) - G(t,t)$ then

$$y'(t) = Q(t)y(t) + \frac{d}{dt}\int_0^t G(t,s)y(s)\,ds + p(t). \tag{5.2.11}$$

We further require that there is a $\Gamma > 0$ with

$$\int_0^t e^{\int_s^t Q(u)\,du}\,ds \leq \Gamma \qquad \text{and} \qquad \int_0^t Q(s)\,ds \to -\infty \tag{5.2.12}$$

as $t \to \infty$, that there is a $J > 0$ such that $0 \leq t_0 \leq t$ implies that

$$\int_{t_0}^t Q(s)\,ds \leq J, \tag{5.2.13}$$

and for each fixed $T > 0$ we have

$$\int_0^T |G(t,v)|\,dv \to 0 \tag{5.2.14}$$

as $t \to \infty$.

In condition (5.2.15) below, if $Q(t) \leq 0$ then (5.2.15) can be replaced by

$$2\sup_{t\geq 0}\int_0^t |G(t,v)|\,dv \leq \alpha. \tag{5.2.15a}$$

Theorem 5.2.1. *Let (5.2.12), (5.2.13), and (5.2.14) hold and let G be defined in (5.2.10). Suppose there is an $\alpha < 1$ such that*

$$\int_0^t [|G(t,v)| + |Q(v)|e^{\int_v^t Q(u)\,du} \int_0^v |G(v,u)|\,du]\,dv \leq \alpha \tag{5.2.15}$$

for $t \geq 0$. Then every solution of (5.2.3) is bounded for every bounded and continuous function p. Moreover, (5.2.5), (5.2.6), and (5.2.7) hold.

Proof. There are four conclusions given and these are proved with one basic argument. For the sake of clarity we will separate them.

PROOF OF THE BOUNDEDNESS. Starting with (5.2.11) we apply the variation of parameters formula and for a given $y(0)$ write

$$\begin{aligned}
y(t) &= e^{\int_0^t Q(s)\,ds}y(0) + \int_0^t e^{\int_s^t Q(u)\,du} \frac{d}{ds}\int_0^s G(s,u)y(u)\,du\,ds \\
&+ \int_0^t e^{\int_s^t Q(u)du}p(s)\,ds \\
&= e^{\int_0^t Q(s)\,ds}y(0) + e^{\int_s^t Q(u)\,du}\int_0^s G(s,u)y(u)\,du\Big|_0^t \\
&+ \int_0^t e^{\int_s^t Q(u)\,du}Q(s)\int_0^s G(s,u)y(u)\,du\,ds \\
&+ \int_0^t e^{\int_s^t Q(u)\,du}p(s)\,ds.
\end{aligned}$$

or

$$\begin{aligned}
y(t) =& e^{\int_0^t Q(s)\,ds}y(0) + \int_0^t G(t,u)y(u)\,du \\
&+ \int_0^t e^{\int_s^t Q(u)\,du}Q(s)\int_0^s G(s,u)y(u)\,du\,ds \\
&+ \int_0^t e^{\int_s^t Q(u)du}p(s)\,ds.
\end{aligned} \tag{5.2.16}$$

Let $(\mathcal{X}, \|\cdot\|)$ be the complete metric space of bounded continuous functions $\phi : [0, \infty) \to R$ with the supremum metric, satisfying $\phi(0) = y(0)$. Define a mapping $P : \mathcal{X} \to \mathcal{X}$ by $\phi \in \mathcal{X}$ and $t \geq 0$ implies

$$\begin{aligned}(P\phi)(t) =& e^{\int_0^t Q(s)\,ds} y(0) + \int_0^t G(t,u)\phi(u)\,du \\ &+ \int_0^t e^{\int_s^t Q(u)\,du} Q(s) \int_0^s G(s,u)\phi(u)\,du\,ds \\ &+ \int_0^t e^{\int_s^t Q(u)\,du} p(s)\,ds. \end{aligned} \tag{5.2.17}$$

By (5.2.12), (5.2.15), and the assumed boundedness of p, if ϕ is bounded so is $P\phi$. To see that P is a contraction, if $\phi, \eta \in \mathcal{X}$ then using (5.2.15) we have

$$\begin{aligned}|(P\phi)(t) - (P\eta)(t)| \leq& \int_0^t |G(t,u)||\phi(u) - \eta(u)|\,du \\ &+ \int_0^t e^{\int_s^t Q(u)\,du} |Q(s)| \int_0^s |G(s,u)||\phi(u) - \eta(u)|\,du\,ds \\ &\leq \alpha\|\phi - \eta\|. \end{aligned}$$

This gives a fixed point representing a bounded solution.

PROOF OF (5.2.5). We now write the solution of (5.2.3) as

$$y(t) = R(t,0)y_0 + \int_0^t R(t,s)p(s)\,ds. \tag{5.2.18}$$

For $y_0 = 0$ the argument just given shows that y is bounded for every bounded and continuous p; hence, $\int_0^t R(t,s)p(s)ds$ is bounded for every bounded and continuous p. By Perron's (1930) theorem (or Burton (1985; p. 114) this says that $\int_0^t |R(t,s)|\,ds$ is bounded, so (5.2.5) holds.

PROOF OF (5.2.6). To see that (5.2.6) is true, we invert (5.2.4) for an arbitrary $\tau \geq 0$. Letting z denote $R(t,\tau)$ in the scalar case, we write

$$z'(t) = A(t)z(t) + \int_\tau^t C(t,s)z(s)\,ds, \qquad z(\tau) = 1, \tag{5.2.19}$$

and using (5.2.10) we have

$$\begin{aligned}z'(t) &= [A(t) - G(t,t)]z(t) + \frac{d}{dt}\int_\tau^t G(t,s)z(s)\,ds \\ &= Q(t)z(t) + \frac{d}{dt}\int_\tau^t G(t,s)z(s)\,ds \end{aligned}$$

so that by the variation of parameters formula we get

$$\begin{aligned}
z(t) &= e^{\int_\tau^t Q(s)\,ds} \\
&\quad + \int_\tau^t e^{\int_s^t Q(u)\,du} \frac{d}{ds} \int_\tau^s G(s,u)z(u)\,du\,ds \\
&= e^{\int_\tau^t Q(s)ds} + e^{\int_s^t Q(u)\,du} \int_\tau^s G(s,u)z(u)\,du\Big|_\tau^t \\
&\quad + \int_\tau^t e^{\int_s^t Q(u)\,du} Q(s) \int_\tau^s G(s,u)z(u)\,du\,ds
\end{aligned}$$

or

$$\begin{aligned}
z(t) &= e^{\int_\tau^t Q(s)\,ds} + \int_\tau^t G(t,u)z(u)\,du \\
&\quad + \int_\tau^t e^{\int_s^t Q(u)\,du} Q(s) \int_\tau^s G(s,u)z(u)\,du\,ds.
\end{aligned} \tag{5.2.20}$$

Let $(\mathcal{M}, \|\cdot\|)$ be the space of bounded continuous functions $\phi : [\tau, \infty) \to R$ with $\phi(\tau) = 1$ and having the supremum metric. Use (5.2.20) to define a new mapping P on $\mathcal{M}$ by

$$\begin{aligned}
(P\phi)(t) &= e^{\int_\tau^t Q(s)\,ds} + \int_\tau^t G(t,u)\phi(u)\,du \\
&\quad + \int_\tau^t e^{\int_s^t Q(u)\,du} Q(s) \int_\tau^s G(s,u)\phi(u)\,du\,ds.
\end{aligned}$$

The mapping P will be a contraction if there is an $\alpha < 1$ with

$$\int_\tau^t |G(t,u)|\,du + \int_\tau^t e^{\int_s^t Q(u)\,du} |Q(s)| \int_\tau^s |G(s,u)|\,du\,ds \leq \alpha. \tag{5.2.21}$$

Clearly, if (5.2.15) holds, so does (5.2.21).

Moreover, if $P\phi = \phi$ then

$$\|\phi\| \leq e^{\int_\tau^t Q(s)\,ds} + \alpha\|\phi\|$$

or by (5.2.13) we have

$$\|\phi\| \leq \frac{e^{\int_\tau^t Q(s)\,ds}}{1-\alpha} \leq \frac{e^J}{1-\alpha}, \tag{5.2.22}$$

a uniform bound on $|R(t, \tau)|$, so (5.2.6) holds.

PROOF OF (5.2.7). Finally, if we add to the space $\mathcal{M}$ the condition that $\phi(t) \to 0$, then we can show that $P\phi \to 0$ and so $R(t,\tau) \to 0$ as $t \to \infty$ and (5.2.7) holds. Here are the details. If $\phi(t) \to 0$ as $t \to \infty$, then consider $t > T > \tau$ and for G not identically zero write

$$\left| \int_\tau^t G(t,u)\phi(u)\,du \right| \le \int_\tau^T |G(t,u)\phi(u)|\,du + \int_T^t |G(t,u)||\phi(u)|\,du.$$

For a given $\varepsilon > 0$, for fixed ϕ, and with (5.2.21) in mind find T so that $t \ge T$ implies that $|\phi(t)| \le \varepsilon/[2\alpha]$, leaving the last term in the above display bounded by $\varepsilon/2$. Then find t so large that, by use of (5.2.14), we have

$$\|\phi\| \int_\tau^T |G(t,u)|\,du < \varepsilon/2.$$

Next, with (5.2.12) and (5.2.15) in mind and $t > T > \tau$ write

$$\begin{aligned}
&\left| \int_\tau^t e^{\int_s^t Q(u)\,du} Q(s) \int_\tau^s G(s,u)\phi(u)\,du\,ds \right| \\
&\le \left| \int_\tau^T e^{\int_s^T Q(u)du} Q(s) \int_\tau^s G(s,u)\phi(u)\,du\,ds \right| e^{\int_T^t Q(u)\,du} \\
&+ \left| \int_T^t e^{\int_s^t Q(u)\,du} Q(s) \int_\tau^s G(s,u)\phi(u)\,du\,ds \right| \\
&\le \alpha\|\phi\| e^{\int_T^t Q(u)du} + \alpha \sup_{t \ge T} |\phi(t)|
\end{aligned}$$

which can be made small by taking T large and then t large. This completes the proof.

Remark 5.2.1. *At this point it is possible to explain the significance of using the Becker resolvent. In the proof of (5.2.6) and (5.2.7) we treated the resolvent equation, (5.2.19), as we did (5.2.3). The stability condition (5.2.15) worked on (5.2.19) as it did on (5.2.3). By contrast, the classical resolvent (see Burton (1983; p. 64) or Burton (1985; p. 101]) is given by*

$$\frac{\partial R(t,s)}{\partial s} = -R(t,s)A(s) - \int_s^t R(t,u)C(u,s)\,du, \quad R(t,t) = I,$$

and it is totally unclear to us how we can invert that equation and apply (5.2.15). This entire discussion rests on Becker's form of the resolvent. That form of the resolvent has also been used effectively in Becker, Burton, and Krisztin (1988), Eloe, Islam, and Zhang (2000), Hino and Murakami (1996), and Zhang (1997).

Remark 5.2.2. *We need to keep in mind that one of the important features of analysis by contraction mappings, as compared with Liapunov's direct method, is that forcing functions often simply drop out of the contraction condition, while in Liapunov's direct method they remain as a product with partial derivatives of the Liapunov function.*

We return to (5.2.1) which we write as

$$x'(t) = A(t)x(t) + \int_0^t C(t,s)x(s)\,ds$$
$$- A(t)[x(t) - h(x(t))] - \int_0^t C(t,s)[x(s) - g(x(s))]\,ds. \tag{5.2.23}$$

Theorem 5.2.2. *Let the conditions of Theorem 5.2.1 and condition (5.2.2) hold. Suppose there is a $\Lambda > 0$ with*

$$|A(t)| + \int_0^t |C(t,s)|\,ds \le \Lambda \tag{5.2.24}$$

for $t \ge 0$. Then all solutions of (5.2.1) are bounded.

Proof. Let x_0 be given and identify the last two terms of (5.2.23) as an inhomogeneous term in (5.2.3). We can then use the resolvent discussed in Theorem 5.2.1 to write the solution of (5.2.23) as

$$x(t) = R(t,0)x_0 + \int_0^t R(t,s)A(s)[h(x(s)) - x(s)]\,ds$$
$$+ \int_0^t R(t,s)\int_0^s C(s,u)[g(x(u)) - x(u)]\,du\,ds. \tag{5.2.25}$$

The standard existence theorem (see Burton (5; p. 187)) assures us that there is a solution, while this formula shows that it is bounded. In fact, for a given x_0 by (5.2.2), (5.2.5), and (5.2.24) we have

$$\begin{aligned} |x(t)| &\le |R(t,0)||x_0| + \int_0^t |R(t,s)|K\Lambda\,ds + \int_0^t |R(t,s)|\Lambda K\,ds \\ &\le |R(t,0)||x_0| + 2K\Lambda \int_0^t |R(t,s)|\,ds \\ &\le Z|x_0| + 2K\Lambda W \\ &=: J^*. \end{aligned} \tag{5.2.26}$$

This completes the proof.

We will now show that there are periodic solutions. Theorem 5.2.1 can supply properties for periodic equations which have been very illusive in the literature.

Consider again

$$y'(t) = A(t)y(t) + \int_0^t C(t,s)y(s)\,ds + p(t) \tag{5.2.27}$$

with A, p continuous on $(-\infty, \infty)$ and C continuous for $-\infty < s \leq t < \infty$. If the conditions of Theorem 5.2.1 hold then every solution of (5.2.27) is bounded. If $y_1(t)$ is a fixed solution of (5.2.27) and if $y_2(t)$ is any other solution, then $y(t) =: y_1(t) - y_2(t)$ solves Equation (5.2.1a)

$$y'(t) = A(t)y(t) + \int_0^t C(t,s)y(s)\,ds$$

and every solution of this equation tends to zero by Theorem 5.2.1. Hence, if $y_1(t)$ is any fixed solution of (5.2.27) then every other solution converges to it. Can we say that $y_1(t)$ has "special properties?" For example, if A and p are periodic, might $y_1(t)$ be periodic under some periodic assumptions on C?

This problem is discussed in Burton (1985; pp. 92-94). In fact, $y_1(t)$ can be periodic only if there are very special orthogonal relationships between p and C. The reason for this is that for an arbitrary continuous T-periodic function ϕ, then

$$A(t)\phi(t) + \int_0^t C(t,s)\phi(s)\,ds + p(t)$$

is not periodic even if $C(t+T, s+T) = C(t,s), A(t+T) = A(t)$, and $p(t+T) = p(t)$. By contrast, under the same conditions

$$A(t)\phi(t) + \int_{-\infty}^t C(t,s)\phi(s)\,ds + p(t)$$

is T-periodic. We can discover special properties of $y_1(t)$ by examining $y_1(t+nT)$ for $n = 1, 2, \ldots$. The details are found in Becker, Burton, and Krisztin (1988) and Burton (1985; pp. 101-107, especially p. 102). In that work we assumed there that there exists $T > 0$ such that:

$$A(t+T) = A(t), \qquad p(t+T) = p(t), \qquad C(t+T, s+T) = C(t,s) \tag{5.2.28}$$

and that there is a number $J > 0$ such that for each $[a, b]$ if $t \in [a, b]$ then

$$\lim_{n\to\infty} \int_{-nT}^t |C(t,s)|\,ds = \int_{-\infty}^t |C(t,s)|\,ds \leq J \tag{5.2.29}$$

and $\int_{-\infty}^t |C(t,s)|\,ds$ is continuous.

Under these conditions, in the aforementioned references we showed that

$$R(t+T, s+T) = R(t,s)$$

and that if (5.2.27) *has a bounded solution* then

$$x(t) = \int_{-\infty}^{t} R(t,s)p(s)\,ds \tag{5.2.30}$$

is a periodic solution of

$$x'(t) = A(t)x(t) + \int_{-\infty}^{t} C(t,s)x(s)\,ds + p(t). \tag{5.2.31}$$

The sequence $y_1(t+nT)$ has a subsequence converging to this periodic solution. We remarked in Burton (1985; p. 107) that the major difficulty lay in proving that (5.2.27) has a bounded solution. Our Theorem 5.2.1 gives a simple solution to the boundedness question. We formally state this as follows.

Theorem 5.2.3. *Let the conditions of Theorem 5.2.1 hold, as well as (5.2.28) and (5.2.29). Then (5.2.31) has a T-periodic solution given by (5.2.30). Moreover, if y is any solution of (5.2.27) then there is a subsequence of $\{y(t+nT)\}$ converging to that T-periodic solution and the convergence is uniform on compact subsets of $(-\infty, \infty)$.*

This theorem is just Theorem 1.5.1 (iii) of Burton (1985; p. 102) in which we have replaced the hypothesis that (5.2.27) have a bounded solution with the assumption that the conditions of Theorem 5.2.1 hold.

Theorem 5.2.1 can reduce stability questions to a triviality and it can supply very illusive conditions in difficult stability problems.

Equation (5.2.3) conceals much of the situation. Often, we are interested in many solutions besides those starting at $t = 0$. In the general case, there is a given $t_0 \geq 0$ and a given continuous initial function $\phi : [0, t_0] \to R$. Moreover, we take $f(t) = 0$ and we will see a new forcing function emerge. Under our continuity conditions there is then a unique continuous solution $y(t, t_0, \phi)$ which agrees with ϕ on the initial interval $[0, t_0]$ and satisfies

$$y'(t) = A(t)y(t) + \int_{t_0}^{t} C(t,s)y(s)\,ds + \int_{0}^{t_0} C(t,s)\phi(s)\,ds \tag{5.2.32}$$

on $[t_0, \infty)$.

If the conditions of Theorem 5.2.1 are satisfied then the resolvent satisfies (5.2.5), (5.2.6), and (5.2.7). Becker (1979; p. 22) notes that we can express the solution of (5.2.32) as

$$y(t,t_0,\phi) = R(t,t_0)\phi(t_0) + \int_{t_0}^{t} R(t,u) \int_0^{t_0} C(u,s)\phi(s)\,ds\,du. \quad (5.2.33)$$

From this we can immediately derive a number of results on stability.

Theorem 5.2.4. *If the conditions of Theorem 5.2.1 are satisfied and if for each $t_0 \geq 0$*

$$\int_0^{t_0} |C(t,s)|\,ds \quad (5.2.34)$$

is bounded for $t_0 \leq t < \infty$, then the zero solution of (5.2.32) is stable.

Proof. For a given $\varepsilon > 0$ and $t_0 \geq 0$, we find $M > 0$ with $\int_0^{t_0} |C(t,s)|\,ds \leq M$ for $t \geq t_0$. For δ yet to be determined and for $|\phi(t)| < \delta$, from (5.2.5), (5.2.6), and (5.2.33) we have

$$\begin{aligned} |y(t,t_0,\phi)| &\leq |R(t,t_0)|\delta + M\delta \int_{t_0}^{t} |R(t,u)|\,du \\ &\leq Z\delta + M\delta W \\ &< \varepsilon \end{aligned}$$

provided that

$$\delta < \varepsilon/[Z + MW].$$

Theorem 5.2.5. *If the conditions of Theorem 5.2.1 are satisfied and if there is a uniform bound on (5.2.34) for all t_0 and all $t \geq t_0$, then the zero solution of (5.2.32) is uniformly stable.*

Proof. In the proof of Theorem 5.2.4 we have M independent of t_0 and so the uniform δ needs to satisfy the relation just given.

A classical theorem states that the convolution of an L^1-function with a function tending to zero does, itself, tend to zero. We have given details in previous chapters. Our proof of (5.2.7) in the proof of Theorem 5.2.1 was a type of extension of this. We need a theorem of this kind to show that the last term in (5.2.33) tends to zero under the conditions of Theorem 5.2.1.

Lemma 5.2.1. *Let $R(t,s)$ satisfy (5.2.5), (5.2.6), and (5.2.7). Let $q : [0,\infty) \to R$ be continuous and let $q(t) \to 0$ as $t \to \infty$. Then $\int_0^t R(t,s)q(s)\,ds \to 0$ as $t \to \infty$.*

Proof. For $0 < T < t$ we have

$$\begin{aligned}\left|\int_0^t R(t,s)q(s)\,ds\right| &\le \left|\int_0^T R(t,s)q(s)\,ds\right| + \left|\int_T^t R(t,s)q(s)\,ds\right| \\ &\le \|q\| \int_0^T |R(t,s)|\,ds + \int_T^t |R(t,s)|\,ds \|q\|_{[T,\infty)}.\end{aligned}$$

where $\|\cdot\|$ will represent the supremum on $[0,\infty)$, while $\|\cdot\|_{[T,\infty)}$ will represent the supremum on $[T,\infty)$. Use (5.2.5) and fix T so that for a given $\varepsilon > 0$ we have $\|q\|_{[T,\infty)} < \frac{\varepsilon}{2W}$, making the last term bounded by $\varepsilon/2$. Then use (5.2.7) and take t so large that $\|q\| \int_0^T |R(t,s)|ds < \varepsilon/2$. This completes the proof.

With this lemma and (5.2.33) we can immediately prove the following result.

Theorem 5.2.6. *If the conditions of Theorem 5.2.1 are satisfied and if for each $t_0 \ge 0$*

$$\int_0^{t_0} |C(t,s)|\,ds \to 0 \quad \text{as} \quad t \to \infty \tag{5.2.35}$$

then the zero solution of (5.2.32) is asymptotically stable.

Proof. We choose $\gamma = \delta$ in the proof of Theorem 5.2.4. Then by (5.2.33) and Theorem 5.2.4 we have

$$|y(t,t_0,\phi)| \le \delta|R(t,t_0)| + \delta \int_0^t |R(t,u)| \int_0^{t_0} |C(u,s)|\,ds\,du.$$

Apply (5.2.7), (5.2.5), (5.2.35), and the lemma to see that $y(t,t_0,\phi) \to 0$.

For uniform asymptotic stability we refer the reader to several formulations in Zhang (1997; pp. 326-7). Some of the formulations are his, some are formulations of Hino and Murakami (1996), and some are a combination. But all of them require some form of (5.2.5). Thus, the conditions of Theorem 5.2.1 can be used in place of (5.2.5) in any of those formulations as sufficient conditions for uniform asymptotic stability. The following is an example of what can be deduced under the conditions of Theorem 5.2.1.

It is taken directly from Zhang (1997) Theorem 2.1 on p. 341 where we replace his assumption of (5.2.5) with our assumption that the conditions of Theorem 5.2.1 hold. Several other results of this type are readily formulated in conjunction with Zhang's paper.

Theorem 5.2.7. *Suppose that $p(t) = 0$ and that the conditions of Theorem 5.2.1 hold. In addition, assume that:*

(i) $\sup_{t\geq s\geq 0}\{|A(t)| + \int_0^t |C(t,s)|\,ds\} < \infty$,

(ii) for any $\sigma > 0$, there exists an $S = S(\sigma) > 0$ such that

$$\int_0^{t-S} |C(t,s)|\,ds < \sigma \quad \textit{for all} \quad t \geq S,$$

and

(iii) $A(t)$ and $C(t,t+s)$ are bounded and uniformly continuous in $(t,s) \in \{(t,s) \in R^+ \times K : -t \leq s \leq 0\}$ for any compact set $K \subset R^-$.
Then the zero solution of (5.2.32) is uniformly asymptotically stable.

But (5.2.5), (5.2.6), and (5.2.7) are derived conditions. They follow from (5.2.12)-(5.2.15). So for difficult problems we would always expect to do more with (5.2.12)-(5.2.15) than we can ever do with (5.2.5), (5.2.6), and (5.2.7), even though the latter are the standard conditions discussed extensively in the literature.

Returning to (5.2.32), a translation $x(t) = y(t + t_0)$ results in

$$\begin{aligned} x'(t) =& y'(t+t_0) = A(t+t_0)x(t) + \int_{t_0}^{t_0+t} C(t_0+t,s)y(s)\,ds \\ &+ \int_0^{t_0} C(t_0+t,s)\phi(s)\,ds \end{aligned}$$

or

$$\begin{aligned} x'(t) =& A(t+t_0)x(t) + \int_0^t C(t_0+t,s+t_0)x(s)\,ds \\ &+ \int_0^{t_0} C(t_0+t,s)\phi(s)\,ds \end{aligned} \tag{5.2.36}$$

which we write as

$$x'(t) = A^*(t)x(t) + \int_0^t C^*(t,s)x(s)\,ds + f^*(t). \tag{5.2.37}$$

But this raises a very big question. Will the stability assumptions for (5.2.37) be the same as those in Theorem 5.2.1 for (5.2.3) and (5.2.11)?

If this part of the stability investigation is to be optimal, we need to know that the conditions which held for (5.2.3) and (5.2.11) will also hold for (5.2.37) when we transform it into the counterpart of (5.2.11). We will check this under the simpler condition (5.2.15a).

Starting with (5.2.37), define

$$G^*(t,s) = -\int_t^\infty C(t_0+u, s+t_0)\,du \tag{5.2.38}$$

and write (5.2.37) as

$$x'(t) = [A(t+t_0) - G^*(t,t)]x(t) + \frac{d}{dt}\int_0^t G^*(t,s)x(s)\,ds + f^*(t) \tag{5.2.39}$$

Proposition 5.2.1. *If for (5.2.3) and (5.2.11) we have $Q(t) \le 0$ and if (5.2.15a) holds, then for (5.2.37) and (5.2.39) we have $Q^*(t) \le 0$ and (5.2.15a) holds for G^*.*

Proof. Since we assume that

$$Q(t) = A(t) - G(t,t) = A(t) + \int_t^\infty C(u,t)\,du \le 0$$

we now have

$$\begin{aligned} Q^*(t) &= A(t+t_0) - G^*(t,t) = A(t+t_0) + \int_t^\infty C(t_0+u, t+t_0)\,du \\ &= A(t+t_0) + \int_{t+t_0}^\infty C(u, t+t_0)\,du \\ &= Q(t+t_0) \le 0. \end{aligned}$$

Next, for (5.2.15a) to hold we have

$$\begin{aligned} \sup_{t\ge 0}\int_0^t |G^*(t,v)|\,dv &= \sup_{t\ge 0}\int_0^t \left|\int_t^\infty C(t_0+u, v+t_0)\,du\right| dv \\ &= \sup_{t\ge 0}\int_0^t \left|\int_{t+t_0}^\infty C(u, v+t_0)du\right| dv \\ &= \sup_{t\ge 0}\int_{t_0}^{t_0+t} \left|\int_{t+t_0}^\infty C(u,v)\,du\right| dv. \end{aligned}$$

If we set $q = t + t_0$, then the last equality can be written as

$$\begin{aligned} \sup_{q\ge t_0}\int_{t_0}^q \left|\int_q^\infty C(u,v)\,du\right| dv &\le \sup_{q\ge 0}\int_0^q \left|\int_q^\infty C(u,v)\,du\right| dv \\ &\le \alpha < 1. \end{aligned}$$

This completes the proof.

Next, we need to examine (5.2.12), (5.2.13), and (5.2.14) for (5.2.37) since these were assumptions in Theorem 5.2.1.

Proposition 5.2.2. *If $Q(t) \le 0$ and if (5.2.12) and (5.2.14) hold for (5.2.3) and (5.2.11), then $Q^*(t) \le 0$ and (5.2.12) and (5.2.14) hold for (5.2.37) and (5.2.39).*

Proof. Assume that (5.2.12) holds for (5.2.3) and (5.2.11). We have

$$Q(t) = A(t) - G(t,t) = A(t) + \int_t^\infty C(u,t)\,du$$

and from the proof of Proposition 5.2.1 we also have

$$Q^*(t) = Q(t + t_0)$$

so

$$\int_0^t Q^*(s)\,ds \to -\infty.$$

Moreover,

$$\begin{aligned}\int_0^t e^{\int_s^t Q^*(u)\,du}\,ds &= \int_0^t e^{\int_s^t Q(u+t_0)\,du}\,ds \\ &= \int_0^t e^{\int_{s+t_0}^{t+t_0} Q(u)\,du}\,ds \\ &\le \int_0^{t+t_0} e^{\int_s^{t+t_0} Q(u)\,du}\,ds \\ &\le \Gamma.\end{aligned}$$

This means that (5.2.12) holds for (5.2.37) and (5.2.39).

Assume that (5.2.14) holds for (5.2.3) and (5.2.11). Then we can say that

$$\int_0^{T+t_0} |G(t,v)|\,dv = \int_0^{T+t_0} \left|\int_t^\infty C(u,v)\,du\right| dv \to 0$$

as $t \to \infty$.

Hence,

$$\begin{aligned}\int_0^T |G^*(t,v)|\,dv &= \int_0^T \left|\int_t^\infty C(t_0+u, v+t_0)\,du\right| dv \\ &= \int_{t_0}^{T+t_0} \left|\int_{t+t_0}^\infty C(u,v)\,du\right| dv \\ &\le \int_0^{T+t_0} \left|\int_{t+t_0}^\infty C(u,v)\,du\right| dv \to 0\end{aligned}$$

as $t \to \infty$. This proves Proposition 5.2.2.

Theorem 5.2.1 applies to important second order classical problems.

All of the work here is for scalar equations. Any linear n–th order normalized Volterra equation can be written as a scalar first order equation. Many systems of Volterra equations can be reduced to a single n–th order Volterra equation. But (5.2.15) may not be natural for equations under such reduction. On the other hand, it is routine to do everything here for a system instead of a scalar equation. We did not do so because the straightforward derivation of (5.2.15) for systems yields a condition depending on the dimension. It is a reasonable conjecture that a clever norm can be constructed which will yield a result compatible with (5.2.15) and be, largely, independent of the dimension. Thus, we leave this as an open problem.

But many higher order equations of considerable interest can be reduced to (5.2.1). Study of stability, boundedness, and periodicity properties of differential equations by fixed point theory depends on exactly one thing: we must invert the equation in such a way that the natural mapping defined by that inversion will map a set containing only functions of the type of the desired solution into itself. Here is an example of this for the Liénard equation with memory in the restoring force which we write as

$$x'' + f(x)x' = -\int_0^t C(t,s)g(x(s))\,ds + e(t). \tag{5.4.40}$$

We suppose that there are positive constants D, H, K such that

$$|g(x) - Hx| \le K, \quad \left|\int_0^x f(s)\,ds - Dx\right| < K \quad \text{for all} \quad x \tag{5.2.41}$$

and a differentiable function k such that

$$C(t,s) = \frac{d}{dt}k(t,s), \quad k(s,s) = 0 \tag{5.2.42}$$

and

$$\int_t^\infty k(u,s)\,du \quad \text{is continuous.} \tag{5.2.43}$$

Remark 5.2.3. *Clever choices for k are rewarded. But the pedestrian choice is*

$$k(t,s) = \int_{t-s}^\infty C(u+s,s)\,du - \int_0^\infty C(u+s,s)\,du,$$

provided the integrals exist. Then (5.2.42) is satisfied. With that choice for $k(t,s)$, then to satisfy (5.2.43) we need

$$\begin{aligned}\int_t^\infty k(u,s)\,du &= \int_t^\infty \left[\int_{u-s}^\infty C(v+s,s)\,dv - \int_0^\infty C(v+s,s)\,dv\right] du \\ &= -\int_t^\infty \int_0^{u-s} C(v+s,s)\,dv\,du\end{aligned}$$

to be continuous. Examples are readily constructed.

To relate our work here to Theorem 5.2.1 we set

$$Q(t) = -D - H\int_t^\infty k(u,t)\,du \tag{5.2.44}$$

and

$$G(t,s) = H\int_t^\infty k(u,s)\,du. \tag{5.2.45}$$

Define

$$F(x) := \int_0^x f(s)\,ds \quad \text{and} \quad E(t) := \int_0^t e(s)\,ds. \tag{5.2.46}$$

Theorem 5.2.8. *If (5.2.41)-(5.2.43) hold, if $E(t)$ is bounded, and if Q and G defined by (5.2.44)-(5.2.45) satisfy (5.2.12)-(5.2.15), then (5.2.40) has a bounded solution for each $x(0), x'(0)$. If $e(t) = 0$ and if $x'(0) + F(x(0)) = 0$, then the linearization of (5.2.4) (see (5.2.49)) has a solution tending to zero.*

Proof. Write (5.2.40) as

$$\begin{aligned} x'' + f(x)x' &= -\int_0^t C(t,s)g(x(s))\,ds + e(t) \\ &= -\frac{d}{dt}\int_0^t \int_0^{t-s} C(u+s,s)\,du\; g(x(s))\,ds \; + e(t) \\ &= -\frac{d}{dt}\int_0^t \int_0^{t-s} \frac{d}{du}k(u+s,s)\,du\; g(x(s))\,ds + e(t) \end{aligned}$$

or

$$x'' + f(x)x' = -\frac{d}{dt}\int_0^t k(t,s)g(x(s))\,ds + e(t). \tag{5.2.47}$$

Thus,

$$x'(t) + F(x) = -\int_0^t k(t,s)g(x(s))\,ds + E(t) + x'(0) + F(x(0)). \tag{5.2.48}$$

To make clear how Theorem 5.2.2 is applied we trivially point out that (5.2.48) can be written as

$$x'(t) + D\frac{F(x)}{D} = -\int_0^t Hk(t,s)\frac{g(x(s))}{H}\,ds + E(t) + x'(0) + F(x(0)).$$

This last equation is parallel to (5.2.1) and we write the equation parallel to (5.2.3) as

$$\begin{aligned} x'(t) &= -Dx(t) - H\int_0^t k(t,s)x(s)\,ds + E(t) + x'(0) + F(x(0)) \\ &= -Dx(t) - H\int_t^\infty k(u,t)\,du\; x(t) \\ &+ \frac{d}{dt}\int_0^t \int_t^\infty Hk(u,s)\,du\; x(s)\,ds + E(t) + x'(0) + F(x(0)). \end{aligned} \tag{5.2.49}$$

Thus, for each fixed $x(0), x'(0)$ the conditions of Theorem 5.2.1 are satisfied and Equation (5.2.49) has a bounded solution. By Theorem 5.2.2, (5.2.40) has a bounded solution. If $e(t) = 0$ and $x'(0) + F(x(0)) = 0$ then we can conclude that there is a solution of (5.2.49) tending to zero by using the same argument as in the proof of (5.2.7) in Theorem 5.2.1.

Our Equation (5.2.40) includes the case of negative damping, $f(0) < 0$, which is known to give rise to relaxation oscillations. Much has been written about the Liénard equation with no delay or with a pointwise delay in the restoring force. We refer the reader to Sansone and Conti(1964), Burton and Townsend (1968, 1971), Graef (1972), Omari, Villari, and Zanolin (1988), Zhang (1992, 1993, 1996, 1997), and the references therein. We do not expect periodic solutions of (5.2.40) because the delay is on $[0, t]$, instead of on $[-\infty, t]$; this is exactly the difficulty cited concerning (5.2.27). If we rewrite (5.2.40) with the infinite delay and with periodic assumptions on e and C then it is possible that conditions very much like those of Theorem 5.2.1 could yield a periodic solution. We have not investigated that possibility. The equations with pointwise delay in the Zhang papers just cited do not share this lack of natural periodicity property which occurs in equations with a distributed delay on $[0, t)$.

The reader will find fascinating examples of relaxation oscillation problems in Haag (1962; pp. 147-171) of both mechanical and electrical type (briefly repeated in Burton (1985; pp. 144-148)). The introduction of a distributed delay goes back to Volterra (1959) and even to Picard in 1907. An interesting discussion and references is given in Davis (1962; pp. 112-113). It tells of the debate over the question of whether distorted material has a memory, as we have indicated, or if there simply needs to be more variables. Both views can be right as one will see in our Equation (5.2.40) by selecting $C(t, s) = e^{t-s}$. For with such a kernel, we can differentiate (5.2.40) again and eliminate the integral, at the cost of having a higher order equation. Volterra (1959; pp. 138-154) gives a detailed discussion of memory effects on elastic materials and their representation by integrals. If our problem is to represent the mechanical system of Haag, then $g(x)$ would represent the restoring force of the spring and the integral takes into account the effects of previous distortions of the spring.

Chapter 6

Open Problems, Global Nonlinearities

6.1 Open Problems

We have presented what may be thought of as a bare bones suggestion of a stability theory for functional differential equations based on fixed point theory. It is driven by examples and there are two main features. First, locally there is as much nonlinearity as one could possibly desire. But globally we have asked for a Lipschitz condition which makes the global theory almost linear in nature. That is certainly the area in which the theory is in greatest need for improvement and we start that study in the next section.

But there are other problems which are very attractive and highly accessible. Virtually all of Chapter 2 was made possible by the work of Cooke and Yorke concerning equations in which every constant is a solution and every solution approaches a constant. That led us to the conjecture that such functions constituted harmless perturbations of stable systems. Moreover, what was then done in Chapter 2 is surely just a sample of what awaits the investigator who considers the long line of papers which were inspired by the Cooke-Yorke paper. It is most reasonable to expect almost every one of those papers to inspire as much innovation as was seen in Chapter 2.

All of the next chapter concerns extensions of these techniques to stohastic problems. This does, indeed, appear to be a marvelous direction for extensions.

In a new vein, one can hardly read about these ideas without wanting to try them on semi-group problems. We have seen that a problem need not have a shred of linearity about it and, yet, through ideas like exact linearization and borrowing they can be treated with a variation of parameters type formula. This should certainly be a large and fruitful area.

There is a Volterra-Levin type problem not discussed here of the form

$$x' = -\int_{t-r(t)}^{t} C(t,s)g(s,x(s))\,ds$$

which was treated in Becker and Burton (2005) when $t - r(t)$ is increasing. So much more needs to be said. Immediate progress can be made by separating the integral into two pieces of the form

$$\int_{t-L}^{t} + \int_{t-r(t)}^{t-L}$$

where L is cleverly chosen.

There are many functional differential equations with application in mathematical biology which can fruitfully be attacked by the methods of this book. These are so important in that realistic conditions must be averages.

We have sought throughout the book to contrast and to unite Liapunov theory and fixed point theory. The matter of global nonlinearities can be studied by first using a Liapunov functional to obtain boundedness and then using fixed point theory with a weighted norm either with contractions or Schauder's theorem.

In most of our problems studied by means of contractions we have first shown that the mapping maps a bounded set into itself before we ever bring up the topic of contractions. We then assume a Lipschitz condition. Many of these problems can fruitfully be studied again by dropping the Lipschitz condition and using a Schauder-type fixed point theorem.

We have used Schaefer's fixed point theorem and the idea of *a priori* bounds in a limited way. This fixed point theory can be greatly developed in the stability investigations.

6.2 Nonlinearities

Early in our investigations we looked at questions of global large nonlinearities, Here, we will offer a few simple ideas from Burton and Furumochi (2001a). This is the place where we feel that much effort needs to be directed.

Consider the scalar equation

$$x'(t) = -a(t)x^3 + g(t, x^2, x) \tag{6.2.1}$$

where

$$a(t) \geq 0, \quad \int_0^\infty a(t)dt = \infty, \quad g(t, y, 0) = 0,$$

and

$$|g(t, y, x) - g(t, y, w)| \leq b(t)|y||x - w|. \tag{6.2.2}$$

Now we come to a recurring problem. We require that for each bounded continuous function $z^2(t)$ with $z^2(t) \geq c$ for some $c > 0$, there is an $\alpha < 1$ with

$$\int_0^t e^{-\int_s^t a(u)z^2(u)du} b(s)z^2(s)\, ds \leq \alpha, \ t \geq 0. \tag{6.2.3*}$$

An obvious sufficient condition is that $a(t) \geq kb(t)$ and $a(t) \geq 0$ for all t and some $k > 1$. But that is much too severe. We would like for a to be zero on long intervals when b is nonzero.

But here we come to a real difficulty. In these fully nonlinear problems we will use the unknown exact solution as part of the mapping. Thus, we need to rely on additional information to ensure that solutions exist on $[t_0, \infty)$. An example of (6.2.1) is

$$x' = -a(t)x^3 + b(t)x^3.$$

If $a(t) < b(t)$ and $b(t) > 0$ on any interval $[t_0, t_1]$, however short, there are solutions with finite escape time. Hence, it will be necessary to work with particular initial times t_0 for which

$$\int_{t_0}^t [-a(s) + b(s)]\, ds \leq 0 \text{ for all } t \geq t_0. \tag{*}$$

Obviously, if $a(t) \geq b(t)$ for all t, this would hold for any t_0.

Lemma 6.2.1. *Let (*) hold and let $x_0 \in R$. Then $x(t, t_0, x_0)$ is defined for all $t \geq t_0$.*

Proof. Let $x(t) = x(t, t_0, x_0)$ be a solution of (6.2.1) with maximal interval of definition $[t_0, t_1)$. It is known that $t_1 = \infty$ or $\lim_{t\to t_1} |x(t)| = \infty$. Then define a Liapunov function

$$V(x) = |x|$$

so that along the solution we have

$$V'(x(t)) \le -a(t)|x(t)|^3 + b(t)|x(t)|^3 = [-a(t) + b(t)]V^3.$$

If we separate variables and integrate we obtain

$$-V(t)^{-2} + V(t_0)^{-2} \le 2\int_{t_0}^{t} [-a(s) + b(s)]\, ds \le 0$$

so that $|x(t)|^3 \le |x(t_0)|^3$, a contradiction to the finite escape time.

We now assume that for a given x_0 and a given $c < |x_0|$, there is an $\alpha < 1$ such that if $z^2(t)$ is continuous and $x_0^2 \ge z^2(t) \ge c$, then

$$\int_{t_0}^{t} e^{-\int_s^t a(u)z^2(u)\,du} b(s)z^2(s)\, ds \le \alpha \text{ for } t \ge t_0. \tag{6.2.3}$$

Next, it will greatly sharpen the results if we also know that solutions either tend to 0 or are bounded strictly away from 0. If we are willing to ask that $a(t) \ge b(t)$ then the above argument with V will yield uniform stability and that will suffice. But we next indicate that we can get the result with fewer assumptions and more work.

Lemma 6.2.2. *Suppose that if $t_n \to \infty$ and if*

$$K_n = \sup_{t \ge t_n} \int_{t_n}^{t} [-a(s) + b(s)]\, ds, \text{ then } \lim_{n\to\infty} K_n = 0. \tag{**}$$

If $x(t)$ is a solution of (6.2.1) on $[t_0, \infty)$ either $x(t) \to 0$ or there is a $c > 0$ with $|x(t)| \ge c$.

Proof. Suppose there is a sequence $t_n \to \infty$, a sequence $s_n > t_n$, and a $c > 0$ with $|x(t_n)| \to 0$ and $|x(s_n)| = c$. Rename indices so that $|x(t_n)| < c/2$ for all n. Using the V as in the proof of Lemma 6.2.1 and taking

$$k_n = \int_{t_n}^{s_n} [-a(s) + b(s)]\, ds$$

upon integration of V' we have the relation

$$-V^{-2}(s_n)) + V^{-2}(t_n)) \le 2k_n.$$

If some $k_n \leq 0$, then $V(s_n) \leq V(t_n)$, a contradiction. Hence, $k_n \geq 0$ for all n and $k_n \leq K_n \to 0$ as $n \to \infty$ so

$$V^{-2}(t_n)) - 2k_n \leq c^{-2},$$

a contradiction since $V(t_n) \to 0$.

In particular, from this result, for each x_0 there is a solution $x(t, 0, x_0) =: z(t)$ defined on $[0, \infty)$. If we strengthen the Lipschitz condition in (6.2.2) it is unique. We will now outline the method to be used on these problems.

1. We shall assume or prove that there is a t_0 and for each x_0 there is a unique solution $x(t, t_0, x_0) =: z(t)$ on $[t_0, \infty)$. When (*) and (**) hold we have shown in the lemmas how this might be done in a particular problem.

2. Hence, $z(t)$ is the unique solution of

$$x' = -a(t)x^2(t)x + g(t, z^2(t), x), \; x(t_0) = x_0. \tag{6.2.4}$$

3. PROBLEM. In what space does $z(t)$ lie? We want to show that it lies in

$$\mathcal{S} = \{\phi : [t_0, \infty) \to R \,|\, \phi(t_0) = x_0, \; \phi(t) \to 0 \text{ as } t \to \infty, \; \phi \in \mathcal{C}\}. \tag{6.2.5}$$

Here, $\|\cdot\|$ will denote the supremum metric.

The unique solution of (6.2.4) is

$$x(t) = x_0 e^{-\int_{t_0}^{t} a(s)z^2(s)\,ds} + \int_{t_0}^{t} e^{-\int_s^t a(u)z^2(u)\,du} g(s, z^2(s), x(s))\,ds. \tag{6.2.6}$$

4. Define $P : \mathcal{S} \to \mathcal{S}$ by

$$(P\phi)(t) = x_0 e^{-\int_{t_0}^{t} a(s)z^2(s)\,ds} + \int_{t_0}^{t} e^{-\int_s^t a(u)z^2(u)\,du} g(s, z^2(s), \phi(s))\,ds. \tag{6.2.7}$$

5. If P has a fixed point, it is z, and so $z \in \mathcal{S}$ which means that $z(t) \to 0$.

6. In this example, when (*) and (**) hold we know from the lemmas that either:

a) $z(t) \to 0$, so there is nothing to prove, or

b) $|z(t)| \geq c > 0$ so $\int_0^\infty a(t)z^2(t)\,dt = \infty$.

Thus, we assume that b) holds.

7. Clearly, $(P\phi)(t_0) = x_0$ and $P\phi \in \mathcal{C}$. We now show that $(P\phi)(t) \to 0$ as $t \to \infty$ and we take $t_0 = 0$ for brevity.

Let $\varepsilon > 0$ and $\phi \in S$ be given and let $c > 0$ be found. Find t_1 so that $|\phi(t)| < \varepsilon/2$ if $t \geq t_1$. Then using (6.2.3) we obtain

$$\begin{aligned}
&\int_0^t e^{-\int_s^t a(u)z^2(u)\,du}|g(s,z^2(s),\phi(s))|\,ds \\
&\leq \int_0^{t_1} e^{-\int_s^t a(u)z^2(u)\,du} b(s)z^2(s)|\phi(s)|\,ds \\
&+ \int_{t_1}^t e^{-\int_s^t a(u)z^2(u)\,du} b(s)z^2(s)|\phi(s)|\,ds \\
&\leq e^{-\int_{t_1}^t a(u)z^2(u)\,du} \int_0^{t_1} e^{-\int_s^{t_1} a(u)z^2(u)\,du} b(s)z^2(s)\|\phi\|\,ds \\
&+ (\varepsilon/2)\int_0^t e^{-\int_s^t a(u)z^2(u)\,du} b(s)z^2(s)\,ds \\
&\leq \alpha\|\phi\|e^{-\int_{t_1}^t a(u)z^2(u)\,du} + (\varepsilon/2)\alpha.
\end{aligned}$$

The first term tends to zero as $t \to \infty$ and the second term can be made as small as we please.

8. To see that we have a contraction, for $\phi, \psi \in S$ we have

$$\begin{aligned}
&|(P\phi)(t) - (P\psi)(t)| \\
&\leq \int_{t_0}^t e^{-\int_s^t a(u)z^2(u)\,du}|g(s,z^2(s),\phi(s)) - g(s,z^2(s),\psi(s))|\,ds \\
&\leq \int_{t_0}^t e^{-\int_s^t a(u)z^2(u)\,du} b(s)z^2(s)|\phi(s) - \psi(s)|\,ds \\
&\leq \alpha\|\phi - \psi\|.
\end{aligned}$$

Thus, P does have a fixed point and it is in $\mathcal{S}$.

Let a and b be continuous, $h > 0$, and consider the equation

$$x'(t) = -a(t)x^3(t) + b(t)x(t-h)x^2(t) \tag{6.2.8}$$

This is a difficult problem even if we ask that $a(t) > |b(t)|$ since $b(t)x(t-h)x^2(t)$ can dominate $-a(t)x^3(t)$ along a solution in a neighborhood of zero. But under that condition it is possible to find a mildly unbounded Liapunov function which will show that all solutions can be continued for all future time, and that is crucial for the fixed point method we use. In this problem any t_0 will work and we simply let $t_0 = 0$. Here are the details.

Lemma 6.2.3. *Let $\psi : [-h, 0] \to R$ be a given continuous initial function and let $z(t)$ be the unique solution determined by ψ. Then z exists on $[0, \infty)$.*

Proof. The conditions of the example ensure that there is a unique local solution. It is well known that the only way in which a solution can fail to be defined for all future time is for there to exist $T > 0$ with $\limsup_{t\to T^-} |z(t)| = +\infty$. In particular, then $z(t-h)$ is a bounded continuous function on $[0,T]$. Now $a(t) > 0$ so there is an $A > 0$ with $a(t) > A$ on $[0,T]$ and there is a $B > 0$ with $B > |b(t)||z(t-h)|$ on that same interval. If we define

$$V(t,x) = (1+|x|)e^{-Kt}$$

then the derivative of V along any solution $x(t)$ of (6.2.8) satisfies

$$\begin{aligned} V' &\le e^{-Kt}[-a(t)|x|^3 + |b(t)x(t-h)||x|^2 - K - K|x|] \\ &\le e^{-Kt}[-A|x|^3 + Bx^2 - K] \le 0 \end{aligned}$$

for large enough K. Thus, $V(t,x(t))$ is bounded on $[0,T]$ and so it can not happen that $\limsup_{t\to T^-} |x(t)| = \infty$.

There will, of course, be problems in which we will be unable to find such a Liapunov function. In such cases we must state that solutions defined for all future time tend to zero.

We are not going to do the work of Lemma 6.2.2 here. Instead, for brevity we ask that

$$|b(t)| \le ka(t), k < 1. \tag{6.2.9}$$

Theorem 6.2.1. *Let (6.2.9) hold and let $a(t) \ge c > 0$. For a given initial function $\eta : [-h,0] \to R$, either:*
a) $x^2(t,0,\eta) \in L^1[0,\infty)$ or
b) $x(t,0,\eta) \to 0$ as $t \to \infty$.

Proof. We let $z(t)$ be the unique solution of (6.2.8) and consider

$$x' = -a(t)z^2(t)x + b(t)z^2(t)x(t-h) \tag{6.2.10}$$

so that the solution can be written as

$$x(t) = \eta(0)e^{-\int_0^t a(s)z^2(s)\,ds} + \int_0^t e^{-\int_s^t a(u)z^2(u)\,du} b(s)z^2(s)x(s-h)\,ds.$$

Let

$$\mathcal{S} = \{\phi : [-h,\infty) \to R \,|\, \phi(t) = \eta(t) \text{ on } [-h,0], \phi(t) \to 0 \text{ as } t \to \infty, \phi \in \mathcal{C}\}$$

and define the mapping P by $(P\phi)(t) = \eta(t)$ if $-h \le t \le 0$ and

$$(P\phi)(t) = \eta(0)e^{-\int_0^t a(s)z^2(s)\,ds} + \int_0^t e^{-\int_s^t a(u)z^2(u)\,du} b(s)z^2(s)\phi(s-h)\,ds$$

for $t \ge 0$. Our condition (6.2.9) readily yields that this is a contraction. Now, if z^2 is not in $L^1[0,\infty)$, then $\int_0^\infty a(t)z^2(t)dt = \infty$ and we can verify that $P : \mathcal{S} \to \mathcal{S}$. Hence, there is a unique fixed point in $\mathcal{S}$, which is the unique solution z, and it goes to zero. This completes the proof.

For completeness of our discussion of (E), we are also going to consider

$$x'(t) = -a(t)x^3(t) + b(t)x^2(t-h)x(t) \tag{6.2.11}$$

even though its solution does not involve fixed point theory. We assume that

$$a(t) \geq c > 0 \tag{6.2.12}$$

and that

$$a(t-h) \geq k|b(t)|, \quad k > 1. \tag{6.2.13}$$

From the argument in the previous section one readily concludes that for each continuous initial function, a unique solution exists on $[0, \infty)$. The next result is proved directly from the variation of parameters formula.

Theorem 6.2.2. *Let (6.2.12) and (6.2.13) hold. For a given continuous $\eta : [-h, 0] \to R$ either:*
a) $x^2(t, 0, \eta) \in L^1[0, \infty)$ or
b) $x(t, 0, \eta) \to 0$ as $t \to \infty$.

We return to (E) which we write as

$$x' = -a(t)x^3(t) + b(t)x^3(t-h) \tag{6.2.14}$$

with a view to showing that solutions tend to 0 without requiring that $a(t)$ be bounded. For brevity we ask that

$$|b(t+h)| \leq \gamma a(t),\ a(t) > 0,\ \gamma < 1. \tag{6.2.15}$$

(With more work we could ask that there is a t_0 with $\int_{t_0}^t [-a(s) + |b(s + h)|]\, ds \leq 0$.) Under this assumption we can use the Liapunov function

$$V(t, x_t) = |x| + \int_{t-h}^t |b(s+h)||x^3(s)|\, ds$$

and obtain

$$V' \leq [-a(t) + |b(t+h)|]|x^3|$$

which yields stability of the zero solution. This means that if $z(t)$ is the unique solution of (6.2.14) with continuous initial function $\eta : [-h, 0] \to R$, we can make z as small as we please by taking $|\eta|$ small. Thus, we ask that there exists $M > 0$ with

$$\int_{t-h}^{t} |b(u+h)|\, du \le M. \tag{6.2.16}$$

Then find $\beta > 0$ with

$$2\beta + \gamma =: \alpha < 1 \tag{6.2.17}$$

and finally take η so small that

$$\int_{t-h}^{t} |b(u+h)| x^2(u,0,\eta)\, du \le \beta. \tag{6.2.18}$$

While this does place an integral bound on b, there is no bound on a.

Theorem 6.2.3. *Let (6.2.15) to (6.2.18) hold with η selected and let $a(t) \ge c > 0$. Either:*
a) $x^2(t,0,\eta) \in L^1[0,\infty)$ or
b) $x(t,0,\eta) \to 0$ as $t \to \infty$.

Proof. Let $z(t) = x(t,0,\eta)$ which is the unique solution of

$$x' = -a(t)z^2(t)x + b(t)z^2(t-h)x(t-h), \quad x_0 = \eta. \tag{6.2.19}$$

Define

$$\mathcal{S} = \{\phi : [-h,\infty) \to R \,|\, \phi(t) = \eta(t) \text{ on } [-h,0], \phi \in \mathcal{C}, \phi(t) \to 0 \text{ as } t \to \infty\}$$

and define $(P\phi)(t) = \eta(t)$ for $-h \le t \le 0$ and for $t \ge 0$ define

$$\begin{aligned}(P\phi)(t) &= \eta(0)e^{-\int_0^t a(s)z^2(s)\,ds} \\ &\quad + \int_0^t e^{-\int_s^t a(u)z^2(u)\,du} b(s)z^2(s-h)\phi(s-h)\, ds.\end{aligned}$$

If a) fails, then $\int_0^t a(s)z^2(s)\,ds \to \infty$ and we can show that $P : S \to S$. Note that

$$\begin{aligned}&b(s)z^2(s-h)\phi(s-h) \\ &= b(s+h)z^2(s)\phi(s) - b(s+h)z^2(s)\phi(s) + b(s)z^2(s-h)\phi(s-h) \\ &= b(s+h)z^2(s)\phi(s) - (d/ds)\int_{s-h}^{s} b(u+h)z^2(u)\phi(u)\, du.\end{aligned}$$

Hence,

$$
\begin{aligned}
(P\phi)(t) &= \eta(0)e^{-\int_0^t a(s)z^2(s)\,ds} \\
&\quad + \int_0^t e^{-\int_s^t a(u)z^2(u)\,du} b(s+h)z^2(s)\phi(s)\,ds \\
&\quad - \int_0^t e^{-\int_s^t a(u)z^2(u)\,du} (d/ds) \int_{s-h}^{s} b(u+h)z^2(u)\phi(u)\,du.
\end{aligned}
$$

We integrate that last term by parts to obtain

$$
\begin{aligned}
& - e^{-\int_s^t a(u)z^2(u)\,du} \int_{s-h}^{s} b(u+h)z^2(u)\phi(u)\,du\Big|_0^t \\
& + \int_0^t e^{-\int_s^t a(u)z^2(u)\,du} a(s)z^2(s) \int_{s-h}^{s} b(u+h)z^2(u)\phi(u)\,du\,ds \\
& = - \int_{t-h}^{t} b(u+h)z^2(u)\phi(u)\,du \\
& + e^{-\int_0^t a(u)z^2(u)\,du} \int_{-h}^{0} b(u+h)z^2(u)\eta(u)\,du \\
& + \int_0^t e^{-\int_s^t a(u)z^2(u)\,du} a(s)z^2(s) \int_{s-h}^{s} b(u+h)z^2(u)\phi(u)\,du\,ds.
\end{aligned}
$$

It follows readily that we have a contraction.

We will now look at two quick examples. Since we want to cover the obvious perturbations of (E) we look at two of those which are very simple. First, let a and b be continuous and consider

$$x'(t) = -a(t)x^2(t-h)x + b(t)x^3(t-h). \tag{6.2.20}$$

It is readily argued that solutions can be defined for all future time, regardless of the signs of the functions. Let $\eta : [-h, 0] \to R$ be a given continuous initial function and let $z(t) = x(t, 0, \eta)$ which is the unique solution of

$$x' = -a(t)z^2(t-h)x + b(t)z^2(t-h)x(t-h),\ x_0 = \eta. \tag{6.2.21}$$

Define S as before and define

$$
\begin{aligned}
(P\phi)(t) &= \eta(0)e^{-\int_0^t a(s)z^2(s-h)\,ds} \\
&\quad + \int_0^t e^{-\int_s^t a(u)z^2(u-h)\,du} b(s)z^2(s-h)\phi(s-h)\,ds.
\end{aligned}
$$

We shall ask that

$$|b(t)| \le k|a(t)|, \quad k < 1. \tag{6.2.22}$$

Theorem 6.2.4. *If (6.2.22) holds and if $a(t) \geq c > 0$, either $x^2(t,0,\eta) \in L^1[0,\infty)$ or $x(t,0,\eta) \to 0$ as $t \to \infty$.*

Finally, let $a(t) > 0$ be a continuous function and consider

$$x' = -a(t)x(t-h)x^2(t).$$

If $a(t) \geq c > 0$, then solutions can have finite escape time. Thus, let $\eta : [-h,0] \to (0,\infty)$ and let $z(t) = x(t,0,\eta)$. Notice that

$$x' = -[a(t)z(t-h)z(t)]x(t)$$

is linear with negative coefficient so the solution decreases monotonically. It is never zero by the uniqueness theorem. We have

$$x(t) = \eta(0)e^{-\int_0^t a(s)z(s-h)z(s)\,ds},$$

and, unless $z(s-h)z(s) \in L^1$, then $x(t) \to 0$. Indeed, $x(t)$ always tends to zero if $\eta > 0$.

This will acknowledge that much of the material presented here was first published in *Dynamic Systems and Applications*, as detailed in Burton and Furumochi (2001a), which is published by Dynamic Publishers.

Chapter 7

Appleby's Stochastic Perturbations

7.0 Acknowledgment

This chapter is entirely the contribution of Dr. John A. D. Appleby of the School of Mathematical Sciences, Dublin City University, Dublin 9, Ireland with e-mail john.appleby@dcu.ie. It was written for this book and is included with his permission.

7.1 Introduction

Earlier, we saw that the unique solution of Levin's equation

$$x'(t) = -\int_{t-L}^{t} p(s-t)g(x(s))\,ds, \quad t > 0 \tag{7.1.1}$$

with $x = \psi$ on $[-L, 0]$ obeys

$$\lim_{t\to\infty} x(t) = 0$$

under some conditions on p and g. It is now reasonable to ask: if this deterministic problem is asymptotically stable, is the zero solution of (7.1.1) stable with respect to stochastic perturbations? For simplicity, we will concentrate on just one type of perturbed stochastic functional differential equation of Itô type, which will take the form

$$dX(t) = -\left(\int_{t-L}^{t} p(s-t)g(X(s))\,ds\right)dt + \sigma(t)\,dB(t). \tag{7.1.2}$$

In this equation B is a scalar standard Brownian motion, and σ is a deterministic function. Roughly speaking, we will show that when σ tends to zero sufficiently quickly, then stability is preserved.

On the other hand, we also saw in an earlier chapter that all solutions of the Cooke–Yorke equation

$$x'(t) = g(x(t)) - g(x(t-L), \quad t > 0 \tag{7.1.3}$$

with $x = \psi$ on $[-L, 0]$ obeys

$$\lim_{t\to\infty} x(t) = k$$

where k is a constant uniquely defined by g and the initial data ψ. We show in this chapter that the stochastic equation

$$dX(t) = (g(X(t)) - g(X(t-L)))\,dt + \sigma(t)\,dB(t) \tag{7.1.4}$$

enjoys the same property of having asymptotically constant solutions, almost surely, provided σ is square integrable. In this case, the limiting solution is a random variable depending on the sample path of the Brownian motion B, σ, g and ψ.

The shape of the chapter is as follows; first we present a section on elements of stochastic analysis that will be used herein. The next section studies the stability of the stochastic version of Levin's equation (7.1.2). This material has appeared in Appleby (2005a). We give in precise terms the problem to be studied, ensuring that the solution of (7.1.2) is unique. We also indicate what we mean by asymptotic stability for these stochastic equations, and give some intuition as to the likely interpretation of the equation being studied. We show that when σ decays sufficiently quickly, and the conditions used to ensure deterministic stability are imposed, then all solutions of the stochastic equation also converge. Finally, we establish that the condition on the decay rate of σ imposed cannot be easily relaxed, while still ensuring the convergence of solutions to zero.

The last section of the chapter considers the asymptotic behaviour of solutions of the stochastic Cooke–Yorke equation (7.1.4). The material in that section appears in Appleby (2005b).

7.2 Stochastic preliminaries

7.2.1 Apologia and reading list

This is not a textbook on stochastic analysis, so our aim here is to simply list some important definitions, concepts and results from probability theory and stochastic analysis that we find useful in what follows. The aim is to give a reasonably self–contained list of the principal results used. Experts in stochastic analysis may find this section distracting, not to say superficial; but they are not the intended audience, and they can safely skip to the next section.

This is a book about application of fixed point theory to the stability theory of functional differential equations, so we do not want to burden the reader with a lengthy list of the many excellent textbooks on probability theory, stochastic analysis, or the stability theory of stochastic functional differential equations that have recently appeared; rather we mention the merest handful.

Among those at an introductory graduate level, we refer to Karatzas and Shreve (1991) for results on Brownian motion and martingales; to Øksendal (2003) for stochastic differential equations; and to Mao (1997) for stochastic delay differential equations, though all touch upon the topics introduced here. Readers wishing to brush up on introductory probability could do worse than to turn to Jacod and Protter (2003), or to Brezeźniak and Zastawniak (1998) for basic stochastic processes.

A textbook which gives a comprehensive perspective on research on stability theory in stochastic functional differential equations up to the year 2000 is Kolmanovskii and Myskhis (1999). We bring it to attention here for two reasons. First, it contains many interesting applications of Liapunov functional arguments to stability; second, it contains a well–compiled bibliography of activity in the stability theory of stochastic functional equations up to that date.

7.2.2 Stochastic processes and Brownian motion

The standard set–up for a random experiment is a probability triple $(\Omega, \mathcal{F}, \mathbb{P})$, where Ω is the collection of all possible outcomes, and $\mathcal{F}$ is the set of all events A to which a probability $\mathbb{P}[A]$ can be attached. $\mathcal{F}$ is called a σ-algebra, and $\mathbb{P}$ a probability measure. We are often very interested in events $A \in \mathcal{F}$ which are realized for almost all $\omega \in \Omega$. Such events are called *almost sure* events, and A is an almost sure event if $\mathbb{P}[A] = 1$. We will often use the abbreviation a.s. to stand for almost sure or almost surely.

$\mathcal{G}$ is said to be a *sub-σ-algebra* of $\mathcal{F}$ if $\mathcal{G}$ is a σ-algebra and for each $A \in \mathcal{G}$ we have $A \in \mathcal{F}$, and we write $\mathcal{G} \subset \mathcal{F}$. A $\mathcal{G}$–measurable *random variable* X is a mapping $X : \Omega \to \mathbb{R}$ such that the event $\{\omega : X(\omega) \leq x\} \in \mathcal{G}$ for all $x \in \mathbb{R}$. We call the function $F_X : \mathbb{R} \to [0,1]$ defined by $F(x) = \mathbb{P}[X \leq x]$ the *distribution function* of X. The *expectation* of the random variable X is given by

$$\mathbb{E}[X] = \int_\Omega X(\omega)\, d\mathbb{P}(\omega),$$

provided $\int_\Omega |X(\omega)|\, d\mathbb{P}(\omega) < \infty$. In terms of the distribution function F_X we have

$$\mathbb{E}[X] = \int_\mathbb{R} x\, dF_X(x)$$

provided $\int_\mathbb{R} |x|\, dF_X(x) < \infty$. The *conditional expectation* of a $\mathcal{F}$–measurable random variable, relative to a sub-σ-algebra $\mathcal{G}$, is a $\mathcal{G}$–measurable random variable denoted by $\mathbb{E}[X|\mathcal{G}]$, and which satisfies

$$\mathbb{E}[(X - \mathbb{E}[X|\mathcal{G}])Y] = 0, \quad \text{for all } \mathcal{G}\text{–measurable } Y.$$

If X obeys $\mathbb{E}[X^2] < \infty$ (we say that X is square integrable), the conditional expectation $\mathbb{E}[X|\mathcal{G}]$ can be viewed as the orthogonal projection of X onto the space of $\mathcal{G}$-measurable random variables which are square integrable. In other words, it is the best approximation of X among all $\mathcal{G}$-measurable random variables, or more intuitively, the best guess for $X(\omega)$, given that ω is contained in a set in $\mathcal{G}$.

Two random variables with the same distribution function are said to have the same distribution. A *standard normal* random variable has distribution function

$$\Phi(x) = \frac{1}{\sqrt{2\pi}} \int_{-\infty}^{x} e^{-\frac{1}{2}u^2}\, du, \quad x \in \mathbb{R}. \tag{7.2.1}$$

X is a normal random variable with mean μ and variance $\sigma^2 > 0$ if the random variable $Z = (X - \mu)/\sigma$ is a standard normal random variable.

The events A and B are said to be *independent* if $\mathbb{P}[A \cap B] = \mathbb{P}[A]\mathbb{P}[B]$. This is consistent with the intuitive notion of "physical" independence, namely that the probability that the event B occurs given that the event A occurs, is the same as the chances of the event B occurring, without any knowledge of whether A occurs or not. The event A is said to be independent of the σ-algebra $\mathcal{G}$ if A is independent of every event $B \in \mathcal{G}$. The random variable X is said to be independent of the σ-algebra $\mathcal{G}$ if, for each $x \in \mathbb{R}$, the event $\{X \leq x\}$ is independent of $\mathcal{G}$.

In this chapter, we consider the solution of a random equation. At every instant in time the solution at that time is a random variable, and the solution as a whole is a collection of random variables, or stochastic process. In modelling a real-world phenomenon, it is reasonable that these variables should all be defined on the same probability space. Towards this end, *stochastic process* $X = \{X(t) : t \geq 0\}$ is a collection of random variables all defined on the same probability triple, and is a mathematical model for the evolution of a single random phenomenon. Since it is often the case that information about the process (or the environment in which the process evolves) increases as time increases, we introduce an increasing family of sub-σ-algebras of $\mathcal{F}$ which are indexed by time. Such a family $\mathcal{F}(t)_{t\geq 0}$ is called a *filtration* and has the property that $\mathcal{F}(s) \subset \mathcal{F}(t)$ for all $0 \leq s < t$ and $\mathcal{F}(t) \subset \mathcal{F}$. A stochastic process $X = \{X(t) : t \geq 0\}$ is said to be *adapted* to the filtration $\{\mathcal{F}(t)\}_{t\geq 0}$ if $X(t)$ is $\mathcal{F}(t)$ measurable, for all $t \geq 0$. This simply reflects the fact that, while the value of the process in the future is unknown, its values up to and including time t $\{X(s) : 0 \leq s \leq t\}$ should be observable at time t. Such a process may be denoted by $X = \{X(t) : \mathcal{F}(t); t \geq 0\}$ to signify the adaptedness to $\{\mathcal{F}(t)\}_{t\geq 0}$. A process $X = \{X(t) : t \geq 0\}$ is said to be *deterministic* if $X(t)$ is $\mathcal{F}(0)$-measurable for all $t \geq 0$. In this terminology, the solution of an ordinary (non–stochastic) differential equation is deterministic, as its value at time t is, in principle at least, known at time 0.

Every time the "experiment" is run, the differing random input leads to a different random output, or *realization*, of the process. The *realization* or *sample path* of the process X for the outcome $\omega \in \Omega$ is the function $X(\omega, \cdot)$ defined by $t \mapsto X(t, \omega)$.

We say that a process is *continuous* if for almost all $\omega \in \Omega$ the function $t \mapsto X(t, \omega)$ is continuous on $[0, \infty)$. It is an *increasing process* if for almost all $\omega \in \Omega$ $t \mapsto X(t, \omega)$ is non–negative, non–decreasing and right–continuous on $[0, \infty)$.

In this chapter, the randomness is supplied by a *standard one–dimensional* $\mathcal{F}(t)$*–Brownian motion* B. It is a stochastic process $B = \{B(t); \mathcal{F}(t); t \geq 0\}$ with $B(0) = 0$, $\mathbb{E}[B(t)] = 0$, $\mathbb{E}[B^2(t)] = t$ for each $t \geq 0$, and moreover possesses stationary and independent increments and continuous sample paths. Mathematically, this means that $B(t+s) - B(t)$ has the same distribution as $B(s)$ for all $t \geq 0$, $s \geq 0$, that $B(t+s) - B(t)$ is independent of $\mathcal{F}(s)$ for all $t \geq 0$, $s > 0$, and that $B(\omega) : [0, \infty) \to \mathbb{R} : t \mapsto B(t, \omega)$ is a continuous function for all $\omega \in \Omega^*$, where $\Omega^* \in \mathcal{F}$ is an a.s. event. The process has been defined in terms of its properties, rather than constructively; however, it turns out that a process with these properties does exist, and is moreover uniquely determined by these properties. In fact, it turns out $B(t)$ is a normally distributed random variable for each $t > 0$.

The properties are a mathematical invocation of pure randomness, once we require that the path traced out by the process is continuous. The independence of increments means that a future change in the process cannot be predicted by studying the past behaviour of the process; the stationarity simply means that changes occurring over any fixed interval of time have the same statistical properties, no matter when that interval starts. The conditions $B(0) = 0$, $\mathbb{E}[B(t)] = 0$, $\mathbb{E}[B^2(t)] = t$ are simply to "standardize" or normalize the process.

Although B has continuous sample paths, these sample paths are nowhere differentiable, a.s., i.e.,

$$\mathbb{P}[\{\omega : t \mapsto B(t,\omega) \text{ is differentiable at some } s \geq 0\}] = 0.$$

Furthermore, the sample paths of B are not of finite variation a.s. It is this property which lies behind many of the complexities of dynamical systems driven by an underlying Brownian motion, and which precludes direct analysis by traditional calculus.

7.2.3 Asymptotic behaviour of martingales

In this chapter, we will often need to use asymptotic results about an important class of stochastic processes called martingales. Such processes arise naturally in the study of the dynamical systems studied here.

A real–valued $\mathcal{F}(t)$–adapted process $M = \{M(t) : t \geq 0\}$ is a *local martingale* if

$$\mathbb{E}[M(t)|\mathcal{F}(s)] = M(s), \quad \text{a.s. for all } 0 \leq s < t < \infty.$$

The *quadratic variation* of a local martingale M is the unique continuous adapted and increasing process $\langle M\rangle = \{\langle M\rangle(t) : t \geq 0\}$, such that the process $M^2 - \langle M\rangle$ is a local martingale vanishing at 0.

The following two results show that the asymptotic behaviour of the quadratic variation determine the asymptotic behaviour of the local martingale. We do not give proofs for either result.

Lemma 7.2.1 (Law of the iterated logarithm for martingales). *Let M be a real–valued continuous local martingale vanishing at 0, with quadratic variation $\langle M\rangle$. If A is a set of positive probability, and*

$$\lim_{t\to\infty} \langle M\rangle(t) = \infty, \quad a.s. \text{ on } A,$$

then

$$\limsup_{t\to\infty} \frac{|M(t)|}{\sqrt{2\langle M\rangle(t) \log\log\langle M\rangle(t)}} = 1, \quad a.s. \text{ on } A.$$

Lemma 7.2.2 (Martingale convergence theorem). *Let M be a real-valued continuous local martingale vanishing at* 0, *with quadratic variation* $\langle M \rangle$. *If A is a set of positive probability, and*

$$\lim_{t\to\infty} \langle M \rangle(t) < \infty, \quad \textit{a.s. on } A,$$

then

$$\lim_{t\to\infty} M(t) \quad \textit{exists a.s. on } A \textit{ and is a.s. finite on } A.$$

Conversely, if there is a set A of positive probability and $\lim_{t\to\infty} M(t)$ *exists a.s. on A and is a.s. finite on A, then*

$$\lim_{t\to\infty} \langle M \rangle(t) \quad \textit{exists a.s. on } A \textit{ and is a.s. finite on } A.$$

7.2.4 The Itô integral and Itô processes

The stochastic dynamical systems studied in this chapter are not automatically amenable to many of the techniques of classical dynamical systems, or may only be analyzed when these techniques have been appropriately modified. One of the major reasons for this is that Brownian motion, which is the source of this randomness, does not have differentiable sample paths. As a consequence, the solutions of the equations studied here are also nowhere differentiable. Therefore, it is not possible to rigorously formulate the problems in this chapter as "differential" equations: instead we must write them as integral equations. Processes which solve such equations that are driven by Brownian motion are important instances of Itô processes, which we introduce here. In order to study such processes, we need to mention a little about integration with respect to Brownian motion, and it is this topic to which we turn first.

Such an integration theory is not completely straightforward. Due to the fact that a standard Brownian motion B is neither differentiable nor of finite variation means that for any $(\mathcal{F}(t))_{t\geq 0}$–adapted process X and fixed $t > 0$ the integral

$$I_X(t) = \int_0^t X(s)\, dB(s)$$

cannot be interpreted as a Stieltjes integral. A consistent interpretation can be ascribed to such integrals, but we cannot hope to give all the details here. The above integral, called the *Itô integral* of X, can be given explicitly when X is a simple process, which is stochastic generalization of a step function. The Itô integral of more general processes can be defined by expressing them as the limit of a sequence of integrals of simple processes. The method of constructing the Itô integral therefore bears certain similarities to the construction of the Riemann integral in conventional calculus.

We record at this point two important properties of Itô integrals. Firstly, Itô integrals give rise to local martingales. For example, if H is a continuous $\mathcal{F}(t)$–adapted process, then M defined by

$$M(t) = \int_0^t H(s)\,dB(s), \quad t \geq 0$$

is a local martingale, null valued at zero, with quadratic variation

$$\langle M \rangle(t) = \int_0^t H^2(s)\,ds, \quad t \geq 0.$$

Secondly, Itô integrals with deterministic integrands are normally distributed. More precisely, if $h \in C([0,\infty);\mathbb{R})$ is a deterministic function, then the local martingale

$$M(t) = \int_0^t h(s)\,dB(s), \quad t \geq 0$$

has the property that, for each $t > 0$, $M(t)$ is normally distributed with zero mean and variance $\int_0^t h^2(s)\,ds$.

Let B be a one–dimensional standard Brownian motion on $(\Omega, \mathcal{F}, \mathbb{P})$. The process $X = \{X(t); \mathcal{F}(t); t \geq 0\}$ is an *Itô process* which has the representation

$$X(t) = X(0) + \int_0^t U(s)\,ds + \int_0^t V(s)\,dB(s), \quad t \geq 0, \tag{7.2.2}$$

where U and V are $\mathcal{F}(t)$–adapted. Here

$$\int_0^t |U(s)|\,ds < \infty, \quad \int_0^t V^2(s)\,ds < \infty, \quad 0 \leq t < \infty, \text{ a.s.}$$

There is a convenient differential shorthand for (7.2.2) which is regularly employed, which is

$$dX(t) = U(t)\,dt + V(t)\,dB(t).$$

Manipulation of Itô processes can be achieved by formulae analogous to the chain rule (called Itô's rule) and integration by parts. We give a special case of the integration by parts result here.

Lemma 7.2.3 (Stochastic integration by parts). *Let X be an Itô process with representation* (7.2.2), *and suppose that Y is a differentiable deterministic function. Then*

$$X(t)Y(t) = X(0)Y(0) + \int_0^t X(s)\,dY(s) + \int_0^t Y(s)\,dX(s), \quad t \geq 0.$$

In this chapter, we need to be able to solve the stochastic differential equation

$$dY(t) = -Y(t)\,dt + \sigma(t)\,dB(t)$$

where σ is a continuous and deterministic function. By analogy with a perturbed deterministic equation, we seek a solution of the form $Y(t) = e^{-t}Z(t)$, where Z is an as yet undetermined Itô process. Rewriting this as $Z(t) = e^t Y(t)$, we may use the integration by parts formula to get

$$dZ(t) = d(e^t Y(t)) = e^t Y(t)\,dt + e^t\,dY(t) = e^t \sigma(t)\,dB(t),$$

which, bearing in mind that $Z(0) = Y(0)$, has equivalent integral form

$$Z(t) = Y(0) + \int_0^t e^s \sigma(s)\,dB(s).$$

Therefore

$$Y(t) = Y(0)e^{-t} + e^{-t}\int_0^t e^s \sigma(s)\,dB(s).$$

In stochastic differential equations and stochastic functional equations the integrand in the Itô integral is called the *diffusion coefficient*, and the integrand of the Riemann integral the *drift coefficient.*

7.3 Stochastic version of Levin's problem

The main purpose of this section is to determine conditions under which solutions of (7.3.4) tend to a unique non-stochastic equilibrium as $t \to \infty$.

7.3.1 Formulation of problem and existence and uniqueness of solution

In this subsection, we consider the existence and uniqueness of solutions of the stochastic functional differential equation (7.3.4). Here, σ represents the intensity of the stochastic perturbation. We first give a precise formulation of this problem in order to ensure that there exists a unique solution to this equation.

From this point on, we work on a given complete filtered probability space $(\Omega, \mathcal{F}, (\mathcal{F}(t))_{t\geq 0}, \mathbb{P})$, where the filtration $(\mathcal{F}(t))_{t\geq 0}$ is the natural filtration of a scalar standard Brownian motion $B = \{B(t); 0 \leq t < \infty; \mathcal{F}^B(t)\}$, viz., $\mathcal{F}(t) = \mathcal{F}^B(t) := \sigma\{B(s) : 0 \leq s \leq t\}$. In the following, we denote by $C([-L,0];\mathbb{R})$ the family of all continuous functions $\varphi : [-L,0] \to \mathbb{R}$ with norm $\|\varphi\| = \sup_{-L\leq\theta\leq 0} |\varphi(\theta)|$. For I an interval of $\mathbb{R}$, we use the notation $C(I;\mathbb{R})$ to denote the family of continuous functions $\varphi : I \to \mathbb{R}$. We denote by $C_0([0,\infty);\mathbb{R})$ the family of continuous functions $\varphi : [0,\infty) \to \mathbb{R}$ such that $\varphi(t) \to 0$ as $t \to \infty$.

Let $L > 0$ and suppose

$$\sigma \in C([0,\infty);\mathbb{R}). \tag{7.3.1}$$

Let $g \in C(\mathbb{R};\mathbb{R})$ satisfy the following global Lipschitz condition:

There exists $K > 0$ such that for all $x, y \in \mathbb{R}$

$$|g(x) - g(y)| \leq K|x - y|. \tag{7.3.2}$$

We suppose also that p obeys

$$p \in C([-L,0];\mathbb{R}). \tag{7.3.3}$$

We denote by $X_t = \{X(t+\theta) : -L \leq \theta \leq 0\}$ the history of the stochastic process X; we may view X_t as a $C([-L,0];\mathbb{R})$–valued stochastic process.

In order to define solutions of the stochastic functional equation, we define an initial function on $[-L,0]$. We ask that

$$X_0 = \psi \in C([-L,0];\mathbb{R}).$$

The following result on the existence and uniqueness of a continuous adapted process X which satisfies the stochastic equation

$$X(t) = \psi(0) + \int_0^t \left\{ -\int_{s-L}^s p(u-s)g(X(u))\,du \right\} ds \tag{7.3.4a}$$
$$+ \int_0^t \sigma(s)\,dB(s), \quad t \geq 0, \quad \text{a.s.}$$

$$X(t) = \psi(t), \quad t \in [-L,0] \tag{7.3.4b}$$

arises from an application of standard results for stochastic functional differential equations.

Theorem 7.3.1. *Let $L > 0$. If g obeys (7.3.2), p obeys (7.3.3), $\sigma : [0, \infty) \to \mathbb{R}$ is continuous, and $\psi \in C([-L, 0]; \mathbb{R})$, then there is a unique continuous $\mathcal{F}(t)$–adapted process $(X_t)_{t \geq 0}$ which obeys (7.3.4). The process is unique in the sense that any other process $\tilde{X}$ which obeys (7.3.4) is indistinguishable from X, viz.,*

$$\mathbb{P}[\{\omega \in \Omega : X(t, \omega) = \tilde{X}(t, \omega) \text{ for all } t \geq -L\}] = 1.$$

Statements and proofs of results which imply this may be found in e.g., Mao (1997).

With questions of existence and uniqueness put to one side, we now may consider the stability of solutions. In this chapter, we are interested in the circumstances under which almost all sample paths of the solution of (7.3.4) converge as $t \to \infty$ to the zero equilibrium of the underlying deterministic equation (7.1.1). However since zero is not a solution of (7.3.4), we speak of convergence to zero, rather than of stability, though we can view this type of convergence as stability with respect to a (stochastic) perturbation. We say that the solution X of (7.3.4) is *almost surely convergent to zero* if

$$\mathbb{P}[\{\omega \in \Omega : \lim_{t \to \infty} X(t, \omega) = 0\}] = 1. \tag{7.3.5}$$

Another way of writing this is

$$\lim_{t \to \infty} X(t) = 0, \quad \text{a.s.}$$

There are many types of convergence which could also be studied, for example, convergence in mean-square

$$\lim_{t \to \infty} \mathbb{E}[X(t)^2] = 0$$

or convergence in probability

$$\lim_{t \to \infty} \mathbb{P}[|X(t)| > \epsilon] = 0, \quad \text{for each } \epsilon > 0,$$

but we do not consider these, or other types of convergence, here. Each type of convergence has its own merit, and while some types are stronger than others (both mean square and almost sure convergence always imply convergence in probability for general stochastic processes), general implications cannot always be found (almost sure convergence does not always imply mean square convergence, nor does mean square convergence always imply almost sure convergence for general processes). However, almost sure behaviour is interesting to study because it records faithfully the behaviour of the random "experiment" (almost) every time that it is run. This means that if a process X which obeys (7.3.5) models a random real–world phenomenon, the trajectory of that real–world phenomenon we observe will tend to zero as $t \to \infty$, except in exceptional circumstances (i.e., for a zero probability set of trajectories).

In this case, we can view (7.3.4) as giving the behaviour of a stochastically perturbed version of (7.1.1). In particular, this perturbation is of a genuinely external nature, as the random contribution is independent of the state (i.e., the diffusion coefficient is independent of X). In this case, we say that the noise is state–independent. By analogy with deterministic equations, we expect that the stability of the equilibrium of the unperturbed deterministic equation can be assured only if the external perturbation dies away in some suitable manner as $t \to \infty$. Our job here is to show that this is the case, and to given a sharp and quantifiable definition (in terms of a critical rate of decay of σ) of the "suitable manner" in which the external perturbation fades.

In some ways, the study of equations which are stochastically perturbed in this way is easier than that of equations in which the noise is state–dependent. Although we do not use fixed point analysis to determine the asymptotic stability of solutions of equations with state–dependent noise in this book, the method of rewriting equations by the introduction of a neutral term employed by the fixed point approach used here can be exploited to determine asymptotic stability conditions for equations like

$$dX(t) = -\left(\int_{t-L}^{t} p(t-s)g(X(s))\,ds\right)\,dt + \sigma X(t)\,dB(t)$$

for σ sufficiently small and p and g satisfying the conditions of this chapter.

7.3.2 Sufficient conditions for asymptotic stability

This section is devoted to establishing sufficient conditions for all solutions of (7.3.4) to obey (7.3.5). To do this, we firstly decompose the solution of (7.3.4) into the solution of a random functional differential equation (whose paths obey perturbed versions of (7.1.1) and may be studied individually by means of fixed point theory), and a linear stochastic differential equation (whose asymptotic behaviour is determined using the martingale limit theorems quoted earlier in this chapter). The other two parts of this section each deal with the asymptotic behaviour of these auxiliary processes.

Reduction of the stability problem

In this subsection, we show that the problem of determining the almost sure stability of solutions of (7.3.4) can be reduced to solving two simpler problems:

(i) Determining the asymptotic stability of the zero solution of (7.1.1) with respect to perturbations $q(t)$ such that $q(t) \to 0$ as $t \to \infty$, where

$$x'(t) = -\int_{t-L}^{t} p(s-t)g(x(s))\,ds + q(t), \quad t > 0, \tag{7.3.6a}$$

$$x(t) = \psi(t), \quad t \in [-L, 0]. \tag{7.3.6b}$$

(ii) Determining the asymptotic stability of the solution of the stochastic differential equation

$$dY(t) = -Y(t)\,dt + \sigma(t)\,dB(t) \tag{7.3.7}$$

In what follows, we request that (7.3.3) and (7.3.2) hold as well as

$$g(0) = 0. \tag{7.3.8}$$

The following high level properties will be required to hold:

$$\left\{ \begin{array}{c} \text{For all } \psi \in C([-L,0];\mathbb{R}),\ q \in C_0([0,\infty);\mathbb{R}), \text{ the} \\ \text{unique solution } x \text{ of (7.3.6) obeys } x(t) \to 0 \text{ as } t \to \infty \end{array} \right. \tag{7.3.9}$$

and

$$\text{The unique solution } Y \text{ of (7.3.7) obeys } \lim_{t\to\infty} Y(t) = 0, \text{ a.s.} \tag{7.3.10}$$

Theorem 7.3.2. *Let $L > 0$. Suppose g obeys* (7.3.2), (7.3.8), *p obeys* (7.3.3), *$\psi \in C([-L,0];\mathbb{R})$, and $\sigma \in C([0,\infty);\mathbb{R})$. Let X be the unique continuous $\mathcal{F}(t)$-adapted process which obeys* (7.3.4). *If Properties* (7.3.9) *and* (7.3.10) *hold, then*

$$\lim_{t\to\infty} X(t) = 0, \quad a.s. \tag{7.3.11}$$

We defer the proof of Theorem 7.3.2 for a moment. Our purpose in writing the result in this form is to show that determining stability of the stochastic equation (7.3.4) can be broken into determining the stability of two related problems, which are easier to study.

(7.3.4) is a stochastic functional differential equation with nowhere differentiable sample paths, so there are three sources of difficulty in the analysis: randomness, lack of regularity of solutions, and the presence of functional dependence. However, by introducing the auxiliary problems (7.3.6) and (7.3.7), we can divide and conquer these difficulties.

(7.3.6) is a functional equation, but is not random, and has C^1 solutions. Indeed, it is quite similar to equations studied elsewhere in this book by fixed point methods. The equation (7.3.7) is certainly still stochastic, and has nowhere differentiable sample paths: however, it is a linear differential equation, and, not possessing complicated path dependence, or awkward nonlinearities, is much more straightforward to analyze.

Proof of Theorem 7.3.2. Let $Y = \{Y(t); -L \leq t < \infty; \mathcal{F}(t)\}$ be the process defined by $Y(t) = \psi(t)$, $t \in [-L, 0]$ and (7.3.7). Note by Property (7.3.10) we have that $Y(t) \to 0$ as $t \to \infty$, a.s. Let

$$\Omega_2 = \{\omega \in \Omega : X(\omega) \text{ is the realization of the unique continuous adapted process obeying (7.3.4), } Y \text{ defined by (7.3.7) obeys } Y(t,\omega) \to 0 \text{ as } t \to \infty\}.$$

Then $\mathbb{P}[\Omega_2] = 1$. For each $\omega \in \Omega_2$ we may define

$$x(t,\omega) = X(t,\omega) - Y(t,\omega), \quad t \geq -L.$$

Clearly $x(t,\omega) = 0$ for $t \in [-L, 0]$. Now $x(\omega) \in C^1((0,\infty); \mathbb{R})$ and indeed

$$x'(t) = -\int_{-L}^{0} p(s)g(X(t+s))\,ds + Y(t), \quad t > 0$$

so

$$x'(t) = -\int_{-L}^{0} p(s)g(x(t+s))\,ds + q(t), \quad t > 0$$

where q is defined by

$$q(t) = -\int_{-L}^{0} p(s)[g(x(t+s)+Y(t+s)) - g(x(t+s))]\,ds + Y(t), \quad t \geq 0.$$

Now, we restrict ω to Ω_2. Then, by (7.3.2),

$$|q(t,\omega)| \leq K\int_{-L}^{0} |p(s)||Y(t+s,\omega)|\,ds + |Y(t,\omega)|.$$

Since p and g are continuous, and x and Y have continuous sample paths, $t \mapsto q(t,\omega)$ is continuous on $[0,\infty)$. Also, because $Y(t,\omega) \to 0$ as $t \to \infty$

for each $\omega \in \Omega_2$, we have $\lim_{t\to\infty} q(t,\omega) = 0$. Now, we invoke Property (7.3.9) so that it follows that

$$\lim_{t\to\infty} x(t,\omega) = 0, \quad \omega \in \Omega_2.$$

Therefore, as $Y(t,\omega) \to 0$ as $t \to \infty$ for each $\omega \in \Omega_2$, we conclude that

$$\lim_{t\to\infty} X(t,\omega) = 0, \quad \omega \in \Omega_2,$$

which implies $X(t) \to 0$ as $t \to \infty$, a.s.

Our task now is to show that Properties (7.3.10) and (7.3.9) hold under conditions on the data. We will show in the next subsection that Property (7.3.10) holds for the auxiliary process Y when σ tends to zero sufficiently quickly. In the subsection following that, we will show that Property (7.3.9) holds under the conditions required of the deterministic problem (7.1.1) in earlier chapters, and can be established by means of a fixed point proof.

Stability of the auxiliary stochastic differential equation

In this subsection, we set out to establish Property (7.3.10) for the solution Y of the stochastic differential equation (7.3.7). As might be expected, the solution of (7.3.7) tends to zero if $\sigma(t) \to 0$ sufficiently quickly. A pointwise decay condition on σ which suffices is

$$\lim_{t\to\infty} \sigma^2(t) \log t = 0. \tag{7.3.12}$$

Later, we show that this condition cannot easily be improved. The result which follows is studied in the finite dimensional case in Appleby and Riedle (2005).

Proposition 7.3.3. *Let Y be a solution of* (7.3.7). *If σ satisfies* (7.3.12) *then Y obeys*

$$\lim_{t\to\infty} Y(t) = 0 \quad a.s.$$

Proof. To prove that $Y(t) \to 0$ as $t \to \infty$, a.s., we note that Y has the representation $Y(t) = Y(0)e^{-t} + U(t)$ where

$$U(t) = e^{-t} \int_0^t e^s \sigma(s)\, dB(s) \qquad \text{for } t \geq 0 \text{ a.s.}$$

Denote the local martingale $(M(t) : t \geq 0)$ by

$$M(t) = \int_0^t e^s \sigma(s)\, dB(s).$$

Then $U(t) = e^{-t} M(t)$.

In the case when $\int_0^\infty e^{2s}\sigma^2(s)\,ds < \infty$, the martingale convergence theorem implies that $M(t) = e^t U(t)$ tends to an almost surely finite random variable almost surely. Hence $U(t) \to 0$ as $t \to \infty$, a.s.

In the case that $\int_0^\infty e^{2s}\sigma^2(s)\,ds = \infty$, then the quadratic variation $(\langle M\rangle(t) : t \geq 0)$ given by

$$\langle M\rangle(t) = \int_0^t e^{2s}\sigma^2(s)\,ds, \quad t \geq 0$$

obeys $\langle M\rangle(t) \to \infty$ as $t \to \infty$, and by the law of the iterated logarithm for martingales we have

$$\limsup_{t\to\infty} \frac{M^2(t)}{2\langle M\rangle(t)\log\log\langle M\rangle(t)} = 1 \quad \text{a.s.}$$

Hence

$$\limsup_{t\to\infty} \frac{U^2(t)}{I(t)} = 1 \quad \text{a.s.,}$$

where $I(t) := 2\int_0^t e^{-2(t-s)}\sigma^2(s)\,ds \log\log \int_0^t e^{2s}\sigma^2(s)\,ds$. If we can now show that $I(t) \to 0$ as $t \to \infty$, the result has been proven. Since σ is bounded, it follows that

$$\limsup_{t\to\infty} \frac{\log\log \int_0^t e^{2s}\sigma^2(s)\,ds}{\log t} \leq 1.$$

Also, as $\sigma^2(t)\log t \to 0$ as $t \to \infty$, for every $\epsilon > 0$ there is a $T = T(\epsilon) > 0$ such that for $t > T$ we have $\sigma^2(t) < \epsilon/\log t$. Hence for $t > T$

$$\begin{aligned}\log t \cdot \int_0^t e^{-2(t-s)}\sigma^2(s)\,ds &\leq e^{-2t}\log t \int_0^T e^{2s}\sigma^2(s)\,ds \\ &\qquad + \epsilon e^{-2t}\log t \int_T^t e^{2s}\frac{1}{\log s}\,ds.\end{aligned}$$

By L'Hôpital's rule, we have

$$\lim_{t\to\infty} \frac{\int_T^t \frac{e^{2s}}{\log s}\,ds}{\frac{e^{2t}}{\log t}} = \frac{1}{2}.$$

Thus

$$\begin{aligned}\limsup_{t\to\infty} I(t) &= \limsup_{t\to\infty} 2\int_0^t e^{-2(t-s)}\sigma^2(s)\,ds \cdot \log t \frac{\log\log \int_0^t e^{2s}\sigma^2(s)\,ds}{\log t} \\ &\leq \epsilon.\end{aligned}$$

Since $\epsilon > 0$ was chosen arbitrarily, $I(t) \to 0$ as $t \to \infty$, and the result follows.

We comment that the condition $\sigma \in L^2([0,\infty);\mathbb{R})$ is also sufficient to ensure that $Y(t) \to 0$ as $t \to \infty$, a.s., so that the stability theorem for the solution of (7.3.4) quoted later can also be stated with the decay condition (7.3.12) replaced by $\sigma \in L^2([0,\infty);\mathbb{R})$.

Fixed point result for the perturbed deterministic equation

We now need to establish Property (7.3.9) for the perturbed equation 7.3.6; the proof of this result will require the imposition of some new additional conditions on g and p:

$$xg(x) \geq 0, \text{and } \gamma := \lim_{x\to 0} \frac{g(x)}{x} \text{ exists.} \tag{7.3.13}$$

$$\text{There exists } \beta > 0 \text{ such that } \frac{g(x)}{x} \geq \beta. \tag{7.3.14}$$

Let $L > 0$ and

$$\int_{-L}^{0} p(s)\, ds = 1. \tag{7.3.15}$$

$$\text{With } K \text{ as in (7.3.2)}, \quad \alpha := 2K \int_{-L}^{0} |p(s)s|\, ds < 1. \tag{7.3.16}$$

We consider the deterministic equation

$$x'(t) = -\int_{t-L}^{t} p(s-t)g(x(s))\, ds, \quad t > 0 \tag{7.3.17}$$

where $x(t) = \psi(t)$, $t \in [-L,0]$ and $\psi \in C([-L,0];\mathbb{R})$.

Under these conditions, it has been shown that the solution of (7.3.17) is asymptotically stable.

Theorem 7.3.4. *Suppose that g obeys (7.3.2), (7.3.8), (7.3.13), (7.3.14), and p obeys (7.3.3), (7.3.15), (7.3.16). If $\psi \in C([-L,0];\mathbb{R})$, then the solution of (7.3.17) obeys*

$$\lim_{t\to\infty} x(t) = 0.$$

The corresponding result for the stochastic equation (7.3.4) is as follows.

Theorem 7.3.5. *Suppose that g obeys (7.3.2), (7.3.8), (7.3.13), (7.3.14), and p obeys (7.3.3), (7.3.15), (7.3.16). If $\psi \in C([-L,0];\mathbb{R})$, and σ obeys (7.3.1), (7.3.12), then the unique continuous adapted process which satisfies (7.3.4) obeys*

$$\lim_{t\to\infty} X(t) = 0, \quad a.s.$$

By virtue of (7.3.12) and Proposition 7.3.3, the solution Y of (7.3.7) obeys Property (7.3.10). Therefore, Theorem 7.3.5 is a consequence of Theorem 7.3.2, once we can show that Property (7.3.9) holds, namely that the solutions of the equation (7.3.6) obey $x(t) \to 0$ as $t \to \infty$ whenever $q(t) \to 0$ as $t \to \infty$, and q is a continuous function.

We now show that (7.3.9) holds. To do this, we mimic the proof of Theorem 3.1 in Burton (2005b).

Since q is continuous and $q(t) \to 0$ as $t \to \infty$, there is a unique continuous solution of (7.3.6). We denote this solution by x. Since

$$\frac{d}{dt}\int_{-L}^{0} p(s)\int_{t+s}^{t} g(x(u))\,du\,ds = g(x(t)) - \int_{t-L}^{t} p(u-t)g(x(u))\,du$$

we have

$$x'(t) = -g(x(t)) + \frac{d}{dt}\int_{-L}^{0} p(s)\int_{t+s}^{t} g(x(u))\,du\,ds + q(t), \quad t > 0.$$

By (7.3.8), we may define a continuous function $a : [0,\infty) \to [0,\infty)$ by

$$a(t) = \begin{cases} \frac{g(x(t))}{x(t)}, & x(t) \neq 0 \\ \gamma, & x(t) = 0. \end{cases}$$

Thus

$$x'(t) = -a(t)x(t) + \frac{d}{dt}\int_{-L}^{0} p(s)\int_{t+s}^{t} g(x(u))\,du\,ds + q(t), \quad t > 0.$$

Using variation of parameters we obtain

$$\begin{aligned} x(t) &= e^{-\int_0^t a(s)\,ds}\psi(0) + \int_0^t e^{-\int_s^t a(u)\,du} q(s)\,ds \\ &\quad + \int_0^t \left\{ e^{-\int_v^t a(s)\,ds} \frac{d}{dv}\left(\int_{-L}^{0} p(s)\int_{v+s}^{v} g(x(u))\,du\right)\right\} dv, \end{aligned}$$

and then integrating by parts as in Theorem 3.1 of Burton (2005b), we find that

$$\begin{aligned} x(t) &= e^{-\int_0^t a(s)\,ds}\psi(0) + \int_0^t e^{-\int_s^t a(u)\,du} q(s)\,ds \\ &+ \int_{-L}^{0} p(s)\int_{t+s}^{t} g(x(u))\,du\,ds - e^{-\int_0^t a(s)\,ds}\int_{-L}^{0} p(s)\int_{s}^{0} g(\psi(u))\,du\,ds \\ &\quad - \int_0^t e^{-\int_v^t a(s)\,ds} a(v)\int_{-L}^{0} p(s)\int_{v+s}^{v} g(x(u))\,du\,ds\,dv. \end{aligned}$$

Let $M = \{\phi : [-L, \infty) \to \mathbb{R} : \phi_0 = \psi, \phi \in C, \phi \text{ bounded }, \phi(t) \to 0 \text{ as } t \to \infty\}$. If $\phi \in M$ and $\|\phi\|_M := \sup_{t\geq 0} |\phi(t)|$ then $(M, \|\cdot\|_M)$ is a complete normed space. Define $P : M \to M$ using the above equation in x. For $\phi \in M$ define $(P\phi)(t) = \psi(t)$ if $t \in [-L, 0]$. If $t \geq 0$ then define

$$(P\phi)(t) = e^{-\int_0^t a(s)\,ds}\psi(0) + \int_0^t e^{-\int_s^t a(u)\,du} q(s)\,ds$$
$$+ \int_{-L}^0 p(s) \int_{t+s}^t g(\phi(u))\,du\,ds - e^{-\int_0^t a(s)\,ds} \int_{-L}^0 p(s) \int_s^0 g(\psi(u))\,du\,ds$$
$$- \int_0^t e^{-\int_v^t a(s)\,ds} a(v) \int_{-L}^0 p(s) \int_{v+s}^v g(\phi(u))\,du\,ds\,dv. \quad (7.3.18)$$

Since $a(t) \geq \beta$, $t \geq 0$, we have that

$$e^{-\int_s^t a(u)\,du} \leq e^{-\beta(t-s)}, \quad 0 \leq s \leq t.$$

We now show that $\phi \in M$ implies $P\phi \in M$. Note that the righthand side of (7.3.18) is clearly continuous when ϕ is continuous; since q is continuous and $q(t) \to 0$ as $t \to \infty$, q is bounded. Therefore the second term on the righthand side of (7.3.18) is bounded and tends to zero as $t \to \infty$. The same is true for the third term, as ϕ is bounded and $\phi(t) \to 0$ as $t \to \infty$, and similarly for the last term, because the function

$$b(t) = a(t) \int_{-L}^0 p(s) \int_{t+s}^t g(\phi(u))\,du\,ds$$

is bounded and $b(t) \to 0$ as $t \to \infty$.

To see that P is a contraction, if $\phi_1, \phi_2 \in M$, we have

$$|(P\phi_1)(t) - (P\phi_2)(t)| \leq \int_{-L}^0 |p(s)| \int_{t+s}^t |g(\phi_1(u)) - g(\phi_2(u))|\,du\,ds$$
$$+ \int_0^t e^{-\int_v^t a(s)\,ds} a(v) \int_{-L}^0 |p(s)| \int_{v+s}^v |g(\phi_1(u)) - g(\phi_2(u))|\,du\,ds\,dv$$

so

$$|(P\phi_1)(t) - (P\phi_2)(t)| \leq 2K \int_{-L}^0 s|p(s)|\,ds \sup_{t\geq 0} |\phi_1(t) - \phi_2(t)|$$
$$= \alpha\|\phi_1 - \phi_2\|_M.$$

Therefore P is a contraction on the complete normed space $(M, \|\cdot\|_M)$. This completes the proof of Theorem 7.3.5.

By Theorems 7.3.5, 7.3.4, we see that the conditions (7.3.2), (7.3.8), (7.3.13), (7.3.14) on g, and (7.3.3), (7.3.15), (7.3.16) on p which guarantee the asymptotic stability of solutions of (7.3.17), also guarantee the asymptotic convergence of solutions of (7.3.4), under the additional condition on the decay of σ, (7.3.12).

We remark that the analysis here can be extended to deal with general functional differential equations, and in particular, to studying the unbounded and infinite Volterra equations studied in earlier parts of this book. See Appleby (2005a) for further details.

7.3.3 Necessary conditions for stability

Next, we want to show that the condition (7.3.12) on σ is difficult to improve upon if we want solutions of (7.3.4) to tend to zero almost surely.

By almost the same argument as used to prove Proposition 7.3.3, we can show that when

$$\text{There exists } c > 0 \text{ such that } \lim_{t\to\infty} \sigma^2(t)\log t = c, \tag{7.3.19}$$

then the solution of (7.3.7) obeys

$$\limsup_{t\to\infty} |Y(t)| = \sqrt{c}, \quad \text{a.s.}$$

Even though (7.3.19) represents only a very slight relaxation on the decay condition (7.3.12), it transpires that we cannot have asymptotically stable solutions, even with positive probability.

Theorem 7.3.6. *Suppose that X is the unique continuous $\mathcal{F}(t)$–adapted process which obeys (7.3.4), and $\sigma \in C([0,\infty);\mathbb{R})$ obeys (7.3.19). If p and g are continuous, then*

$$\mathbb{P}[\{\omega \in \Omega : \lim_{t\to\infty} X(t,\omega) = 0\}] = 0.$$

It is also interesting to note that convergence is not possible, whatever conditions are placed upon g and p; in this case, when the noise fades sufficiently slowly, the deterministic forces which restore the solution to equilibrium in (7.1.1) cannot stabilize the solution, *no matter how strong those forces might be.*

Proof of Theorem 7.3.6. We define

$$\Omega_1 = \{\omega \in \Omega : X(t,\omega) \to 0 \text{ as } t \to \infty,\ X(\omega) \text{ obeys (7.3.4)}\}$$

and suppose that $\mathbb{P}[\Omega_1] > 0$. Define G by

$$G(t) = -\int_{t-L}^{t} p(t-s) g(X(s))\, ds, \quad t \geq 0.$$

Stochastic integration by parts now gives

$$e^t X(t) = \psi(0) + \int_0^t (e^s G(s) + e^s X(s))\, ds + \int_0^t e^s \sigma(s)\, dB(s).$$

Define $e_1(t) = e^{-t}$ for all $t \geq 0$, and the process $Y = \{Y(t); 0 \leq t < \infty; \mathcal{F}(t)\}$ by

$$Y(t) = e^{-t} \int_0^t e^s \sigma(s)\, dB(s), \quad t \geq 0. \tag{7.3.20}$$

Then, if we define

$$F(t) = G(t) + X(t), \quad t \geq 0 \tag{7.3.21}$$

we have

$$X(t) = e^{-t}\psi(0) + (e_1 * F)(t) + Y(t), \quad t \geq 0. \tag{7.3.22}$$

Now, since $X(t) \to 0$ on Ω_1, and $g(X(t,\omega)) \to 0$ as $t \to \infty$ for all $\omega \in \Omega_1$ we have

$$\lim_{t\to\infty} G(t)(\omega) = 0,$$

so obviously, as $X(t,\omega) \to 0$ as $t \to \infty$, we have that $F(t,\omega) \to 0$ as $t \to \infty$. Since $e_1 \in L^1(0,\infty)$, it follows that $(e_1 * F)(t,\omega) \to 0$ as $t \to \infty$ for all $\omega \in \Omega_1$. Thus

$$\lim_{t\to\infty} Y(t,\omega) = 0, \quad \omega \in \Omega_1,$$

or $Y(t) \to 0$ as $t \to \infty$ on a set of positive probability. But Y is a process which obeys the stochastic differential equation (7.3.7), and by (7.3.19) we have $\limsup_{t\to\infty} |Y(t)| = \sqrt{c} > 0$, a.s. This contradiction implies that $\mathbb{P}[\Omega_1] = 0$, as required.

The importance of the decay condition (7.3.12) is further emphasized by a result in Appleby (2005a). There, it is shown that the condition

$$\lim_{t\to\infty} \sigma^2(t)\log t = \infty$$

is sufficient to ensure that all solutions of (7.3.4) are almost surely unbounded viz., they satisfy

$$\mathbb{P}[\{\omega : \limsup_{t\to\infty} |X(t,\omega)| = \infty\}] = 1.$$

Finally, it turns out that when $t \mapsto \sigma^2(t)$ is decreasing, the condition (7.3.12) on the decay rate of the intensity of the stochastic perturbation σ is necessary if solutions of (7.3.4) are still to tend to zero with probability one. The following result in this direction appears in Appleby (2005a).

Theorem 7.3.7. *Suppose that g obeys* (7.3.2), (7.3.8), (7.3.13), *and* (7.3.14), *and p obeys* (7.3.3), (7.3.15), *and* (7.3.16). *Suppose that σ obeys* (7.3.1). *Let $t \mapsto \sigma^2(t)$ be a decreasing function, $\psi \in C([-L,0];\mathbb{R})$. If X is the unique continuous $\mathcal{F}(t)$–adapted process which obeys* (7.3.4) *then the following are equivalent:*

(a) $\lim_{t\to\infty} \sigma^2(t)\log t = 0$;

(b) $\lim_{t\to\infty} X(t) = 0$, *a.s.*

A result in this direction for finite dimensional linear Volterra equations is presented in Appleby and Riedle (2005). A result for a scalar stochastic delay differential equation whose coefficients are globally polynomial is presented in Appleby and Rodkina (2005). The condition (7.3.12) was first shown to be necessary and sufficient for the asymptotic stability of nonlinear stochastic differential equations in Chan and Williams (1989).

7.4 Stochastic Cooke–Yorke problem

The main purpose of this section is to determine conditions under which solutions of the stochastic functional differential equation (7.1.4) are asymptotically constant a.s. Once again, σ represents the intensity of the stochastic perturbation.

7.4.1 Formulation of problem and existence and uniqueness of solution

We first give a precise formulation of (7.1.4) in order to ensure that it possesses a unique solution. As in the last section, we work on the complete filtered probability space $(\Omega, \mathcal{F}, (\mathcal{F}(t))_{t\geq 0}, \mathbb{P})$.

Let $L > 0$ and suppose as before that σ obeys (7.3.1). We also ask that $g \in C(\mathbb{R};\mathbb{R})$ satisfy the global Lipschitz condition (7.3.2). In order to define solutions of the stochastic functional equation, we define an initial function on $[-L, 0]$. We ask that $X_0 = \psi \in C([-L, 0];\mathbb{R})$.

The following result on the existence and uniqueness of a continuous adapted process X which satisfies the stochastic equation

$$X(t) = \psi(0) + \int_0^t \{g(X(s)) - g(X(s-L))\}\, ds \qquad (7.4.1a)$$

$$+ \int_0^t \sigma(s)\, dB(s), \quad t \geq 0, \quad \text{a.s.}$$

$$X(t) = \psi(t), \quad t \in [-L, 0] \qquad (7.4.1b)$$

arises from an application of standard results for stochastic functional differential equations.

Theorem 7.4.1. *Let $L > 0$. If g obeys (7.3.2), σ obeys (7.3.1), and $\psi \in C([-L, 0];\mathbb{R})$, then there is a unique continuous $\mathcal{F}(t)$–adapted process $(X_t)_{t\geq 0}$ which obeys (7.4.1). The process is unique in the sense that any other process $\tilde{X}$ which obeys (7.4.1) is indistinguishable from X, viz.,*

$$\mathbb{P}[\{\omega \in \Omega : X(t,\omega) = \tilde{X}(t,\omega) \text{ for all } t \geq -L\}] = 1.$$

(7.4.1) may be viewed as a stochastic analogue of the deterministic Cooke–Yorke equation

$$x'(t) = g(x(t)) - g(x(t-L)), \quad t > 0, \qquad (7.4.2)$$

where $x(t) = \psi(t)$, $t \in [-L, 0]$ and $\psi \in C([-L, 0];\mathbb{R})$; this is particularly evident when we write (7.4.2) in the equivalent integral form

$$x(t) = \psi(0) + \int_0^t \{g(x(s)) - g(x(s-L))\}\, ds, \quad t \geq 0,$$

and compare with (7.4.1). Once again, the perturbation in (7.4.1) has an intensity which is time dependent, but not state dependent; in other words it represents a stochastic influence from outside the system whose intensity varies over time.

7.4.2 Discussion of results for the stochastic Cooke–Yorke problem

We now consider the connection between the asymptotic behaviour of (7.4.1) and (7.4.2). We suppose, in addition to (7.3.2), that g obeys

$$\text{With } K \text{ given as in (7.3.2), there is } \alpha \in (0,1) \text{ such that } \alpha := KL. \tag{7.4.3}$$

Under these conditions, it has been shown that the solution of (7.4.2) is asymptotically constant.

Theorem 7.4.2. *If g obeys (7.3.2), (7.4.3) and $\psi \in C([-L,0];\mathbb{R})$, then*

there exists a unique constant k satisfying

$$k = \psi(0) + g(k)L - \int_{-L}^{0} g(\psi(s))\,ds \tag{7.4.4}$$

such that the unique continuous solution of (7.4.2) obeys

$$\lim_{t\to\infty} x(t,\psi) = k. \tag{7.4.5}$$

We now show that the properties (7.4.4) and (7.4.5) enjoyed by solutions of (7.4.2) are also enjoyed by solutions of (7.4.1), under some slight amendments which allow for randomness. It turns out that in order for solutions to be asymptotically constant, it is sufficient for the intensity $\sigma(t)$ to tend to zero as $t \to \infty$ so quickly that it obeys

$$\sigma \in L^2([0,\infty);\mathbb{R}). \tag{7.4.6}$$

The corresponding result for the stochastic equation (7.4.1) is as follows.

Theorem 7.4.3. *Suppose that g obeys (7.3.2), (7.4.3). If $\psi \in C([-L,0];\mathbb{R})$ and σ obeys (7.3.1), (7.4.6), then*

there is a unique a.s. finite $\mathcal{F}(\infty)$–measurable random variable

$$\kappa \text{ such that } \quad \kappa = \psi(0)+g(\kappa)L-\int_{-L}^{0} g(\psi(s))\,ds+\int_{0}^{\infty} \sigma(s)\,dB(s) \tag{7.4.7}$$

such that X the unique continuous $\mathcal{F}(t)$–adapted process which obeys (7.4.1) obeys

$$\lim_{t\to\infty} X(t,\psi) = \kappa, \quad a.s. \tag{7.4.8}$$

Before stating converse results or giving a proof of this result, we give an indication as to why the condition that σ is in $L^2([0,\infty);\mathbb{R})$ is reasonable. To do this first consider the following perturbed version of (7.4.2):

$$x'(t) = g(x(t)) - g(x(t-L)) + f(t), \quad t > 0.$$

This has integrated form given by

$$x(t) = \psi(0) + \int_0^t \{g(x(s)) - g(x(s-L))\}\, ds + \int_0^t f(s)\, ds, \quad t \geq 0.$$

In the case where g is linear, by a result of Krisztin and Terjéki (1988), we know that f in $L^1([0,\infty);\mathbb{R})$ suffices to ensure that x tends to a non–trivial limit. In other words, when the perturbed term in the integrated form of the equation tends to a limit, then so does the solution of the equation itself.

Comparing the integrated form of this deterministically perturbed equation with (7.4.1), the above discussion indicates that if we want the solution to tend to a nontrivial limit, we should request that the stochastic perturbation in this integrated form tend to a limit, i.e., $\lim_{t\to\infty} \int_0^t \sigma(s)\, dB(s)$ exists a.s. The martingale convergence theorem guarantees this in the case that σ is in $L^2([0,\infty);\mathbb{R})$, motivating our choice of this hypothesis here.

We will also show that the condition (7.4.6) is well–chosen; indeed it is necessary for (7.4.6) to hold in order for solutions of (7.4.1) to be asymptotically constant.

Theorem 7.4.4. *Suppose that g obeys (7.3.2) and $\psi \in C([-L,0];\mathbb{R})$ and let σ obey (7.3.1). If the unique continuous $\mathcal{F}(t)$ adapted process X which obeys (7.4.1) obeys (7.4.8) then σ obeys (7.4.6) and the limit κ obeys (7.4.7).*

We can therefore distill the results of Theorem 7.4.3 and 7.4.4 into a single result.

Theorem 7.4.5. *Suppose that g obeys (7.3.2) and (7.4.3), $\psi \in C([-L,0];\mathbb{R})$ and σ obeys (7.3.1). Let X be the unique continuous $\mathcal{F}(t)$ adapted process which obeys (7.4.1). Then the statements*

a) There exists a $\mathcal{F}(\infty)$–measurable and almost surely finite random variable κ such that

$$\lim_{t\to\infty} X(t,\psi) = \kappa, \quad a.s.$$

b) $\sigma \in L^2([0,\infty);\mathbb{R})$,

are equivalent. Furthermore, κ satisfies (7.4.7).

We note that Theorem 7.4.3 is very similar to Theorem 7.4.2 but has some important differences; in Theorem 7.4.2 the limit of each solution is known precisely, once the initial function is known. In Theorem 7.4.4, however, the stochastic perturbation also influences the limiting value, so that this limit cannot be known a priori. However, the distribution of the limiting random value can be known, because the function σ is deterministic. This statement is made precise in the following theorem.

Theorem 7.4.6. *Suppose that g obeys (7.3.2), (7.4.3), $\psi \in C([-L,0];\mathbb{R})$ and σ obeys (7.3.1), (7.4.6). If κ is the random variable defined by (7.4.7), then κ has the same distribution as the random variable $G^{-1}(U)$, where $G \in C(\mathbb{R};\mathbb{R})$ is defined by $G(x) = x - g(x)L$, and U is a normally distributed random variable with mean*

$$\psi(0) - \int_{-L}^{0} g(\psi(s))\,ds$$

and variance $\int_0^\infty \sigma^2(s)\,ds$.

More specifically, the distribution function of κ is given by

$$F_\kappa(x) = \Phi\left(\frac{G(x) - \psi(0) + \int_{-L}^{0} g(\psi(s))\,ds}{\sqrt{\int_0^\infty \sigma^2(s)\,ds}}\right), \quad x \in \mathbb{R}, \tag{7.4.9}$$

where Φ is given by (7.2.1).

Proof. By Theorem 7.4.3, κ is a $\mathcal{F}(\infty)$–measurable random variable defined by (7.4.7). Since

$$U = \psi(0) - \int_{-L}^{0} g(\psi(s))\,ds + \int_0^\infty \sigma(s)\,dB(s)$$

is normally distributed with mean $\psi(0) - \int_{-L}^{0} g(\psi(s))\,ds$ and variance $\int_0^\infty \sigma^2(s)\,ds$, we see that $G(\kappa) = U$, and the result holds provided that G is increasing. Note that (7.3.2), (7.4.3) imply for $x > y$ that $|L(g(x)-g(y))| \leq \alpha(x-y)$, and so, as $\alpha \in (0,1)$,

$$G(x) - G(y) = x - y + L(g(y) - g(x)) \geq x - y - \alpha(x-y) > 0,$$

which shows G is increasing.

To obtain the formula (7.4.9), because κ has the same distribution as $G^{-1}(U)$, we have

$$F_\kappa(x) = \mathbb{P}[\kappa \le x] = \mathbb{P}[G^{-1}(U) \le x] = \mathbb{P}[U \le G(x)].$$

But, as U is normally distributed with mean $\psi(0) - \int_{-L}^{0} g(\psi(s))\,ds$ and variance $\int_0^\infty \sigma^2(s)\,ds$, we have

$$\mathbb{P}[U \le G(x)] = \Phi\left(\frac{G(x) - \psi(0) + \int_{-L}^{0} g(\psi(s))\,ds}{\sqrt{\int_0^\infty \sigma^2(s)\,ds}}\right),$$

as required.

Theorem 7.4.6 enables us to show how the initial function and noise perturbation effect the central location and spread of the distribution of κ.

First, there is a strong connection between the average value of the random limit κ for (7.4.1) and the limiting value k of (7.4.2) when both problems have the same initial data ψ. In fact k is the median of κ, i.e.,

$$\mathbb{P}[\kappa \le k] = \mathbb{P}[\kappa > k] = \frac{1}{2}.$$

To see this, note that as k is defined by (7.4.4), we have $G(k) = \psi(0) - \int_{-L}^{0} g(\psi(s))\,ds$. Remembering that the median of a standard normal random variable is zero, we see that $\Phi(0) = 1/2$, and so, by (7.4.9), we get

$$\mathbb{P}[\kappa \le k] = \Phi\left(\frac{G(k) - \psi(0) + \int_{-L}^{0} g(\psi(s))\,ds}{\sqrt{\int_0^\infty \sigma^2(s)\,ds}}\right) = \Phi(0) = \frac{1}{2}.$$

Also, we can see that the dispersion of the random limit κ from the deterministic limit is greater the larger the cumulative intensity of the noise perturbation (as measured by $\int_0^\infty \sigma^2(s)\,ds$). This can readily be established from the formula (7.4.9) in the following form. Suppose that X_1 is the solution of (7.4.1) with $\sigma = \sigma_1$ and X_2 is the solution of (7.4.1) with $\sigma = \sigma_2$, and let

$$\int_0^\infty \sigma_1^2(s)\,ds < \int_0^\infty \sigma_2^2(s)\,ds.$$

If κ_1 and κ_2 are the limiting values of X_1 and X_2 respectively, then we have

$$\mathbb{P}[k \le \kappa_1 \le x] \ge \mathbb{P}[k \le \kappa_2 \le x], \quad x > k,$$
$$\mathbb{P}[\kappa_1 \le x] \le \mathbb{P}[\kappa_2 \le x], \quad x < k.$$

7.4.3 Proofs of Theorems 7.4.3–7.4.4

The key to the proof of both theorems is the same as exploited throughout this book; that is, the solution of (7.4.1) is expressed in "integral form" (which is the only mathematically legitimate form for (7.4.1)), and the integral

$$\int_0^t \{g(X(s)) - g(X(s-L))\}\, ds$$

is telescoped. A fixed point proof then establishes the asymptotic constancy of solutions; however, this must be shown path by path for the solution of (7.4.1).

Proof of Theorem 7.4.3. Since $\sigma \in L^2(0,\infty)$, the martingale convergence theorem assures that

$$\lim_{t\to\infty} \int_0^t \sigma(s)\, dB(s) \quad \text{exists a.s. and is an a.s. finite random variable.}$$

We denote this limit by $\int_0^\infty \sigma(s)\, dB(s)$.

Now, let

$$\Omega^* = \{\omega \in \Omega : \lim_{t\to\infty} (\int_0^t \sigma(s)\, dB(s))(\omega) \quad \text{exists,}$$

$$\text{and } X \text{ is the solution of (7.4.1)}\}.$$

Then $\mathbb{P}[\Omega^*] = 1$. By (7.4.1), we can deduce that X obeys

$$X(t) = \psi(0) - \int_{-L}^0 g(\psi(s))\, ds + \int_{t-L}^t g(X(s))\, ds + \int_0^t \sigma(s)\, dB(s), \quad t \geq 0. \tag{7.4.10}$$

Now, fix $\omega \in \Omega^*$, and let $X(\cdot,\omega)$ be the realisation of X, which obeys

$$X(t,\omega) = \psi(0) - \int_{-L}^0 g(\psi(s))\, ds + \int_{t-L}^t g(X(s,\omega))\, ds + \left(\int_0^t \sigma(s)\, dB(s)\right)(\omega), \quad t \geq 0. \tag{7.4.11}$$

Note also that $m_\infty(\omega) := \lim_{t\to\infty} \left(\int_0^t \sigma(s)\, dB(s)\right)(\omega)$ exists. Write

$$m(t,\omega) = \left(\int_0^t \sigma(s)\, dB(s)\right)(\omega), \quad t \geq 0.$$

Next, we show that there is a unique $\kappa = \kappa(\omega)$ such that

$$\kappa(\omega) = \psi(0) - \int_{-L}^{0} g(\psi(s))\, ds + g(\kappa(\omega))L + m_\infty(\omega). \tag{7.4.12}$$

To do this, we define a mapping $Q_\omega : \mathbb{R} \to \mathbb{R}$ by

$$Q_\omega \kappa = \psi(0) + g(\kappa)L - \int_{-L}^{0} g(\psi(s))\, ds + m_\infty(\omega).$$

If κ_1, $\kappa_2 \in \mathbb{R}$, by (7.3.2), we have

$$|Q_\omega \kappa_1 - Q_\omega \kappa_2| = |L(g(\kappa_1) - g(\kappa_2))| \leq LK|\kappa_1 - \kappa_2| = \alpha|\kappa_1 - \kappa_2|.$$

By (7.4.3), we have that Q_ω is a contraction on the complete normed space $(\mathbb{R}, |\cdot|)$. Thus Q_ω has a unique fixed point defined by (7.4.12).

Next, let $(M_\omega, \|\cdot\|)$ be the complete normed space of bounded continuous functions $\phi : [-L, 0] \to \mathbb{R}$ with $\phi(t) = \psi(t)$ for $t \in [-L, 0]$ and $\phi(t) \to \kappa(\omega)$ as $t \to \infty$, in such a way that ϕ is the realization of a $\mathcal{F}(t)$–adapted process.

Define $P_\omega : M_\omega \to M_\omega$ for $\phi \in M_\omega$ so that

$$(P_\omega \phi)(t) = \begin{cases} \psi(t), & t \in [-L, 0], \\ \psi(0) - \int_{-L}^{0} g(\psi(s))\, ds + \int_{t-L}^{t} g(\phi(s))\, ds + m(t, \omega), & t \geq 0. \end{cases}$$

We note that if $\phi \in M_\omega$ obeys $P_\omega \phi = \phi$, then ϕ solves (7.4.11).

Clearly, $P_\omega : M_\omega \to M_\omega$; if $\phi(t) \to \kappa(\omega)$ as $t \to \infty$, then the continuity of g and the fact that $m(t, \omega) \to m_\infty(\omega)$ as $t \to \infty$, we have

$$(P_\omega \phi)(t) \to \psi(0) - \int_{-L}^{0} g(\psi(s))\, ds + Lg(\kappa(\omega)) + m_\infty(\omega) = \kappa(\omega),$$

where the fact that $\kappa(\omega)$ is the unique fixed point of Q_ω has been used to deduce the last equality. Clearly $P_\omega \phi$ is continuous, and if ϕ is the realization of a $\mathcal{F}(t)$–adapted process, then so is $P_\omega \phi$. Finally, if $\phi_1, \phi_2 \in M_\omega$, then for any $t \geq 0$, we have

$$\begin{aligned} |(P_\omega \phi_1)(t) - (P_\omega \phi_2)(t)| &\leq \int_{t-L}^{t} |g(\phi_1(s)) - g(\phi_2(s))|\, ds \\ &\leq KL\|\phi_1 - \phi_2\|. \end{aligned}$$

By (7.4.3), P_ω is a contraction, so it has a unique fixed point $X(\cdot, \omega) \in M_\omega$. Therefore, because the construction is valid for each $\omega \in \Omega^*$, and Ω^* is an almost sure event, it is true that the unique continuous adapted solution of (7.4.1) obeys (7.4.7) and (7.4.8).

We remark that no intrinsically stochastic result has been invoked here, but from the perspective of stochastic analysis, this may not be so surprising. Indeed, in the case where the diffusion coefficient of a stochastic differential equation is independent of the state, it is known that the existence of a unique continuous and adapted solution process can be assured with what is principally a standard deterministic argument, but applied on a pathwise basis. See for example Karatzas and Shreve (1991), Chapter 5.2, pp. 294–295.

Proof of Theorem 7.4.4. Since X is the unique continuous $\mathcal{F}(t)$–adapted process which obeys (7.4.1), it also obeys (7.4.10). Now define

$$\Omega^* = \{\omega \in \Omega : \text{there exists a finite } \kappa(\omega) \quad \text{such that } \lim_{t\to\infty} X(t,\omega) = \kappa(\omega)\}.$$

By hypothesis $\mathbb{P}[\Omega^*] = 1$. Rearranging (7.4.10), and fixing $\omega \in \Omega^*$, we have

$$\lim_{t\to\infty}\left(\int_0^t \sigma(s)\,dB(s)\right)(\omega) = \kappa(\omega) - \psi(0) + \int_{-L}^0 g(\psi(s))\,ds - g(\kappa(\omega))L.$$

Since $\kappa = \kappa(\omega)$ is almost surely finite, we have that

$$\lim_{t\to\infty}\int_0^t \sigma(s)\,dB(s) \quad \text{exists and is a.s. finite a.s.}$$

From this, we conclude that $\sigma \in L^2([0,\infty);\mathbb{R})$.

Writing $\int_0^\infty \sigma(s)\,dB(s)$ for this limit, we see that

$$\int_0^\infty \sigma(s)\,dB(s) = \kappa - \psi(0) + \int_{-L}^0 g(\psi(s))\,ds - g(\kappa)L,$$

which rearranges to give (7.4.7). If (7.4.3) also holds, we can see that for a given limit $\int_0^\infty \sigma(s)\,dB(s)$ that the corresponding κ must be unique.

References

Appleby, J. A. D. (2005a). Fixed points, stability and harmless stochastic perturbations. preprint.

Appleby, J. A. D. (2005b). Fixed points and stochastic functional differential equations with asymptotically constant or periodic solutions. preprint.

Appleby, J. A. D. and Riedle, M (2005). Almost sure asymptotic stability of stochastic Volterra integro–differential equations with fading perturbations. *Stochastic Anal. Appl.*, submitted.

Appleby, J. A. D. and Rodkina, A. (2005). On the asymptotic stability of polynomial stochastic delay differential equations. *Funct. Differ. Equ.*, to appear.

Arino, O. and Pituk, M. (1998). Asymptotic constancy for neutral functional differential equations. *Differential Equations Dynam. Systems* **6**, 261-273.

Atkinson, F. and Haddock, J. (1983). Criteria for asymptotic constancy of solutions of functional differential equations. *J. Math. Anal. Appl.* **91**, 410-423.

Atkinson, F., Haddock, J., and Staffans, O. (1982). Integral inequalities and exponential convergence of differential equations with bounded delay. in "Ordinary and Partial Differential Equations: Proceedings, Dundee, Scotland 1982, Lecture Notes in Mathematics: Edited by W. Everitt and B. Sleeman," Springer, New York, pp. 56-68.

Avramescu, Cezar and Vladimirescu, Cristian. (2004). Fixed point theorems of Krasnoselskii type in a space of continuous functions. *Electronic J. Qualitative Theory of Differential Equations* **3**, 1-7.

Banach, S. (1932). "Théorie des Opérations Linéairs" (reprint of the 1932 ed.). Chelsea, New York.

Becker, L. C. (1979). Stability considerations for Volterra integro-differential equations. Ph.D. dissertation. Southern Illinois University, Carbondale, Illinois.

Becker, L. C., Burton, T. A., and Krisztin, T. (1988). Floquet theory for a Volterra equation. *J. London Math. Soc.* **37**, 141–147.

Becker, L. C. and Burton, T. A. (2005). Stability, fixed points, and inverses of delays. *Proc. Roy. Soc. Edinburgh: Section A*, to appear.

Bellman, R. (1953). "Stability Theory of Differential Equations." McGraw-Hill, New York.

Bihari, I. (1956). A generalization of a lemma of Bellman and its application to uniqueness problems of differential equations. *Acta. Math. Sci. Hungar.* **7**, pp. 71-94.

Brezeźniak, Z. and Zastawniak, T. (1998). "Basic Stochastic Processes." Springer, New York.

Burton, T. A. (1965). The generalized Liénard equation. *SIAM J. Control* **3**, 223–230.

Burton, T. A. (1978). Uniform asymptotic stability in functional differential equations. *Proc. Amer. Math. Soc.* **68**, 195–199.

Burton, T. A. (1980). An integrodifferential equation. *Proc. Amer. Math. Soc.* **79**, 393–399.

Burton, T. A. (1983). "Volterra Integral and Differential Equations." Academic Press, Orlando. (Second edition available from Elsevier, Amsterdam, 2005.)

Burton, T. A. (1984). Periodic solutions of nonlinear Volterra equations. *Funkcial. Ekvac.* **27**, 301-317.

Burton, T. A. (1985). "Stability and Periodic Solutions of Ordinary and Functional Differential Equations." Academic Press, Orlando. (Or Dover, New York, 2005.)

Burton, T. A. (1989). Differential inequalities and existence theory for differential, integral, and delay equations. "Comparison Methods and Stability Theory," Xinzhi Liu and David Siegel, editors. Dekker, New York, pp. 35-56.

Burton, T. A. (1996). Integral equations, implicit functions, and fixed points. *Proc. Amer. Math. Soc.* **124**, 2383-2390.

Burton, T. A. (1998). A fixed-point theorem of Krasnoselskii. *Appl. Math. Lett.* **11**, 85-88.

Burton, T. A. (2002). Liapunov functionals, fixed points, and stability by Krasnoselskii's theorem. *Nonlinear Stud.* **9**, 181-190.

Burton, T. A. (2003a). Stability by fixed point theory or Liapunov theory: a comparison. *Fixed Point Theory* **4**, 15–32.

Burton, T. A. (2003b). Perron-type stability theorems for neutral equations. *Nonlinear Anal.* **55**, 285–297.

Burton, T. A. (2004a). Fixed points and stability of a nonconvolution equation. *Proc. Amer. Math. Soc.* **132**, 3679-3687.

Burton, T. A. (2004b). Liénard equations, delays, and harmless perturbations. *Fixed Point Theory* **5**, 209-223.

Burton, T. A. (2004c). Stability and fixed points: addition of terms. *Dynam. Systems Appl.* **13**, 459-478.

Burton, T. A. (2004d). Fixed points and differential equations with asymptotically constant or periodic solutions. *Electron. J. Qual. Theory Differ. Equ.* **11**, 1-31.

Burton, T. A. (2004e). Stability by fixed point methods for highly nonlinear delay equations. *Fixed Point Theory* **5**, 3-20.

Burton, T. A. (2005a). Fixed points, stability, and exact linearization. *Nonlinear Anal.* **61**, 857-870.

Burton, T. A. (2005b). Fixed points, stability, and harmless perturbations. *Fixed Point Theory and Applications* **1**, 35-46.

Burton, T. A. (2005c). Fixed points, Volterra equations, and Becker's resolvent. *Acta Math. Hungar.* **108**, 261-281.

Burton, T. A., and Furumochi, Tetsuo (2001a). Fixed points and problems in stability theory for ordinary and functional differential equations. *Dynam. Systems Appl.* **10**, 89–116.

Burton, T. A. and Furumochi, Tetsuo (2001b). A note on stability by Schauder's theorem. *Funkcial. Ekvac.* **44**, 73-82.

Burton, T. A. and Furumochi, Tetsuo (2002a). Asymptotic behavior of solutions of functional differential equations by fixed point theorems. *Dynam. Systems Appl.* **11**, 499-519.

Burton, T. A. and Furumochi, Tetsuo (2002b). Krasnoselskii's fixed point theorem and stability. *Nonlinear Anal.* **49**, 445-454.

Burton, T. A. and Furumochi, Tetsuo (2005). Asymptotic behavior of nonlinear functional differential equations by Schauder's theorem. *Nonlinear Stud.* **12**, 73-84.

Burton, T. A. and Hatvani, L. (1993). Asymptotic stability of second order ordinary, functional, and partial differential equations. *J. Math. Anal. Appl.* **176**, 261-281.

Burton, T. A. and Kirk, Colleen (1998). A fixed point theorem of Krasnosel'skii-Schaefer type. *Math. Nachr.* **189**,23-31.

Burton, T. A., and Mahfoud, W. E. (1983) Stability criteria for Volterra equations. *Trans. Amer. Math. Soc.* **279**, 143–174.

Burton, T. A., and Mahfoud, W. E. (1985) Stability by decompositions for Volterra equations. *Tohoku Math. J.* **37**, 489–511.

Burton, T. A. and Townsend, C. G. (1968). On the generalized Liénard equation with forcing function. *J. Differential Equations* **4**, 620-633.

Burton, T. A. and Townsend, C. G. (1971). Stability regions of the forced Liénard equation. *J. London Math. Soc.* **3**, 393-402.

Burton, T. A., and Zhang, B. (1990). Uniform ultimate boundedness and periodicity in functional differential equations. *Tohoku Math. J.* **42**, 93–100.

Burton, T. A. and Zhang, Bo (2004). Fixed points and stability of an integral equation: nonuniqueness. *Appl. Math. Lett.* **17**, 839-846.

Caccioppoli, R. (1930) Un teorema generale sull'esistenza di elementi uniti in una trasformazione funzionale. *Rend. Accad. Lincei* **11**, 794-799.

Cesari, L. (1962). "Asymptotic Behavior and Stability Problems in Ordinary Differential Equations." Springer, New York.

Chan, T. and Williams, D. (1989). An "excursion" approach to an annealing problem. *Math. Proc. Camb. Philos. Soc.* **105** (1), 169–176.

Coddington, E. A. and Levinson, N. (1955). "Theory of Ordinary Differential Equations" McGraw-Hill, New York.

Cooke, Kenneth L. and Yorke, James A. (1973). Some equations modelling growth processes and gonorrhea epidemics. *Math. Biosciences* **16**, 75-101.

Davis, H. T. (1962). "Introduction to Nonlinear Differential and Integral Equations." Dover, New York.

De Pascale, E. and De Pascale, L. (2002). Fixed points for non-obviously contractive operators. *Proc. Amer. Math. Soc.* **127**, 3249-3254.

Eloe, P., Islam, M., and Zhang, Bo (2000). Uniform asymptotic stability in linear Volterra integrodifferential equations with application to delay equations. *Dynamic Systems and Applications* **9**, 331-344.

Ergen, W. K. (1954). Kinetics of the circulating-fuel nuclear reactor. *J. Appl. Phys.* **25**, 702–711.

Furumochi, Tetsuo (2004). Asymptotic behavior of solutions of some functional differential equations by Schauder's theorem. *Proc. 7th Coll. Qualitative theory of Diff. Equ.* in *Electron. J.Qual. Theory Differ. Equ.* **10**, 1-11.

Furumochi, Tetsuo (2004). Stabilities in FDEs by Schauder's theorem. *Proc. World Congress on Nonlinear Analysis*, to appear.

Galbraith, A. S., McShane, E. J., and Parrish, G. B. (1965). On the solutions of linear second-order differential equations. *Proc. Nat. Acad. Sci. U.S.A.* **53**, 247-249.

Graef, J. R. (1972). On the generalized Liénard equation with negative damping. *J. Differential Equations* **12**, 34–62.

Graef, J. R., Qian, Chuanxi, and Zhang, Bo (2000). Asymptotic behavior of solutions of differential equations with variable delays. *Proc. London Math. Soc.* **81**, 72-92.

Haag, Jules (1962). " Oscillatory Motions." Wadsworth, Belmont, CA.

Haddock, J. R. (1987). Functional differential equations for which each constant function is a solution: A narrative. "Proc. Eleventh Int. Conf. on Nonlinear Oscillations, M. Farkas, V. Kertesz, and G. Stepan, eds., Janos Bolyai Math. Soc., Budapest." pp. 86-93.

Haddock, J. R., Krisztin, T., Terjéki, J., and Wu, J. H. (1994). An invariance principle of Lyapunov-Razumikhin type for neutral functional differential equations. *J. Differential Equations* **107**, 395-417.

Haddock, J. and Terjéki, J. (1990). On the location of positive limit sets for autonomous functional-differential equations with infinite delay. *J. Differential Equations* **86**, 1-32.

Halanay, A. (1965). On the asymptotic behavior of the solutions of an integrodifferential equation. *J. Math. Anal. Appl.* **10**, 319–324 .

Hale, J. K. (1965) Sufficient conditions for stability and instability of autonomous functional-differential equations. *J. Differential Equations* **1**, 452–482.

Hale, J. K. (1969). "Ordinary Differential Equations." Wiley, New York.

Hartman, P. (1964). "Ordinary Differential Equations." Wiley, New York.

Hatvani, L. (1996). Integral conditions on the asymptotic stability for the damped linear oscillator with small damping. *Proc. Amer. Math. Soc.* **124**, 415-422.

Hatvani, L. (1992). Nonlinear oscillation with large damping. *Dynamic Systems Appl.* **1**, 257-270.

Hatvani, L. (1997b). Annulus arguments in the stability theory for functional differential equations. *Differential and Integral Equations* **10**, 975–1002.

Hatvani, L. (2002). On the asymptotic stability for nonautonomous functional differential equations by Lyapunov functionals. *Trans. Amer. Math. Soc.* **354**, 3555–3571.

Hatvani, L., and Krisztin, T. (1995). Asymptotic stability for a differential-difference equation containing terms with and without a delay. *Acta Sci. Math.* (Szeged) **60**, 371–384.

Hatvani, L. and Krisztin, Tibor (1997). Necessary and sufficient conditions for intermittent stabilization of linear oscillators by large damping. *Differential and Integral Equations* **10**, 265-272.

Hatvani, L., Krisztin, Tibor, and Totik, Vilmos (1995). A necessary and sufficient condition for the asymptotic stability of the damped oscillator. *J. Differential Equations* **119**, 209-223.

Hatvani, L. and Totik, Vilmos (1993). Asymptotic stability of the equilibrium of the damped oscillator. *Differential and Integral Equations* **6**, 835-848.

Hino, Y. and Murakami, S. (1996). Stabilities in linear integrodifferential equations. *Lecture Notes in Numerical and Applied Analysis* **15**, 31-46.

Horn, W. A. (1970). Some fixed point theorems for compact maps and flows in Banach spaces. *Trans. Amer. Math. Soc.* **149**, 391–404.

Jacod, J. and Protter, P. (2003). "Probability Essentials," 2nd. edition. Springer, Berlin.

Jehu, C., Comportement asymptotique des solutions de l'equation $x'(t) = -f(t, x(t)) + f(t, x(t-1)) + h(t)$. *Annales de la Societe Scientifique de Bruxelles* **92**, IV , 263-269.

Kaplan, James L, Sorg, M., and Yorke, James A. (1979). Solutions of $x'(t) = f(x(t), x(t-L))$ have limits when f is an order relation. *Nonlinear Anal.* **3**, 53-58.

Karatzas, I. and Shreve, S. E. (1991) "Brownian Motion and Stochastic Calculus." Volume 113 of *Graduate Texts in Mathematics*. Springer, New York.

Kato, J. (1980). An autonomous system whose solutions are uniformly ultimately bounded but not uniformly bounded. *Tohoku Math. J.* **32**, 499–504.

Kolmanovskii, V. and Myshkis, A. (1999). "Introduction to the Theory and Applications of Functional Differential Equations." Volume 463 of *Mathematics and Its Applications*. Kluwer Academic Publishers, Dordrecht.

Krasnoselskii, M. A. (1958). *Amer. Math. Soc. Transl. (2)* **10**, 345–409.

Krasovskii, N. N. (1963). "Stability of Motion." Stanford Univ. Press.

Krisztin, T. and Terjéki, J. (1988). On the rate of convergence of solutions of linear Volterra equations. *Boll. Un. Mat. Ital. B (7)*, 2 (2), 427–444.

Krisztin, Tibor and Wu, Jianhong (1996). Asymptotic behaviors of solutions of scalar neutral functional differential equations. *Differential Equations Dynam. Systems* **4**, 351-366.

Krisztin, Tibor (1991). On the stability properties for one-dimensional delay-differential equations. *Funkcial. Ekvac.* **34**, 241-256.

Lakshmikantham, V., and Leela, S. (1969). "Differential and Integral Inequalities," Vol. I. Academic Press, New York.

LaSalle, J. and Lefschetz, S. (1961). "Stability by Liapunov's Direct Method." Academic Press, New York.

Lefschetz, S. (1965). "Stability of Nonlinear Control Systems." Academic Press, New York.

Levin, J. J. (1963). The asymptotic behavior of a Volterra equation. *Proc. Amer. Math. Soc.* **14**, 434–451.

Levin, J. J. (1968). A nonlinear Volterra equation not of convolution type. *J. Differential Equations* **4**, 176–186.

Levin, J. J., and Nohel, J. A. (1960). On a system of integrodifferential equations occuring in reactor dynamics. *J. Math. Mechanics* **9**, 347–368.

Levin, J. J. and Nohel, J. A. (1964). On a nonlinear delay equation. *J. Math. Anal. Appl.* **8**, 31–44.

Liénard, A. (1928). Étude des oscillations entretenues. *Revue Générale de l'Électricité* **23**, 901–946.

Lou, B. (1999). Fixed points for operators in a space of continuous functions and applications. *Proc. Amer. Math. Soc.* **127**, 2259-2264.

Lurie, A. I. (1951). "On Some Nonlinear Problems in the Theory of Automatic Control." H. M. Stationery Office, London.

Lyapunov, A. M. (1992). The general problem of the stability of motion. *Int. J. Control.* **55**, 531-773. This is a modern translation by A. T. Fuller of the 1892 monograph.

MacCamy, R. C. and Wong, J. S. W. (1972). Stability theorems for some functional equations. *Trans. Amer. Math. Soc.* **164**, 1-37.

MacCamy, R. C. (1977b). A model for one-dimensional, nonlinear viscoelasticity. *Quart. Appl. Math.* **35**, 21–33.

Mao, X. (1997). "Stochastic Differential Equations and Applications." *Horwood Series in Mathematics and Applications* Horwood, Chichester.

Massera, J. L. (1950). The existence of periodic solutions of a system of differential equations. *Duke Math. J.* **17**, 457–475.

Maynard Smith, J. (1974). "Models in Ecology." Cambridge Univ. Press, London and New York.

Minorsky, Nicholas (1960). "Investigation of Nonlinear Control Systems. Part 1, Continuously Acting Control Systems. Part 2." Office of Naval Research, Washington, D. C.

Minorsky, Nicholas (1962). "Nonlinear Oscillations." D. Van Nostrand, New York.

Morosanu, Gheorghe and Vladimirescu, Cristian (2005a). Stability for a nonlinear second order ODE. *Funkcial. Ekvac* **48**, 49-56.

Morosanu, Gheorghe and Vladimirescu, Cristian (2005b). Stability for a damped nonlinear oscillator. *Nonlinear Anal.* **60**, 303-310.

Omari, P., Villari, G., and Zanolin, F. (1988). A survey of recent applications of fixed point theory to periodic solutions of the Liénard equation. In "Contemporary Mathematics: Fixed Point Theory and Its Applications, Vol. 72." Amer. Math. Soc., Providence, RI., pp. 171 - 178.

Øksendal, B (2003). "Stochastic Differential Equations" 6th. edition. Springer, Berlin.

Perron, O. (1930). Die stabilitatsfrage bei differential-gleichungssysteme. *Math. Z.* **32**, 703–728.

Pielou, E. C. (1969). "An Introduction to Mathematical Ecology." Wiley, New York.

Pucci, P. and Serrin, J. (1994). Asymptotic stability for intermittently controlled nonlinear oscillators. *SIAM J. Math. Anal.* **25**, 815-835.

Rus, Ioan A. (2001). "Generalized Contractions and Applications." Cluj Univ. Press, Cluj-Napoca, Romania.

Rus, Ioan A., Petrusel, Adrian, and Petrusel, Gabriela (2002). "Fixed Point Theory 1950-2000 Romanian Contributions." House of the Book of Science, Cluj-Napoca, Romania.

Schaefer, H. (1955). Über die Methode der a priori-Schranken. *Math. Annalen* **129**, 415-416.

Schauder, J. (1930). Der Fixpunktsatz in Funktionalraümen. *Studia Math.* **2**, 171-80.

Schauder, J. (1932). Über den Zusammenhang zwischen der Eindeutigkeit and Lösbarkeit partieller Differentialgleichungen zweiter Ordung von Elliptischen Typus. *Math. Ann.* **106**, 661–721.

Seifert, G. (1973). Liapunov-Razumikhin conditions for stability and boundedness of functional differential equations of Volterra type. *J. Differential Equations* **14**, 424-430.

Seiji, Saito (1989). Global stability of solutions for quasilinear ordinary differential systems. *Math. Japonica* **34**, 821–829.

Serban, M.-A. (2001). Global asymptotic stability for some difference equations via fixed point technique. Seminar on Fixed Point Theory, Cluj-Napoca **2**, 87–96.

Smart, D. R. (1980). "Fixed Point Theorems." Cambridge Univ. Press, London.

Smith, R. A. (1961). Asymptotic stability of $x'' + a(t)x' + x = 0$. *Quart. J. Math. Oxford ser. (2)* **12**, 123-126.

Somolinos, A. (1977). Stability of Lurie-type functional equations. *J. Differential Equations* **26**, 191–199.

Somolinos, A. (1978). Periodic solutions of the sunflower equation $x'' + (a/r)x' + (b/r)\sin x(t-r) = 0$. *Quart. Appl. Math.* **35**, 465-478.

Volterra, V. (1928). Sur la théorie mathématique des phénomès héréditaires. *J. Math. Pur. Appl.* **7**, 249–298.

Volterra, Vito (1959). "Theory of Functionals and of Integral and Integro-differential Equations. Dover, New York.

Wang, Tingxiu (1992). Equivalent conditions on stability of functional differential equations. *J. Math. Anal. Appl.* **170**, 138–157.

Wang, Tingxiu (1994a). Asymptotic stability and the derivatives of solutions of functional differential equations. *Rocky Mountain J. Math.* **24**, 403–427.

Wang, Tingxiu (1994b). Upper bounds for Liapunov functionals, integral Lipschitz condition and asymptotic stability. *Differential and Integral Equations* **7**, 441–452.

Yoshizawa, T. (1966). "Stability Theory by Liapunov's Second Method." Math. Soc. Japan, Tokyo.

Zhang, Bo (1992). On the retarded Liénard equation. *Proc. Amer. Math. Soc.* **115**, 779-783.

Zhang, Bo (1993). Boundedness and stability of solutions of the retarded Liénard equation with negative damping. *Nonlinear Anal.* **20**, 303-313.

Zhang, Bo (1996). Necessary and sufficient conditions for boundedness and oscillation in the retarded Liénard equation. *J. Math. Anal. Appl.* **200**, 453-473.

Zhang, Bo (1997). Periodic solutions of the retarded Liénard equation. *Ann. Mat. pura appl.* **CLXXII**, 25-42.

Zhang, B. (1995). Asymptotic stability in functional-differential equations by Liapunov functionals. *Trans. Amer. Math. Soc.* **347**, 1375–1382.

Zhang, B. (1997). Asymptotic stability criteria and integrability properties of the resolvent of Volterra and functional equations. *Funkcial. Ekvac.* **40**, 335–351.

Zhang, B. (2001). Formulas of Liapunov functions for systems of linear ordinary and delay differential equations. *Funkcial. Ekvac.* **44**, 253–278.

Zhang, Bo (2004). Contraction mapping and stability in a delay-differential equation. *Dynam. Systems and Appl.* **4**, 183-190.

Zhang, Bo (2005). Fixed points and stability in differential equations with variable delays. *Nonlinear Anal.* **63**, e233-e242.

Zhang, Bo and Gao, Hang (2005). Fixed Points and Controllability in Delay Systems. *Fixed Point Theory and Applications*, to appear.

Author Index

Subject Index

A CATALOG OF SELECTED

DOVER BOOKS

IN SCIENCE AND MATHEMATICS

Chemistry

THE SCEPTICAL CHYMIST: THE CLASSIC 1661 TEXT, Robert Boyle. Boyle defines the term "element," asserting that all natural phenomena can be explained by the motion and organization of primary particles. 1911 ed. viii+232pp. 5⅜ x 8½. 0-486-42825-7

RADIOACTIVE SUBSTANCES, Marie Curie. Here is the celebrated scientist's doctoral thesis, the prelude to her receipt of the 1903 Nobel Prize. Curie discusses establishing atomic character of radioactivity found in compounds of uranium and thorium; extraction from pitchblende of polonium and radium; isolation of pure radium chloride; determination of atomic weight of radium; plus electric, photographic, luminous, heat, color effects of radioactivity. ii+94pp. 5⅜ x 8½. 0-486-42550-9

CHEMICAL MAGIC, Leonard A. Ford. Second Edition, Revised by E. Winston Grundmeier. Over 100 unusual stunts demonstrating cold fire, dust explosions, much more. Text explains scientific principles and stresses safety precautions. 128pp. 5⅜ x 8½. 0-486-67628-5

THE DEVELOPMENT OF MODERN CHEMISTRY, Aaron J. Ihde. Authoritative history of chemistry from ancient Greek theory to 20th-century innovation. Covers major chemists and their discoveries. 209 illustrations. 14 tables. Bibliographies. Indices. Appendices. 851pp. 5⅜ x 8½. 0-486-64235-6

CATALYSIS IN CHEMISTRY AND ENZYMOLOGY, William P. Jencks. Exceptionally clear coverage of mechanisms for catalysis, forces in aqueous solution, carbonyl- and acyl-group reactions, practical kinetics, more. 864pp. 5⅜ x 8½. 0-486-65460-5

ELEMENTS OF CHEMISTRY, Antoine Lavoisier. Monumental classic by founder of modern chemistry in remarkable reprint of rare 1790 Kerr translation. A must for every student of chemistry or the history of science. 539pp. 5⅜ x 8½. 0-486-64624-6

THE HISTORICAL BACKGROUND OF CHEMISTRY, Henry M. Leicester. Evolution of ideas, not individual biography. Concentrates on formulation of a coherent set of chemical laws. 260pp. 5⅜ x 8½. 0-486-61053-5

A SHORT HISTORY OF CHEMISTRY, J. R. Partington. Classic exposition explores origins of chemistry, alchemy, early medical chemistry, nature of atmosphere, theory of valency, laws and structure of atomic theory, much more. 428pp. 5⅜ x 8½. (Available in U.S. only.) 0-486-65977-1

GENERAL CHEMISTRY, Linus Pauling. Revised 3rd edition of classic first-year text by Nobel laureate. Atomic and molecular structure, quantum mechanics, statistical mechanics, thermodynamics correlated with descriptive chemistry. Problems. 992pp. 5⅜ x 8½. 0-486-65622-5

FROM ALCHEMY TO CHEMISTRY, John Read. Broad, humanistic treatment focuses on great figures of chemistry and ideas that revolutionized the science. 50 illustrations. 240pp. 5⅜ x 8½. 0-486-28690-8

Mathematics

FUNCTIONAL ANALYSIS (Second Corrected Edition), George Bachman and Lawrence Narici. Excellent treatment of subject geared toward students with background in linear algebra, advanced calculus, physics and engineering. Text covers introduction to inner-product spaces, normed, metric spaces, and topological spaces; complete orthonormal sets, the Hahn-Banach Theorem and its consequences, and many other related subjects. 1966 ed. 544pp. 6⅛ x 9¼. 0-486-40251-7

ASYMPTOTIC EXPANSIONS OF INTEGRALS, Norman Bleistein & Richard A. Handelsman. Best introduction to important field with applications in a variety of scientific disciplines. New preface. Problems. Diagrams. Tables. Bibliography. Index. 448pp. 5⅜ x 8½. 0-486-65082-0

VECTOR AND TENSOR ANALYSIS WITH APPLICATIONS, A. I. Borisenko and I. E. Tarapov. Concise introduction. Worked-out problems, solutions, exercises. 257pp. 5⅝ x 8¼. 0-486-63833-2

AN INTRODUCTION TO ORDINARY DIFFERENTIAL EQUATIONS, Earl A. Coddington. A thorough and systematic first course in elementary differential equations for undergraduates in mathematics and science, with many exercises and problems (with answers). Index. 304pp. 5⅜ x 8½. 0-486-65942-9

FOURIER SERIES AND ORTHOGONAL FUNCTIONS, Harry F. Davis. An incisive text combining theory and practical example to introduce Fourier series, orthogonal functions and applications of the Fourier method to boundary-value problems. 570 exercises. Answers and notes. 416pp. 5⅜ x 8½. 0-486-65973-9

COMPUTABILITY AND UNSOLVABILITY, Martin Davis. Classic graduate-level introduction to theory of computability, usually referred to as theory of recurrent functions. New preface and appendix. 288pp. 5⅜ x 8½. 0-486-61471-9

ASYMPTOTIC METHODS IN ANALYSIS, N. G. de Bruijn. An inexpensive, comprehensive guide to asymptotic methods–the pioneering work that teaches by explaining worked examples in detail. Index. 224pp. 5⅜ x 8½ 0-486-64221-6

APPLIED COMPLEX VARIABLES, John W. Dettman. Step-by-step coverage of fundamentals of analytic function theory–plus lucid exposition of five important applications: Potential Theory; Ordinary Differential Equations; Fourier Transforms; Laplace Transforms; Asymptotic Expansions. 66 figures. Exercises at chapter ends. 512pp. 5⅜ x 8½. 0-486-64670-X

INTRODUCTION TO LINEAR ALGEBRA AND DIFFERENTIAL EQUATIONS, John W. Dettman. Excellent text covers complex numbers, determinants, orthonormal bases, Laplace transforms, much more. Exercises with solutions. Undergraduate level. 416pp. 5⅜ x 8½. 0-486-65191-6

RIEMANN'S ZETA FUNCTION, H. M. Edwards. Superb, high-level study of landmark 1859 publication entitled "On the Number of Primes Less Than a Given Magnitude" traces developments in mathematical theory that it inspired. xiv+315pp. 5⅜ x 8½. 0-486-41740-9

CALCULUS OF VARIATIONS WITH APPLICATIONS, George M. Ewing. Applications-oriented introduction to variational theory develops insight and promotes understanding of specialized books, research papers. Suitable for advanced undergraduate/graduate students as primary, supplementary text. 352pp. 5⅜ x 8½. 0-486-64856-7

COMPLEX VARIABLES, Francis J. Flanigan. Unusual approach, delaying complex algebra till harmonic functions have been analyzed from real variable viewpoint. Includes problems with answers. 364pp. 5⅜ x 8½. 0-486-61388-7

AN INTRODUCTION TO THE CALCULUS OF VARIATIONS, Charles Fox. Graduate-level text covers variations of an integral, isoperimetrical problems, least action, special relativity, approximations, more. References. 279pp. 5⅜ x 8½. 0-486-65499-0

COUNTEREXAMPLES IN ANALYSIS, Bernard R. Gelbaum and John M. H. Olmsted. These counterexamples deal mostly with the part of analysis known as "real variables." The first half covers the real number system, and the second half encompasses higher dimensions. 1962 edition. xxiv+198pp. 5⅜ x 8½. 0-486-42875-3

CATASTROPHE THEORY FOR SCIENTISTS AND ENGINEERS, Robert Gilmore. Advanced-level treatment describes mathematics of theory grounded in the work of Poincaré, R. Thom, other mathematicians. Also important applications to problems in mathematics, physics, chemistry and engineering. 1981 edition. References. 28 tables. 397 black-and-white illustrations. xvii + 666pp. 6⅛ x 9¼. 0-486-67539-4

INTRODUCTION TO DIFFERENCE EQUATIONS, Samuel Goldberg. Exceptionally clear exposition of important discipline with applications to sociology, psychology, economics. Many illustrative examples; over 250 problems. 260pp. 5⅜ x 8½. 0-486-65084-7

NUMERICAL METHODS FOR SCIENTISTS AND ENGINEERS, Richard Hamming. Classic text stresses frequency approach in coverage of algorithms, polynomial approximation, Fourier approximation, exponential approximation, other topics. Revised and enlarged 2nd edition. 721pp. 5⅜ x 8½. 0-486-65241-6

INTRODUCTION TO NUMERICAL ANALYSIS (2nd Edition), F. B. Hildebrand. Classic, fundamental treatment covers computation, approximation, interpolation, numerical differentiation and integration, other topics. 150 new problems. 669pp. 5⅜ x 8½. 0-486-65363-3

THREE PEARLS OF NUMBER THEORY, A. Y. Khinchin. Three compelling puzzles require proof of a basic law governing the world of numbers. Challenges concern van der Waerden's theorem, the Landau-Schnirelmann hypothesis and Mann's theorem, and a solution to Waring's problem. Solutions included. 64pp. 5⅜ x 8½. 0-486-40026-3

THE PHILOSOPHY OF MATHEMATICS: AN INTRODUCTORY ESSAY, Stephan Körner. Surveys the views of Plato, Aristotle, Leibniz & Kant concerning propositions and theories of applied and pure mathematics. Introduction. Two appendices. Index. 198pp. 5⅜ x 8½. 0-486-25048-2

TENSOR CALCULUS, J.L. Synge and A. Schild. Widely used introductory text covers spaces and tensors, basic operations in Riemannian space, non-Riemannian spaces, etc. 324pp. 5⅜ x 8¼. 0-486-63612-7

ORDINARY DIFFERENTIAL EQUATIONS, Morris Tenenbaum and Harry Pollard. Exhaustive survey of ordinary differential equations for undergraduates in mathematics, engineering, science. Thorough analysis of theorems. Diagrams. Bibliography. Index. 818pp. 5⅜ x 8½. 0-486-64940-7

INTEGRAL EQUATIONS, F. G. Tricomi. Authoritative, well-written treatment of extremely useful mathematical tool with wide applications. Volterra Equations, Fredholm Equations, much more. Advanced undergraduate to graduate level. Exercises. Bibliography. 238pp. 5⅜ x 8½. 0-486-64828-1

FOURIER SERIES, Georgi P. Tolstov. Translated by Richard A. Silverman. A valuable addition to the literature on the subject, moving clearly from subject to subject and theorem to theorem. 107 problems, answers. 336pp. 5⅜ x 8½. 0-486-63317-9

INTRODUCTION TO MATHEMATICAL THINKING, Friedrich Waismann. Examinations of arithmetic, geometry, and theory of integers; rational and natural numbers; complete induction; limit and point of accumulation; remarkable curves; complex and hypercomplex numbers, more. 1959 ed. 27 figures. xii+260pp. 5⅜ x 8½. 0-486-63317-9

POPULAR LECTURES ON MATHEMATICAL LOGIC, Hao Wang. Noted logician's lucid treatment of historical developments, set theory, model theory, recursion theory and constructivism, proof theory, more. 3 appendixes. Bibliography. 1981 edition. ix + 283pp. 5⅜ x 8½. 0-486-67632-3

CALCULUS OF VARIATIONS, Robert Weinstock. Basic introduction covering isoperimetric problems, theory of elasticity, quantum mechanics, electrostatics, etc. Exercises throughout. 326pp. 5⅜ x 8½. 0-486-63069-2

THE CONTINUUM: A CRITICAL EXAMINATION OF THE FOUNDATION OF ANALYSIS, Hermann Weyl. Classic of 20th-century foundational research deals with the conceptual problem posed by the continuum. 156pp. 5⅜ x 8½. 0-486-67982-9

CHALLENGING MATHEMATICAL PROBLEMS WITH ELEMENTARY SOLUTIONS, A. M. Yaglom and I. M. Yaglom. Over 170 challenging problems on probability theory, combinatorial analysis, points and lines, topology, convex polygons, many other topics. Solutions. Total of 445pp. 5⅜ x 8½. Two-vol. set. Vol. I: 0-486-65536-9 Vol. II: 0-486-65537-7